1996

GENETIC ANALYSIS OF PATHOGENIC BACTERIA

GENETIC ANALYSIS OF PATHOGENIC BACTERIA

A LABORATORY MANUAL

Stanley R. Maloy

University of Illinois

Valley J. Stewart

Cornell University

Ronald K. Taylor

Dartmouth Medical School

Cold Spring Harbor Laboratory Press 1996

GENETIC ANALYSIS
OF PATHOGENIC BACTERIA
A Laboratory Manual

© 1996 Cold Spring Harbor Laboratory Press
All rights reserved
Design by Emily Harste

Front Cover: Artistic rendition of *Vibrio cholerae* progressing to and colonizing the epithelial surface of the intestinal mucosa. Once attached, the bacterium multiplies to form microcolonies in the small crypts of the intestine where it secretes exotoxin. (Artwork courtesy of Ginny Togyre.)

Back Cover: Photograph showing replacement of the wild-type *lacZ* gene of *Vibrio cholerae* with an in-vitro-constructed deletion using an *rpsL*-positive selection allelic exchange vector (Skorupski and Taylor 1995). A streptomycin-sensitive merodiploid strain carrying the *lacZ* deletion plasmid integrated at homologous sequences in the chromosome was streaked onto an agar plate containing streptomycin and XGal. Resulting streptomycin-resistant colonies have excised the vector sequences and either retained the wild-type allele (blue) or acquired the deletion (white). (For further details on allelic exchange, see Experiment 14.)

Library of Congress Cataloging-in-Publication Data

Maloy, Stanley R.
    Genetic analysis of pathogenic bacteria : a laboratory manual / by
  Stanley R. Maloy, Valley J. Stewart, Ronald K. Taylor.
       p.   cm.
    Includes bibliographical references and index.
    ISBN 0-87969-452-1 (alk. paper). -- ISBN 0-87969-453-X (alk.
  paper)
    1. Bacterial genetics. 2. Pathogenic bacteria.   I. Stewart,
  Valley Joseph. II. Taylor, Ronald K.   III. Title.
  QH434.M353 1996
  589.9'015--dc20   95-20145
                                                          CIP

The polymerase chain reaction process is covered by certain patent and proprietary rights. Users of this manual are responsible for obtaining any licenses necessary to practice PCR or to commercialize the results of such use. COLD SPRING HARBOR LABORATORY MAKES NO REPRESENTATION THAT THE USE OF THE INFORMATION IN THIS MANUAL WILL NOT INFRINGE ANY PATENT OR OTHER PROPRIETARY RIGHT.

Students and researchers using the procedures in this manual do so at their own risk. Cold Spring Harbor Laboratory makes no representations or warranties with respect to the material set forth in this manual and has no liability in connection with the use of these materials.

All Cold Spring Harbor Laboratory Press publications may be ordered directly from Cold Spring Harbor Laboratory Press, 10 Skyline Drive, Plainview, New York 11803-2500. Phone: 1-800-843-4388 in Continental U.S. and Canada. All other locations: (516) 349-1930. FAX: (516) 349-1946.

This manual is dedicated
to the memory of Max Delbrück
on the fiftieth anniversary
of the first Phage Course
at Cold Spring Harbor Laboratory

Max Delbrück (1906–1981)

Delbrück's Legacy
A BRIEF HISTORY OF THE PHAGE AND BACTERIAL GENETICS COURSES

Max Delbrück organized the first Phage Course in 1945. His goal was to train a diverse group of scientists to attack basic biological questions using phage as simple model systems. The Phage Course shaped the new field of Molecular Biology. It influenced what kinds of questions were asked, how these questions were answered experimentally, and how the results were interpreted. Delbrück was emphatic that biological experiments be answered quantitatively—anything less he considered simply phenomenology or "stamp-collecting." He also believed that scientists should be able to freely give and receive criticism of their work without taking it personally. He shunned pretensions and formalities. Scientists called one another by their first names, regardless of their prestige. As the "guru" of the phage group, Delbrück's ideas became the accepted etiquette for the fields of phage and bacterial genetics that has remained to this day.

The molecular dissection of phage provided many important clues into central biological processes, but there were some questions that could not be answered with phage—questions that required studying cells directly. As insights into molecular genetics of phage accumulated and new genetic tools were developed, it became feasible to begin probing the molecular genetics of more complex organisms—bacteria. Thus, in 1950, Milislav Demerec began an offshoot of the Phage Course that emphasized bacterial genetics. The Bacterial Genetics Course immediately followed the Phage Course, and many students took the two courses sequentially to obtain training in both phage and bacterial genetics.

Over time, many of the concepts and techniques from the Phage Course were integrated into the Bacterial Genetics Course, until the two courses merged into a single course in 1971. Like the Phage Course, the

Bacterial Genetics Course had a dramatic impact on science. In addition to training the many participating scientists, the course engendered a very influential laboratory manual written by Jeffrey Miller, one of the course instructors. Published by Cold Spring Harbor Laboratory in 1972, the *Experiments in Molecular Genetics* manual provided a "cookbook" for a burgeoning group of scientists who wanted to use the "new" tools of molecular genetics but did not know how.

In the mid 1970s, two newly discovered tools—molecular cloning and transposons—began a technological revolution in genetics. Integration of these new tools into the course led to another name change, to Advanced Bacterial Genetics. The *Advanced Bacterial Genetics* laboratory manual written by Ron Davis, David Botstein, and John Roth was published by Cold Spring Harbor Laboratory in 1980. At a time when most scientists realized the power of cloning and transposons but few knew how to use these tools, the impact of this laboratory manual was phenomenal. Until *Molecular Cloning: A Laboratory Manual* written by Tom Maniatis, Ed Fritsch, and Joe Sambrook was published two years later, the manual by Davis, Botstein, and Roth was the primary source for these techniques.

In 1981, Tom Silhavy, Lynn Enquist, and Mike Berman assumed responsibility for the Advanced Bacterial Genetics Course and brought another new technology—the use of gene and operon fusions to analyze gene expression in vivo. This resulted in another very influential laboratory manual published by Cold Spring Harbor Laboratory in 1984, *Experiments with Gene Fusions*. The use of fusions popularized by this manual opened the doors for studying the in vivo expression of any gene, whether or not there was a convenient assay for the gene product.

In 1986, George Weinstock, Russ Maurer, and Peter Berget began teaching the course. They continued to use transposons, cloning, and gene fusions but also incorporated new technologies such as pulsed-field gel electrophoresis. Then in 1991 we began teaching the course. In addition to bringing in new technologies such as use of challenge phage to study DNA-protein interactions and in vivo cloning of genes, we emphasized classical genetic analyses—approaches that still provide essential tools for dissecting the structure and function of genes but have been missed by a generation of students raised on "cut-and-paste" genetics. We have also incorporated a less genetically developed organism into the course (*Vibrio cholerae*), to demonstrate how to approach the genetics of an interesting bacterium from scratch.

Although the name of the course has changed since 1945—from Phage, to Bacterial Genetics, to Advanced Bacterial Genetics—the philosophy and goals of the course have remained true to Delbrück's original intent: training scientists how to use a sophisticated set of state-of-the-art genetic tools, emphasizing critical interpretation of experimental results, and serving a missionary role to disseminate correct genetic thinking to "the masses of poor souls who haven't seen the light."

In many cases, it is now possible to study many complex eukaryotic systems directly, so one might ask: Will phage and bacterial genetics continue to have an important role in science in the future? Many indicators suggest that it will. Phage and bacteria are still useful model systems for understanding cells in general. The genetic tests needed to establish a direct "cause-and-effect" relationship are still easier to carry out in phage and bacteria. Thus, phage and bacteria are likely to continue to provide valuable model systems for many years to come. Furthermore, we can now begin to apply these techniques to the wide variety of bacterial species that have important roles in human health and global ecology. Genetic characterization of diverse bacteria is important because they have practical applications, and because the diversity of these organisms is likely to yield exciting new biological insights. But this very diversity will demand new genetic tools and a new contingent of scientists trained in bacterial genetics.

Contents

EXPERIMENTS

APPENDICES

Preface

This manual is intended for experienced researchers who wish to apply a particular technique at the laboratory bench, for researchers with little experience in bacterial genetics who wish to learn how to apply genetic approaches or develop new genetic techniques for their favorite bacterium or phage, and for instructors preparing advanced laboratory courses on molecular genetics.

The manual evolved from the Advanced Bacterial Genetics Course we taught at Cold Spring Harbor Laboratory from 1991 to 1995. The course was designed to demonstrate how the power of bacterial and phage genetics can be harnessed to answer a variety of biological questions. We soon realized that our largest clientele consisted of scientists working in bacterial pathogenesis, so we have emphasized the application of genetic approaches to study bacterial pathogens.

Although it is now relatively easy to clone and sequence genes from a variety of pathogens, genetic approaches are often needed to identify previously unknown genes involved in virulence and to fulfill the "Molecular Koch's Postulates" to demonstrate that a particular gene has a role in pathogenesis. Efficient application of genetic tools requires knowledge not only of a particular technique, but also of when to use one technique versus another and how to interpret the results. However, due to the influx of scientists from other fields, many people working on bacterial pathogenesis lack sufficient understanding of bacterial genetics to properly apply existing genetic tools or to develop new genetic tools when needed. Therefore, we have attempted in this manual to provide a solid background in bacterial genetics as well as a description of how to apply a variety of powerful genetic tools. Of course, it is not possible to cover the entire spectrum of bacterial pathogens. Instead, the experiments focus on two bacterial pathogens: *Salmonella typhimurium*, an organism with well-developed genetics, and *Vibrio cholerae*, an organism for which the available genetic tools are more typical of those available for the majority of bacterial pathogens. Although many of the specific methods in this manual are presently only applicable in *S. typhimurium*, the concepts are widely applicable. With a little ingenuity, analogous methods could be developed for many other bacteria.

Many people have made important contributions to this manual. We greatly benefited from the helpfulness of the previous course instructors, especially Peter Berget, Russ Maurer, John Roth, Tom Silhavy, and George Weinstock. We were lucky to have outstanding course assistants: Scott Allen, Clark Brown, Li-Mei Chen, Andrew Darwin, Barry Goldman, Melissa Kaufman, Janine Lin, Alicia Muro-Pastor, Paula Ostrovsky, Joel Peek, James Pfau, Ross Rabin, Karen Skorupski, and Thomas Zahrt. The course assistants corrected many errors and provided numerous pragmatic suggestions on the experiments. Their dedication and effort were invaluable. Special thanks are due to Barry Goldman for the Mud-P22 mapping experiment and Tom Zahrt for the PFGE experiment. In addition, the course participants put the experiments through a careful screen that contributed to many insertions, deletions, and other rearrangements, as well as repair of many point mutations in the manual. Each of the course participants deserves our personal thanks: Celia Alpuche, Irmgard Behlau, Amanda Brown, Elena Budrene, Asa Eliasson, Cecil Forsberg, Donald Granger, Reiner Hedderich, Laura Kiessling, Rafael Maldonado, Philippe Moreillon, Robert Poole, Philippe Regnier, Deborah Smith, Jeffrey Stein, Ronald Thune, Soumaya Laalami, James Liao, Maria Sanna, Liselotte Søgaard-Andersen, Trudee Tarkowski, Joyce Velterop, Alberto Villaseñor, Michael Waldor, Philippe Cottagnoud, Ansley Crockford, Amando Flores, Richard Frazee, Valerie Grandjean, Allen Honeyman, Weihong Hsing, Katherine Kezdy, Tamara Barlow, Lisa Barsom, Hein Boot, Kerstin Calia, Angeles Dominguez, Howard Gold, Geeta Gupte, Joseph Hendrick, Ute Hentschel, Alison Holmes, Timothy Hoover, Krystyna Krajewska-Grynkiewicz, Leah Macfayden, Lara Madison, Barbara Troup, Jaap Van Dissel, Paul Anderson, Katherine Andrews-Cramer, Lisa Armitage, Kauline Cipriani, Carmen Collazo-Custodio, Klas Flärdh, Alicia Gil, Marten Hammar, Karin Hansson, Ronald Mackenzie, Patricia Marini, Lalita Ramakrishnan, Cecilia Toro, Renee Tsolis, Anne Vanet, Carol Webber, Joseph Barbieri, Creg Darby, Robert Edwards, Darrell Galloway, Julie-Ann Gavigan, Mary Hondalus, K. Heran Hong, Michael Konkel, Alvydas Mikulskis, Heather Miyagi, Marlena Moors, Mads Nørregaard-Madsen, Ruth Ann Schmitz, Paul Sullam, Friedrich Widdel, and Eva Zydch.

The personnel at Cold Spring Harbor Laboratory also deserve our heartfelt thanks for their services and hospitality during our summers there. The staff of Cold Spring Harbor Laboratory Press merit special appreciation. John Inglis enthusiastically encouraged us to publish this manual, Nancy Ford helped with the conceptual design of the manual, Mary Cozza effectively kept us to a time schedule, Dotty Brown carefully edited the final manuscript, and Maryliz Dickerson and Susan Schaefer ably typeset the manual.

Finally, we are grateful to our wives, Lisa, Janine, and Darlene, for their patience, understanding, and support over the years we taught the Advanced Bacterial Genetic Course and while we prepared this manual.

S.R. Maloy
V.J. Stewart
R.K. Taylor

OVERVIEW

"Geneticists, like other biologists, have a passionate attraction to organisms, but, unlike their colleagues, they are usually monogamous, wedded to one organism for much of their careers. This fidelity is not a manifestation of dreamy romanticism, but rather a consequence of the dedication required to create a standard organism suitable for genetic studies. The emphasis is on 'create' because, contrary to the common perception, good genetic organisms are not found in nature; they are shaped by geneticists. There are, however, intrinsically bad genetic organisms, those that have long life cycles and are difficult to study in the laboratory (whales) or those with no sexual stage (Penicillium). Clearly, one must begin with an organism that is easily cultured in the laboratory and has a tractable sexual cycle. But all the rest is hard work. Mutant strains must be designed so that the biochemistry, physiology, and even the genotype can be manipulated at will. Once the organism has been redesigned so that crosses, complementation, recombination and transformation can be carried out with facility, the modified organism can be used by all biologists. Witness the use of bacteriophage lambda, *Escherichia coli*, yeast and Drosophila in genetic engineering experiments by evolutionists, biophysicists, crystallographers and embryologists. None of these standard organisms exists in nature; all have been painstakingly altered to perform the scientist's bidding.

"The goal of this single-minded devotion is the creation of a standard organism that can be used to reveal themes fundamental to all organisms or groups of organisms. A great deal is known about bacteria because of the millions of laboratory hours that have been invested in a single bacterium, *E. coli*. The encyclopedic knowledge about *E. coli*, its genetics and biochemistry make it an invaluable standard, akin to the meter stick, against which other procaryotes and even eucaryotes can be compared. A system of knowledge based on the *E. coli* standard in no way diminishes our interest in other microorganisms. In fact, it intensifies interest and increases the quality of questions that can be posed. The answer to the question, 'Is this like *E. coli*?' has profound meaning because the quality and quantity of work on this model organism elevate the criteria for comparison. If the answer is 'Yes', then the question has been answered to a first approximation. If the answer is 'No', then a new phenomenon has been uncovered."

Gerald R. Fink
Reprinted, with permission, from Notes of a Bigamous Biologist.
Genetics 118: 549–550 (1988)

Approaching Genetic Analysis of Diverse Bacteria

Interest in understanding the genetics of diverse bacterial species has been stimulated by a variety of factors in recent years. The increasing importance of antibiotic resistance in clinical settings and the rise in the number of immunocompromised patients have both reinforced the critical importance of understanding the biology of pathogenic bacteria. Biotechnology increasingly seeks to engineer diverse bacterial species in more sophisticated ways. The field of microbial ecology has been stimulated by new methods for the study of as yet uncultured bacteria and by the potential for effective bioremediation of polluted sites. Finally, more and more investigators are turning their attention to the study of diverse bacterial species from a range of niches, including hot springs, deep thermal vents, plant and animal symbionts, and even the open ocean. It is argued that the future of microbiology lies in the exploration of its diversity (Woese 1994).

This upsurge in activity has been driven in large part by the ability to employ molecular biology in the study of virtually any biological process. Technological advances in the identification, cloning, and sequencing of genes have expanded molecular analysis far beyond the bounds of such traditional genetic organisms as *Escherichia coli* and *Bacillus subtilis*. Microbiology journals are overwhelmed with papers reporting the cloning and sequence analysis of interesting and important genes from bacterial species of medical, industrial, and environmental importance.

The advances in molecular analysis have led to the view that "extraordinary progress had all but overwhelmed the traditional discipline of genetics" (Lewin 1994). We argue the opposite. Whereas molecular tech-

3

niques are invaluable in identifying and characterizing specific genes of known importance, understanding the in vivo functions of the resultant gene products is not so easily achieved with a purely molecular approach. Analysis of specific and well-characterized mutants is essential for determining the significance of the wild-type phenotype. Identification of previously unknown genes that specify a particular process is best achieved by mutant isolation, followed by genetic analysis and phenotypic characterization. These factors are particularly critical in cases where a truly novel phenomenon is uncovered, and where the genes involved are not related by sequence similarity to other characterized genes of known function. Indeed, genome sequencing projects reveal a large proportion of previously unknown genes that are unrelated in sequence to any other identified genes (Borodovsky et al. 1994). Our aim with this laboratory manual is to provide examples of how formal genetic analysis can be integrated with molecular biological techniques to provide a modern and comprehensive approach to the study of distinct problems not only in pathogenic bacteriology, but also in the analysis of diverse bacterial species in assorted contexts.

We have chosen two pathogenic species to illustrate these concepts. *Salmonella typhimurium* has an exceptionally well-developed genetic system that in many respects is even more sophisticated than that for *E. coli*. The approaches described herein will of course be useful to the many scientists who study *S. typhimurium* and its relatives, but we also hope that they will provide inspiration and motivation for the continued development of genetic techniques for other species as well. *Vibrio cholerae* in contrast has a rather rudimentary genetic system that is not altogether different from those currently available for a wide range of species. Our approaches to the study of *V. cholerae* illustrate how a largely molecular biological approach can nevertheless incorporate the essence and aim of formal genetic analysis and take the investigator far beyond the process of cloning and sequencing and into the realm of functional in vivo analysis. Finally, the P22-based challenge phage system provides a broadly applicable means for the detailed study of the DNA-protein interactions that are so critical for determining the range of phenotypes in all organisms. Challenge phage allows transplanting a specific DNA-protein interaction from any organism into *S. typhimurium*, thereby providing the means for refined and detailed genetic investigation of these interactions and their functional consequences.

This chapter provides a broad view of the considerations and goals that may be employed in approaching genetic analysis of an organism that has only a primitive genetic system. Specific topics are cross-referenced to the following chapters, which provide more in-depth description and detail.

BEFORE YOU START

The quote from G.R. Fink that introduces this section emphasizes the fact that developing a genetic system for a previously unstudied organism demands commitment and much hard work. Such development is also greatly facilitated if workers in other laboratories share a common interest in that particular organism. A solitary investigator faced with the task of developing mutant isolation, gene transfer, and genetic mapping methods for a given species will face much frustration and slow progress. Therefore, a pragmatic initial consideration is to determine the amount of interest in a particular organism and its phenotype. The chosen problem should be of sufficient significance to attract the attention of both funding agencies and colleagues alike. The lack of a critical mass of interested investigators will often signal a project's ultimate doom. A corollary to this is that workers in all laboratories ought to focus their efforts on a single representative strain, so that mutants, genetic and physical mapping data, and other information may be profitably exchanged.

An important aspect of this decision-making process is to find a balance between the amount of work required to develop a system and the overall importance of the problem. For example, *Mycobacterium tuberculosis* is a notoriously difficult organism with which to work, yet it is also increasingly recognized as a serious pathogen of considerable significance. For many, the inherent importance of the problem outweighs the difficulties encountered in its study. Of course, widespread recognition of a problem's significance (and fundability) also helps to attract new workers to the field, thereby building a critical mass and ultimately accelerating the development of diverse methods for genetic analysis.

With an interesting problem in hand, the next essential consideration is to understand where the organism in question is situated with respect to bacterial phylogeny. Often, an organism of interest will be closely related to one that has been well-studied, giving one the hope that methods and approaches might be transplanted to the newly chosen organism with relatively little effort. For example, investigations with *Salmonella typhi*, a significant pathogen with a poorly developed genetic system, benefit from adapting methods already developed for *S. typhimurium* (see, e.g., Zahrt et al. 1994).

By the same token, knowledge of an organism's relatives may allow workers to choose a surrogate organism to facilitate initial studies. Many examples abound. Although *S. typhimurium* is a human pathogen in its own right (causing gastroenteritis), it provides in many ways a model for understanding aspects of *S. typhi* pathogenesis, which leads to a rather more serious disease (typhoid fever). Likewise, the relatively fast-growing *Mycobacterium smegmatis* provides a kind of genetic proving ground for developing and optimizing methods that may ultimately find use with the much more slowly growing *M. tuberculosis*.

A classic example of the use of surrogate hosts comes from Miller and Mekalanos (1985), who cloned the *V. cholerae* regulatory gene *toxR*. The ToxR protein activates expression of the cholera toxin structural gene operon, *ctxAB*. A *ctxA-lacZ* operon fusion is expressed at a low level in *E. coli*, because the *toxR* gene is not present in this organism. To identify the *toxR* gene, a *V. cholerae* gene bank was introduced into an *E. coli ctxA-lacZ* strain; strongly Lac+ colonies harbored clones carrying the *toxR*+ gene. Once identified and isolated, the function of the *toxR* gene in *V. cholerae* was amenable to study (see, e.g., Miller and Mekalanos 1988; see also Experiments 14 and 15).

IMMEDIATE GOALS

Rudimentary genetic analysis precedes sophisticated analysis, and one needs a starting point. The following methods and considerations are important for developing a first-level system for genetic analysis. Most of these focus on genetic exchange, particularly between species (i.e., from *E. coli* into the species of interest). This allows the investigator to manipulate a given gene in *E. coli* and then to reintroduce it into the organism in question in order to probe the function of that particular gene.

Mutagenesis

The essential ingredient for genetic analysis is the ability to generate and recover mutants with informative phenotypes. Isolation of mutants leads to the identification of the genes required for a given process, and analysis of various types of mutations allows gene function to be probed.

Initial genetic analysis often begins with chemical mutagenesis, which can be performed on organisms with even very limited genetic systems. Two critical factors must be addressed. First, a suitable chemical or physical (e.g., UV light) mutagen must be identified. Many pathways of mutagenesis require error-prone repair systems, which are not present in all bacterial species (see Section 4, Mutagenesis). In addition, the mutagen must result in sufficiently high levels of mutagenesis so that individuals of the desired phenotype are recovered without undue difficulty. A very good first-choice mutagen for exploratory studies is *N*-methyl-*N'*-nitro-*N*-nitrosoguanidine (NTG). This powerful and efficient mutagen directly alkylates bases in DNA, resulting in direct mutagenesis through mispairing (i.e., largely independent of error-prone repair systems).

The serious drawback of NTG mutagenesis is the fact that it tends to induce multiple clustered mutations (through its action on the replication fork), potentially complicating subsequent genetic analysis. For this reason, NTG is shunned by many investigators for routine use. However, its power as a mutagen makes it ideal for pilot studies, in which the researcher asks, Can I recover mutants of the desired phenotype at reasonable frequencies? If the answer is yes, the investigator is then encouraged to explore alternate mutagenesis schemes. Various strategies are detailed in Section 4 (Mutagenesis).

The second critical factor is in designing effective selections, screens, or enrichments for recovering the mutants of interest. A positive selection, in which only the desired mutants are able to form colonies, is greatly valued by geneticists, but many forms of mutant recovery rely on screens and/or enrichments instead (see Section 2, Culture Media; Antibiotics, Antibiotic Resistance, and Positive and Negative Selections). An in-depth knowledge of bacterial physiology in relation to the process of interest is essential for designing effective selections and screens (see Section 2, Bacterial Physiology).

Gene Transfer Systems

Conjugation is a method of great utility for transferring genes from one strain (or species) to another. Several broad host range plasmids for gram-negative bacteria (e.g., RP4, R100, and R6) have been engineered for use with diverse species (see Section 7, Plasmids and Conjugation). Conjugal transfer has been facilitated by the development of specialized *E. coli* donor strains, although triparental matings are also effective (see Section 7, Plasmids and Conjugation). Likewise, plasmids initially discovered in *Staphylococcus aureus* have found utility in diverse gram-positive species (Novick 1991).

A second useful means for introducing plasmid DNA into a given organism is through transformation or electrotransformation (see Section 7, Transformation and Electrotransformation). Many species are naturally competent for DNA uptake (e.g., *Acinetobacter calcoaceticus*), although competence often reflects a specific physiological state that must be generated by growth under appropriate conditions (e.g., *Bacillus subtilis*). In some species, natural transformation requires the presence of specific signals within the DNA sequence (e.g., *Haemophilus influenzae*). Alternatively, competence may be induced by artificial means, such as $CaCl_2$ shock (e.g., *E. coli*). The advent of electroporation has greatly broadened the range of species that can be transformed (see Section 7, Transformation and Electrotransformation), and an electroporation apparatus is an essential component of many bacterial genetics laboratories.

Conjugation and transformation both may require considerable optimization to achieve suitably high and reproducible efficiencies. Important parameters with respect to preparing the donor and recipient cultures include the growth phase, composition of the culture medium, and the level of aeration (see Section 2, Bacterial Physiology).

A further consideration in developing plasmid-based genetic exchange is the development of convenient shuttle vectors that replicate both in *E. coli* and in the organism of interest. Broad host range vectors are generally large and unwieldy, and thus more compact, specific shuttle vectors find utility. Often, this entails building a plasmid with two replication origins, one for each organism, as well as selectable markers that function in both hosts.

Conjugation and transformation methods are generally limited to transferring plasmid DNA between strains and are less useful for exchanging specific segments of chromosomal DNA. Generalized transduction is the method of choice for this latter purpose, and generalized transducing systems have been developed for many bacterial species (see Section 3, Generalized Transduction). An available or readily developed generalized transduction system will prove invaluable in the genetic analysis of any bacterial species. However, extensive efforts with some organisms (e.g., *Rhodobacter sphaeroides*) have to date failed to provide workable generalized transduction systems, and thus developing this particular method may be relegated to the category of long-term goals in many cases.

Restriction and Modification Systems

Restriction endonuclease systems can provide a formidable barrier to interspecies gene transfer into many bacterial species. Restriction can reduce gene transfer efficiency by a factor of 100 or more in certain cases. In other cases (e.g., *V. cholerae*), restriction barriers are insignificant and do not hinder genetic exchange. Thus, this aspect must be evaluated on an individual basis. Difficulty with interspecies genetic exchange may be overcome by testing for the presence of restriction endonucleases as a possible inhibiting influence.

Some useful initial approaches rely on in vitro manipulations. Often, donor DNA can be appropriately modified (methylated) by incubation with an EDTA-treated crude extract of the recipient organism. EDTA inhibits the activity of most restriction endonucleases but leaves methylase activity intact. The modified DNA may prove to be a superior substrate for transformation, thereby pinpointing the barrier as restriction, at least in part.

Their commercial utility has led to the discovery and characterization of innumerable restriction endonucleases, and isoschizomers abound. Many suppliers provide gel electrophoresis patterns for specific substrates (e.g., λ DNA) cleaved with various enzymes. In vitro cleavage of such a substrate with an extract of the organism in question might provide a pattern that can be identified in relation to an enzyme of known specificity. With this knowledge in hand, additional steps can be taken to overcome the restriction barrier. For example, many suppliers now also market cognate methylases. If a crude extract from a given organism yields a cleavage pattern corresponding to that for *Bam*HI (for example), then *Bam*HI methylase-treated DNA might be a good substrate for transformation.

Even if these in vitro methods succeed in identifying the restriction barrier, they do not provide convenient long-term solutions and of course cannot be used with conjugation. However, the presence of the cloned methylase gene in the *E. coli* donor strain can result in effective protection of the transferred donor DNA (Haselkorn 1991).

By far the most effective long-term solution is to isolate a restriction-deficient mutant, preferably one that retains modification function ($r^- m^+$ phenotype). Such mutants can often be recognized among individuals of a mutagenized culture that is then subjected to transformation with a plasmid or infection with a lytic bacteriophage. Survivors can be retested for increased transformation frequencies indicative of a defective restriction barrier. Many organisms contain multiple restriction endonucleases, and thus mutational inactivation of each requires considerable effort (Bullas and Ryu 1983; Raleigh et al. 1991; Carlson et al. 1994).

Counterselections

A critical component of any conjugation-based genetic exchange system is a means to select against the donor strain. Most often, this is achieved by introducing an antibiotic resistance marker into the recipient (see Section 2, Antibiotics, Antibiotic Resistance, and Positive and Negative Selections). Commonly used markers include resistance to streptomycin (*rpsL*; ribosomal protein S12), spectinomycin (*rpsE*; ribosomal protein S5), rifampin and related antibiotics (*rpoB*; RNA polymerase β subunit), and nalidixic acid (*gyrA*; DNA gyrase). Isolation of such mutants is straightforward. A concentrated culture is plated on medium containing a relatively high concentration of the antibiotic in question (as much as 1 mg/ml for streptomycin). Spontaneous resistant mutants are generally very rare, owing to the specific nature of the change that yields a functional yet resistant product. Indeed, this low background of resistant mutants is an essential feature for a counterselectable marker.

A second but less convenient form of counterselection involves auxotrophic donor strains. A conjugal donor that requires one or more amino acids for growth will not form colonies on minimal medium; nonreverting mutations such as deletions are preferred. However, this method is only suitable if the recipient grows well on minimal medium, and it may constrain the types of indicators and other screens employed.

INTERMEDIATE GOALS

With workable genetic exchange systems in hand, suicide vectors and methods may be developed for complementation analysis.

Suicide Vectors

Suicide vectors are plasmids or bacteriophages that can replicate in one host but not in another. A simple form of suicide vector is simply a narrow host range plasmid, such as those based on ColE1-like replicons. These plasmids will replicate only in enterobacteria and closely related species. Other types of conditionally replicating vectors have a temperature-sensitive replicon or a nonsense mutation in an essential gene. A widely used series of suicide vectors is based on the R6K replicon, in which the essential π replication protein is provided in trans in the donor host. Recipients do not synthesize the π protein and are thereby unable to maintain the plasmid (see Section 7, Plasmids and Conjugation).

Suicide vectors are essential for two important aspects of genetic analysis. First, they provide the most efficient delivery vehicles for transposons (see Section 5, Uses of Transposons in Bacterial Genetics). Second, suicide vectors are essential for efficient allelic exchange mutagenesis (see Section 4, Broad Host Range Allelic Exchange Systems), wherein a manipulated version of a cloned gene is transplanted into the chromosome in place of the resident wild-type allele.

Complementation Analysis

Providing mutants with a second, wild-type copy of a given gene in the absence of homologous recombination (complementation) serves many purposes: determining the number of genes at a given locus, determining whether a particular mutation acts in cis or trans, confirming that a cloned gene confers the expected phenotype, and inferring the nature of a specific mutation (loss or gain of function). Methods and considerations for conducting complementation analysis are described in Section 4 (Genetic Complementation).

LONG-TERM GOALS

As the genetic system in question becomes more sophisticated, further methods may be developed for more detailed genetic analysis.

Physical Mapping

Genetic maps based on conjugation methods continue to be important for many bacterial species. However, the advent of general methods for physical map construction has greatly expanded the range of organisms for which maps are available (see Section 3, Genetic Mapping of Chromosomal Genes). A genetic or physical map is very useful in the study of multigenic phenomena, allowing the determination of the number of genes and their physical relationship to each other. For example, mutations that confer a novel phenotype can be mapped to determine whether a previously unknown gene has been identified or whether a known gene has a previously unknown function.

The most useful physical maps are those which contain genetic information as well—for example, the locations of auxotrophic transposon insertions. High-resolution maps require a number of genetic or physical markers and therefore rely on a community of workers to supply well-characterized markers. The combined genetic/physical map of *S. typhimurium* provides a good example of rigorous map construction combined with abundant genetic information (Liu et al. 1993a).

Generalized Transduction

Finally, generalized transduction is a technique that sees daily use in laboratories fortunate enough to work with organisms for which this technique is available. For example, the utility of phage-P22-mediated generalized transduction has been a major catalyst in the development of *S. typhimurium* as one of the preeminent bacterial genetic systems. Some strategies for finding and characterizing generalized transducing phage are described in Section 3 (Generalized Transduction).

SECTION 1
Bacterial Pathogenesis

"Mac Edds, a marvelous teacher and research embryologist, coaxed me into working with *Dictyostelium* spp. The *Dictyostelium* experiments were simply feeding these predatory creatures different bacteria as a food source. I felt somewhat like a traitor. In any event, the amoeba form of *Dictyostelium* could clear a petri dish of bacterial growth with extraordinary dispatch prior to aggregating into 'slugs.' However, I was, as one might guess, interested in the bacterial survivors of the phagocytic carnage. The answer played a role in my thinking throughout the rest of my career. The enteric bacterial survivors, *E. coli* and *Salmonella* spp., as well as *Pseudomonas aeruginosa*, were mucoid variants. I had seen remarkably similar strains isolated from the sputum of cystic fibrosis patients. Cystic fibrosis patients had always suffered from serious lung infections. Prior to the antibiotic era, they were largely caused by staphylococci (reportedly quite mucoid variants), but with the advent of antimicrobial therapy in the 1950s, one increasingly saw gram-negative infections, particularly with *E. coli* and later with *P. aeruginosa*. These student experiments with *Dictyostelium* spp., therefore, led me to begin to understand that the bacterial genetic apparatus selected to resist phagocytosis by mammalian phagocytes had its roots much, much earlier in evolution in the early battles against natural microbial predators. It is less difficult for me to think about the determinants of pathogenicity like that of botulinum toxin when I consider that it was likely evolved to act on some microscopic predatory bacteriophagic creature like nematodes in the soil rather than as a deliberate strategy to poison some poor soul who ate a spoiled can of food. Yet, the life of microbes outside of the human disease state gets relatively little space in textbooks, or in conversations with bacteriologists for that matter. I tease my friends who work on *Pseudomonas* spp. that it is not a 'proper' pathogen because it causes disease only in compromised human hosts. Yet, *P. aeruginosa* comes armed with an extensive array of toxins and other virulence attributes and it regulates them exquisitely. What is the real target of of these virulence determinants in the natural aquatic environment inhabited by *P. aeruginosa*?"

Stanley Falkow

Reprinted, with permission, from A Look Through the Retrospectoscope.
In *Molecular Genetics of Bacterial Pathogenesis: A Tribute to Stanley Falkow*
(ed. V.L. Miller et al.), pp. xxiii–xxxix. ASM Press, Washington, DC. (1994)

General Concepts in Bacterial Pathogenesis

Pathogenic bacteria exhibit a wide range of biochemical and physiological processes that contribute directly to the virulence of the organism or possibly to pathogenesis by solicitation of host responses that cause tissue damage in the vicinity of the infection. A number of the attributes of bacterial pathogens are readily approachable by genetic analysis under both in vitro and in vivo growth conditions. Initially, null mutations can be used to identify genes that encode particular virulence properties associated with a pathogenic microorganism. The use of genetics and molecular biology to correlate a particular gene with a virulence property has been coined as Molecular Koch's Postulates (Falkow 1988), which include correlation between knockout of gene function and loss of the virulence trait. The trait should be restored upon isolation of the wild-type gene and its reintroduction into the null mutant (gene replacement) or a surrogate host.

Once null mutations have established the importance of a particular parameter in contributing to the virulence of an organism—association of a gene with a particular virulence phenotype—a more thorough analysis aimed at determining the molecular mechanisms by which a certain function is carried out can be undertaken. A plethora of genetic tools have been designed for these analyses, with considerations as to host range, delivery vehicles, selectable and screenable markers, and methods to perform complementation analyses. Generally, transposons are the method of choice for initial mutagenesis to identify genes with a pathogenic role. Transposons provide the advantages of encoding an easily followed selectable marker and the creation of a null mutation upon insertion into a gene. Tn*phoA* is an example of a transposon that has been useful for enrichment of insertions in genes encoding cell surface or secreted virulence determinants (see, e.g., Section 5, Uses of Transposons in Bacterial Genetics and Operon and Gene Fusions). A more complete genetic analysis can then be carried out after this initial identification.

A general attribute of a bacterial strain that classifies it as pathogenic is its ability to metabolize and multiply in or on host tissue with resulting host compromise or damage. Pathogenic bacteria can generally be distributed into three classes based on their invasive potential (Finlay and Falkow 1989). Extracellular parasites multiply outside of cells and are destroyed when phagocytosed. Facultative intracellular parasites can survive inside or outside of cells and thus have mechanisms to avoid intracellular killing. In addition to this property, obligate intracellular parasites require for growth certain metabolites or energy sources found inside of cells. Additional general properties include the ability to gain access to the host location where favorable conditions such as particular nutrients, temperature, and oxygen levels are attainable, the resistance or avoidance of host immune mechanisms, and a method for eventual dissemination.

The initial step of entry into the host is often a direct consequence of the mode of transmission. Typical modes of microbial transmission involve direct contact between hosts, indirect contact as in the case of an insect vector, droplet infection spread as through coughing, or ingestion of contaminated food or water sources. In the latter two instances, the ability of the pathogen to survive outside of a host is a major factor in its transmission.

One of the first steps in establishing an infection is the ability to adhere to, and perhaps multiply on, a specific host surface—often a mucosal membrane. Several bacterial surface structures can generally be implicated in this step. Fimbriae or pili are long polymeric structures that often have a role in colonization by displaying adhesins that bind to host cell receptors at an extended distance from the bacterial surface. Such a distance from the bacterial surface may overcome electrostatic forces that might tend to repel the bacterium from the host cell surface. The often hydrophobic nature of pili may have a further role in host association as well as promoting agglutination between individual bacteria to facilitate microcolony formation. In addition to these organized macromolecular appendages, certain outer membrane proteins have also been implicated in colonization and may act at a level different from that of pilus-mediated colonization. Another well-characterized surface molecule that promotes bacterial adherence is the lipoteichoic acid (LTA) layer of *Streptococcus pyogenes*. This molecule complexes with M protein on the bacterial surface to form very thin hair-like structures termed fibrillae that mediate binding of the bacteria to fibronectin and perhaps other surface molecules on host cells (Beachey and Courtney 1987). Lipoteichoic acid is a constituent of nearly all gram-positive bacteria and may thus generally promote adherence for a number of species within this group. The capsule layer, which is perhaps most notable for its antiphagocytic role in immune avoidance by many bacterial species, can sometimes have the additional role of promoting adherence, perhaps most significantly for bacteria that colonize the oral cavity.

A number of selections and screens have facilitated genetic study of bacterial surface structures. The utilization of cell surface constituents as receptors by bacteriophage can provide a strong selection for mutants since phage resistance can often be correlated with the absence of the receptor. Genetic screens have been developed to assess the presence of certain structures. For example, the interaction of pili with cell surface receptors is often reflected by their ability to agglutinate red blood cells, providing an easy and quantitative assay. The presence of surface structures may cause resistance to certain dyes or other compounds. The use of seroagglutination with antibodies directed against surface components can also provide a useful screen. A change in the expression of surface determinants can manifest itself in ways as simple as a visible change in the properties of the bacterial culture itself, such as the ability of the bacteria to agglutinate to each other in culture and fall out of suspension (Taylor et al. 1987) or the colony morphology on agar medium (Swanson 1982; Swanson and Barrera 1983).

As colonization is established, the functions associated with persisting and multiplying are usually what mediate host damage and define a particular bacterium as being pathogenic. One way in which host damage arises is due to the toxicity of an organism. Toxins elaborated by bacteria may exert their effect in the vicinity of the organism, at distant sites within the host, or may even be ingested as preformed toxins in contaminated food. Bacterial toxins have a variety of roles in pathogenesis. For example, cytolytic toxins may destroy cells, particularly of the immune system, as in the case of leukotoxins. Hemolysins may provide sources of iron and are essentially a subclass of cytolytic toxins. Toxin action may inhibit the function of certain host cells, such as crippling specialized cells like those involved with peristaltic clearance. Crippled or destroyed cells may also leach out additional nutrients for bacterial growth. Ultimately, some toxins may contribute to the dissemination of the bacteria such as in the case of those that promote fluid secretion in diarrhea. The actions of many toxins have been extensively characterized biochemically, which has provided a number of assays that have served as genetic screens. A few selections for toxin-negative derivatives have been established but are relatively specialized (Bramucci and Holmes 1978). In addition to genetic studies on toxin genes themselves, there is much recent interest regarding the mechanisms by which toxins and other molecules are secreted outside of bacterial cells and targeted to host cell receptors.

The invasive capacity of a pathogen is a second major mechanism contributing to host damage during organism multiplication and persistence. The ability of an organism to invade cells and to disseminate is dependent on its overcoming host antibacterial barriers and avoiding the immune system. It must be able to obtain needed nutrients, and indeed the availability of specific nutrients in certain tissues may be the ultimate determinant of the focus of infection. A key metabolite that is scarce in most hosts is iron. Free iron in the host is chelated by molecules such as transferrin and lactoferrin, making it unavailable for bacterial growth. The ability of pathogens to acquire iron from the host through the elaboration of siderophores such as enterobactin is an important determinant of virulence. In fact, the shortage of iron inside the host is so universal that low-iron conditions often act as a signal for inducing the expression of virulence genes.

A number of strategies are utilized to circumvent immune defenses. These include antigenic variation of surface determinants, as in the case of the hundreds of variants displayed by gonococcal pili, and the major outer membrane variations of *Borrelia* that are responsible for the pattern of relapsing fever. Similar in some ways is the alteration of antigenic patterns due to phase variation, where the presence of surface structures is determined by whether they are in the "on" or "off" phase of expression. An example in *Salmonella* is when the "on" or "off" expression translates into an either/or situation for the two antigenic types of flagellin, only one of which is expressed at a time. Some organisms secrete products specifically designed to impede immune mechanisms, such as IgA proteases.

Another surface structure that has an important role in immune avoidance is the elaboration of a nonimmunogenic surface capsular coating. The capsule is usually composed of polymeric carbohydrate, and in addition to being nonimmunogenic, it is often anti-opsonic, preventing the action of antibodies and the promotion of phagocytosis. Another way to avoid the immune system is by direct intracellular spread between cells without exiting one to reenter another. *Shigella* and *Listeria* provide well-studied examples of this mechanism of spread (Galan et al. 1993; Goldberg and Sansonetti 1993; Cossart and Kocks 1994). The intracellular survival mechanism of these and many other bacteria are being studied genetically using screens for intracellular survival and for expression of gene fusion products such as β-galactosidase using substrates that yield a product that can be visualized during intracellular growth (Wiater et al. 1994). Some bacteria such as *Legionella* can actually grow within phagocytic cells due at least in part to their ability to inhibit phagosome/

lysosome fusion. The study of the initial cell entry is again being approached at the genetic level in a number of laboratories. At least several general mechanisms and combinations appear to mediate this event.

A useful selection to isolate genes involved in the cell entry pathway is to introduce a genomic library from an invasive bacterium into a normally noninvasive strain of *Escherichia coli*, incubate these with a susceptible cultured cell line, and then select for bacteria that have entered cells by virtue of their resistance to a nonpermeable antibiotic such as gentamicin (Isberg and Falkow 1985; Isberg 1991). Bacteria that have attained invasive capacity can be isolated from within the cells by gentle detergent extraction and plating. This selection can then greatly aid in the further genetic analysis of the cloned fragment and be useful for assessing its activity when mutations within it are used to replace the wild-type copy in the original pathogenic strain.

Many of the selections and screens outlined here are done by in vitro means that mimic in vivo conditions as closely as is feasible. Of course, the mutants isolated by these methods can then be studied in an appropriate in vivo system. Yet, a question that has been difficult to answer is how to identify just which genes are specifically turned on in the host and to understand the mechanisms that stimulate their expression. The following approaches are being developed to address this question. One is through the use of subtraction libraries constructed after reverse transcribing RNA isolated from bacteria freshly harvested from in vivo growth as compared to in vitro grown organisms (Plum and Curtiss-Clark 1994). A second approach has been termed IVET for in vivo expression technology (Mahan et al. 1993a) (see Section 5, Operon and Gene Fusions). This system relies on formation of gene fusions via integration of a suicide vector containing random cloned fragments from the host bacterial chromosome. The methodology allows for selection of genes expressed in vivo because promoters activated under these conditions result in an active gene fusion that complements an auxotrophic mutation in the host bacterium that normally prevents growth in the animal model. The vector integration event results in a strain that is merodiploid for the gene containing the fusion, which allows for bacterial survival even if the interrupted gene is essential for in vivo survival. Several derivatives of this approach are being tried in different species. These approaches for identifying genes expressed in vivo would not work if the genes involved with virulence were not regulated in some fashion. A new in vivo approach that does not rely on regulated gene expression uses transposons carrying unique DNA sequence tags to specifically identify each individual bacterium within a large mixed inoculum for a given animal model system (Hensel et al. 1995) (see Section 5, Uses of Transposons in Bacterial Genetics). This technique provides a unique so-called signature or name tag to every input bacterium. DNA hybridization comparing the unique sequences of the isolates that survived in the in vivo system to those in the inoculum pool which did not survive provides a method to retrieve bacteria that are defective for in vivo growth and thus have insertions within potential virulence genes.

Once virulence genes are identified by one method or another, much can be learned about their regulation. Usually, the method of choice for such studies is again to use gene fusions. For example, a gene that is positively regulated may not be expressed well in E. coli. Fusing lacZ to such a gene in a lac deletion background would result in a Lac⁻ phenotype. Introducing a genomic library into such a strain can allow for the isolation of Lac⁺ bacteria that have acquired the positive regulator (Miller and Mekalanos 1985). Perhaps more complex in the regulatory issue is just what in vivo signals modulate virulence gene expression and how these signals are perceived by the bacterial cell. Having the virulence genes and their regulators isolated provides a foundation for such studies.

The examples given here provide just a few strategies for how bacteria can colonize and persist in a host and how investigators can approach an understanding of these mechanisms using genetic analyses. For some excellent reviews of pathogenic mechanisms and volumes dealing with various facets of the molecular biology of pathogenesis, see Finlay and Falkow (1989), Miller et al. (1989), Mekalanos (1992), Clark and Bavoil (1994), Salyers and Whitt (1994), and Mandel et al. (1995).

Paradigms of Bacterial Pathogenesis

SALMONELLA TYPHIMURIUM

Salmonella are members of the family Enterobacteriaceae: gram-negative, facultative aerobes, straight rods with peritrichous flagella. *Salmonella typhimurium* and its close relative *Escherichia coli* have been the major workhorses of bacterial genetics since the early 1950s when Norton Zinder and Joshua Lederberg discovered bacteriophage-P22-mediated generalized transduction with *S. typhimurium* strains LT2 and LT7 (Zinder and Lederberg 1952).

Salmonella Classification[1]

The genus *Salmonella* includes a group of closely related bacteria that cause disease in mammals, birds, and reptiles. Different strains of *Salmonella* can cause a variety of diseases, and they are often very host-specific. For many years, the classification of *Salmonella* was based on host specificity, the presence of specific surface antigens (mostly the lipopolysaccharide O antigens and flagellar H antigens), and sensitivity to phage (which also depends on surface characteristics). This resulted in more than 2100 "species" of *Salmonella* named after their preferred host or the place in which they were originally found. However, such surface features do not necessarily reflect overall genetic relatedness because they are subject to strong selection in nature and hence are quite variable (Selander et al. 1990). Comparison of *Salmonella* by using techniques such as DNA-DNA hybridization, protein isozyme analysis, and DNA sequence similarity indicated that most of these *Salmonella* species do not differ from each other sufficiently to be considered separate species. Thus, it has been recommended that all *Salmonella* be classified as a single species,

---

[1]Any classification that does not express an evolutionary idea is nothing but an arbitrary, if convenient, method of pigeonholing facts (P.B. White 1926).

Salmonella enteritica, with numerous serovars (Le Minor and Popoff 1987). Under this new system, the bacterium previously designated *S. typhimurium* would be correctly designated *S. enteritidis*, subspecies *enterica*, serovar *Typhimurium*. However, this nomenclature is extremely cumbersome, and thus it is still commonly abbreviated to *S. typhimurium*.

Salmonella Infections

Salmonella cause two basic types of infections in humans. Many *Salmonella* cause gastroenteritis, a localized infection of the intestinal tract and regional lymph nodes that results in diarrhea usually lasting a few days. Other *Salmonella* cause a potentially lethal systemic infection called typhoid fever. The type of infection caused depends on the host. For example, *S. typhimurium* causes gastroenteritis in otherwise healthy humans, but it causes serious systemic disease in mice and in immunocompromised humans.

The initial infection with *Salmonella* is similar for both types of diseases. *Salmonella* ingested in contaminated food or water first passes through the stomach where conditions are very acidic.[2] After passage from the stomach into the distal ileum, *Salmonella* associates with the epithelial lining where it adheres to and enters the apical membrane of M cells in the Peyer's patches. Although *Salmonella* preferentially enters M cells, it can also enter other types of epithelial cells. The ligand that *Salmonella* recognizes on the eukaryotic cell surface may vary for different types of host cells. Interaction of the bacteria with the host stimulates a complex pathway of signal transduction in the host cell, causing the eukaryotic cell membrane to form ruffles, and produces dramatic cytoskeletal rearrangements at the point of the bacteria–host cell interaction (Finlay et al. 1991; Bliska et al. 1993; Jones et al. 1993). The ruffling and cytoskeletal rearrangements seem to be essential for bacterial entry because (1) mutants unable to cause ruffling are severely defective for entry into the host cells, (2) inhibitors of cytoskeletal protein rearrangements block *Salmonella* entry, and (3) Ca^{++} chelators and Ca^{++}-channel agonists that prevent the cytoskeletal rearrangements also block *Salmonella* entry. The cytoskeletal rearrangements cause "micropinocytosis" of the bacteria into membrane-enclosed vacuoles in the host cell cytoplasm.

Salmonella strains that pass through the basolateral membrane are engulfed by macrophages into membrane-bound phagosomes. Lysosomes may fuse with these phagosomes to form phagolysosomes that elaborate a

---

[2]*Salmonella* is "...transmitted by the fecal-oral route. This is not a pleasant thought from the (human) host's standpoint, but the physiologic burden upon the pathogen is no less, so to speak, unpleasant" (Bliska et al. 1993).

battery of antibacterial conditions, including oxygen radicals, low pH, toxic peptides, and limiting nutrients. Bacteria that cause enteric fever survive this attack and spread to the mesenteric lymph nodes, spleen, and liver where they persist in fixed phagocytes. Escape of bacteria into the bloodstream results in bacteremia. The sustained high fever and other symptoms associated with typhoid fever are probably caused by the release of cytokines stimulated by the bacterial lipopolysaccharide. The mortality for systemic infection with *Salmonella typhi* is approximately 20–50% unless treated with antibiotics. About 2% of the typhoid fever survivors become asymptomatic chronic carriers, continuing to release bacteria from the gallbladder into the intestine. Such carriers may secrete 10^6 bacteria per gram of feces.

Virulence Mechanisms

Most studies on *Salmonella* virulence mechanisms have been done with virulent strains of *S. typhimurium* because it has a practical animal host and a well-developed genetic system. The chromosome of *S. typhimurium* is about 4800 kb (~100 kb larger than the closely related bacterium *E. coli*) (Liu et al. 1993a). Rough estimates from transposon mutagenesis studies suggest that of the approximately 5000 genes on the *S. typhimurium* chromosome, about 4% encode virulence factors (Groisman and Saier 1990). Many of these genes are only expressed under conditions found during infection of the eukaryotic host. For example, expression of genes required for invasion is regulated by the growth conditions, so stationary phase cells and cells grown with high oxygen do not effectively invade eukaryotic cells (Lee and Falkow 1990).

The ability to cross the intestinal epithelial barrier has a dramatic effect on the virulence of *Salmonella* (for review, see Falkow et al. 1992). Genes that encode proteins required for invasion have been identified by (1) screening for mutants that prevent invasion or survival of *S. typhimurium* in tissue culture cells (Finlay et al. 1988; Gahring et al. 1990), (2) screening for clones from wild-type *S. typhimurium* that complement a noninvasive bacterium for invasion of tissue culture cells (Elsinghorst et al. 1989; Galan and Curtiss 1989), and (3) enriching for constitutive mutants that allow efficient invasion of tissue culture cells even after aerobic growth (Lee et al. 1992). A large number of the virulence genes required for invasion are clustered between 58 and 60 minutes on the *S. typhimurium* linkage map, between *srl* and *mutS*. Null mutations in some, but not all, of these genes dramatically decrease virulence in mice. The function of these gene products remains unknown, but based on DNA sequence analysis, the genes seem to encode a complex surface structure that may interact with the eukaryotic cell (Bliska et al. 1993).

The ability to survive in macrophages is also essential for virulence of *Salmonella* (for review, see Groisman and Saier 1990). Several approaches have been used to identify genes required for survival in macrophages. Fields et al. (1986) isolated a large number of random Tn*10* insertions in *S. typhimurium* and subsequently screened each mutant for survival in macrophages in tissue culture. Four major classes of mutants were obtained: (1) mutants with increased sensitivity to oxygen radicals, (2) mutants with increased sensitivity to cationic peptides (called defensins), (3) auxotrophs with mutations in the *aroA* or *purA* genes, and (4) mutants in the global regulatory loci *cya*, *crp*, or *ompR*. All of the mutants with decreased survival in macrophages also had decreased virulence in mice. One of these loci identified the *phoP/phoQ* genes, encoding a two-component system that regulates expression of about 20 genes in response to environmental conditions such as starvation and low pH (Miller et al. 1989). Certain genes, designated *pag*, are activated by PhoPQ. Other genes, designated *prg*, are repressed by PhoPQ. Null mutations in some of the *pag* and *prg* genes dramatically decrease survival in macrophages in tissue culture and virulence in mice, indicating that virulence requires both expression of certain genes and repression other genes at specific times.

Another major class of mutations that decrease virulence affects lipopolysaccharide (LPS) biosynthesis. The intact O-antigen of the *Salmonella* LPS helps protect the bacteria from host defenses by (1) decreasing sensitivity to complement by preventing the activated complement complex from interacting with the bacterial membrane, (2) stimulating interactions with macrophages, and (3) decreasing sensitivity to cationic peptides released in lysosomes.

In addition to the chromosomal virulence genes, some strains of *Salmonella* (e.g., *S. typhimurium*) carry a large plasmid (between 50 and 100 kb, depending on the strain) required for virulence (Gulig and Curtiss 1987; Guiney et al. 1994). Strains cured of the virulence plasmid and strains with specific mutations in the virulence plasmid have decreased virulence in mice but remain competent to invade tissue culture cells. Thus, the genes on the virulence plasmid may be involved in "serum resistance" and possibly growth and survival in the spleen.

Caveats

Studies on *Salmonella* virulence emphasize some problems that should be considered but are often overlooked when studying bacterial pathogenesis (Salyers and Whitt 1994). First, what is true for tissue culture cells may not be true in the host. For example, *S. typhimurium* can be enticed to invade a variety of types of tissue culture cells, but considerable evidence indicates that invasion is much more restrictive in the host. This may have important implications when interpreting the results of tissue culture experiments because different cell types may respond differently to the invading bacterium (Buchmeier and Heffron 1989; Ginocchio et al. 1994). This may explain why certain mutations that seem to affect infection of tissue culture cells have little effect on virulence in the animal host. Second, what is true for one bacterium-host interaction may not be true for another bacterium-host interaction, even if the two bacteria are closely related. For example, *S. typhi* is strictly a human pathogen that causes typhoid fever, a systemic disease. Relatively little is known about the pathogenic mechanisms of *S. typhi* due to the lack of an adequate animal model and of useful genetic tools. Consequently, *S. typhimurium* has been frequently used to infer aspects of typhoid fever in humans, because it causes a typhoid-like disease in mice. However, *S. typhimurium* may not be an adequate model for *S. typhi* pathogenesis because a number of important differences exist between the systemic diseases caused by *S. typhi* and those caused by *S. typhimurium*. Additionally, *S. typhi* clearly lacks some of the pathogenesis functions (such as genes encoded on the large virulence plasmid) required by *S. typhimurium*.

Salmonella typhimurium LT2

Although the original isolate of *S. typhimurium* LT2 caused gastrointestinal infections in humans, the *S. typhimurium* LT2 strains used for genetic studies are no longer appreciably virulent in humans (Sanderson and Hartman 1978). Nevertheless, it is essential to use good microbiological techniques with *S. typhimurium* because it can cause serious infections in immunocompromised hosts and because many of the strains carry drug resistance plasmids or transposons that could conceivably be transferred to other enteric bacteria. Several common sense safety rules should be followed when working with bacteria: Decontaminate all glassware and media with bleach or autoclave prior to routine washing, and wash hands with a bacteriocidal soap after handling cultures.

VIBRIO CHOLERAE

Vibrio cholerae, the causative agent of cholera, has proven to be a model pathogen in which to analyze virulence because of its amenability to a variety of genetic techniques and the immediate potential application of these findings to improved vaccine development. Although cholera remains a serious epidemic disease in various parts of the world, a

suitable anti-cholera vaccine has not yet been developed. *V. cholerae* is a gram-negative curved rod that is highly motile by means of a single polar flagellum. The *V. cholerae* strains that cause epidemic cholera have, until recently, all been of the O1 serogroup and can be divided into two biotypes, classical and El Tor, based on hemolysin production and several biochemical tests. The classical biotype was responsible for essentially all cholera throughout the world until 1961, when it was replaced by the El Tor biotype. In late 1992, the first recorded non-O1 strain to be associated with epidemic cholera was identified in an epidemic that rapidly spread throughout the Indian subcontinent and into other regions of Asia. This strain, designated O139, appears to have arisen from the El Tor biotype and, after briefly replacing the El Tor biotype, is currently coexistent with it in the region of the initial outbreak. Cholera pathogenesis involves oral ingestion through contaminated water or food, survival during passage through the gastric acid barrier of the stomach, colonization of the upper small intestine with concomitant toxin production, and ultimate dissemination in a watery diarrhea. Throughout this process, the bacterium elaborates a number of proven and potential virulence determinants that help it to reach, adhere to, and colonize the intestinal epithelial layer. Some of the virulence factors, such as cholera toxin, have been extremely well characterized, whereas others, for example, those involved in colonization, have more recently been identified.

Cholera Toxin

The main virulence determinant of *V. cholerae* is the secreted exotoxin. This is suggested by the finding that even a small a dose of toxin results in the full-blown diarrhea characteristic of the disease (Levine et al. 1979). In addition, mutants engineered specifically for loss of the toxin genes or their expression are greatly attenuated (Herrington et al. 1988; Levine et al. 1988). The toxin is composed of one A subunit (27,215 m.w.) and five identical B subunits (11,677 m.w.) (Gill 1976; Lospalluto and Finkelstein 1972). The three-dimensional structure of the toxin revealed that the B subunits are arranged as a pentamer in the shape of a doughnut, with the A subunit extending out of the hole in the center (Sixma et al. 1991; Merritt et al. 1994). The B subunits are required for secretion of the toxin out of the bacterial cell (Hirst et al. 1984) and for its interaction with target cell surface ganglioside GM_1 receptors (Cutrecasas 1973). The pentameric structure also likely participates in entry of the toxin A subunit into the target cell. The A subunit is proteolytically nicked to become active (Mekalanos et al. 1979), and its amino-terminal fragment, termed A_1, catalyzes the transfer of the ADP-ribose moiety from NAD to the mammalian adenylate cyclase regulatory protein G_s, thus stimulating an increase in cyclase activity and concomitant level of cAMP in the target cell (Cassel and Pfeuffer 1978; Gill and Meren 1978). The increased cAMP leads to the secretion of ions and fluid responsible for the severe, watery diarrhea characteristic of cholera (Field 1980).

The toxin (*ctx*) genes are located on the CTX genetic element that is composed of a core region carrying the *ctx* genes with flanking RS1 sequences (Mekalanos et al. 1983). RS1 appears to act as a site-specific transposon (Pearson et al. 1993). This element can be introduced into *V. cholerae* strains that do not carry toxin genes, and it will always insert into the same position of the chromosome in a *recA*-independent manner. Classical biotype strains generally have two copies of the element at two different positions in the chromosome. El Tor biotype strains can also carry multiple copies of the element, but these are arranged as tandem duplications that can amplify or decrease in number by homologous recombination requiring *recA* (Goldberg and Mekalanos 1985). The toxin genes, *ctxA* and *ctxB*, are encoded in an operon with *ctxA* being promoter-proximal. The ratio of A to B subunits is apparently achieved by more efficient translation of the *ctxB* open reading frame (Mekalanos et al. 1983; Lockman et al. 1984).

Additional Toxins

Volunteers who have ingested *ctx* deletion strains often still experience residual symptoms including moderate diarrhea, abdominal cramps, and low-grade fever. A search for additional toxins that might contribute to these symptoms has led to the discovery of the Zot and Ace toxins. Zot (zonal occludens toxin) decreases ileal tissue resistance by affecting intercellular tight junctions as detected in Ussing chambers (Fasano et al. 1991). The *zot* gene is located upstream of the *ctx* genes but not within the same transcriptional unit. Yet another toxin, termed Ace (accessory cholera enterotoxin) was identified with the Ussing chamber technique (Trucksis et al. 1993). This toxin causes increased short-circuit current across ileal tissue and also leads to increased fluid secretion in ligated rabbit ileal loops. The *ace* gene is also located on the CTX element. Ace contains a hydrophobic structure that may implicate its ability to insert into target cell membranes. Interestingly, an additional gene has been identified within the element that shows homology with a fimbrial protein involved in colonization by *Aeromonas*. This gene, called *cep* (core-encoded pilin), may contribute to colonization (Pearson et al. 1993). Thus, the CTX element can be considered as a mobile genetic element that carries a virulence gene cassette.

Colonization Factors

Fimbriae (pili) are bacterial surface appendages that mediate adherence of many gram-negative bacterial species to eukaryotic cell surfaces. Genetic and immunological studies have identified a role for *V. cholerae* fimbriae in colonization. It now appears that the O1 and O139 serotypes of *V. cholerae* can elaborate at least two fimbrial types under different growth conditions. One of these, composed of a 20.5-kD subunit, is expressed under the same growth conditions that elicit the highest levels of toxin production and has thus been termed TCP, for toxin coregulated pilus (Taylor et al. 1987). The amino-terminal 25-amino-acid residue region of the mature subunit pilin protein, termed TcpA, is highly homologous to a group of pilin proteins called type 4 pilins (for review, see Strom and Lory 1993). Type 4 pili are implicated in the adhesion of several bacterial species that colonize or interact with mucosal surfaces. To test the role of TCP in colonization, transposon Tn*phoA* (see Section 5, Transposons and Fusions) was used to create insertions within the *tcpA* structural gene. The wild-type O395 and a *tcpA*::Tn*phoA* mutant provided an isogenic pair of strains with which to test the role of TCP in host colonization by *V. cholerae*. An infant mouse cholera model showed an increase in the LD_{50} by approximately 5 logs for the mutant. This was similar to the LD_{50} seen for *ctx* toxin structural gene deletion mutants, demonstrating the central role of TcpA in *V. cholerae* pathogenesis. Subsequent studies using additional defined *tcpA* mutants have established the role of this organelle in the colonization of human volunteers, as well as its universal presence among O1 and O139 strains (Rhine and Taylor 1994).

Hemagglutination is an assay often used for examining the expression of bacterial factors that might potentially mediate cell attachment because the molecules that could serve as receptors on the surface of red blood cells are generally representative of those found on other cell types within the body. In addition to TCP, *V. cholerae* elaborates both soluble and cell-associated hemagglutinins (Hanne and Finkelstein 1982). Recently, the MSHA (mannose-sensitive hemagglutinin) elaborated by the El Tor biotype has been associated with a pilus structure composed of a 16-kD pilin subunit (Jonson et al. 1991). Antibodies directed against this structure protect against challenge by strains of the El Tor biotype in the infant mouse cholera model (Osek et al. 1994). Thus, there are potentially multiple adhesive organelles elaborated by *V. cholerae* that may differ between biotypes and in their roles in colonization. Proof of a role of MSHA pili in colonization awaits genetic analysis.

Yet another set of genes that appear to have a role in colonization are termed the *acf* genes (accessory colonization factor). These genes were identified by Tn*phoA* fusions that exhibited regulatory properties similar to those of the toxin and TCP genes but still expressed these factors (Peterson and Mekalanos 1988). The *acf* genes are clustered together but are transcribed from three different promoters, one of which is involved in expression of the *tcp* genes (Everiss et al. 1994; Brown and Taylor 1995). Mutations in *acf* genes exhibit a tenfold decrease in colonization of the infant mouse cholera model, but the exact reason for this defect is not yet known.

In addition to specific attachment organelles, accumulating evidence indicates that motility has a role in the ability of the organism to colonize. Motile strains show greater adherence, although some nonmotile mutants are actually more virulent than wild type, and others are less virulent in several animal models (Richardson 1991). In addition, the residual diarrhea that led to the discovery of the Zot and Ace toxins was not relieved when the corresponding mutant strains were assessed in human volunteer studies (Tacket et al. 1993). However, potential vaccine strains determined to be less reactinogenic are also less motile. Thus, there appears to be a correlation between motility and the ability to cause some diarrhea simply through heavy colonization of the small intestine (Mekalanos and Sadoff 1994). There may also be a complex inverse relationship between expression of ToxR-regulated virulence genes and motility (Harkey et al. 1994).

Extracellular Secretion of Virulence Factors

The Tn*phoA* approach was further exploited to identify genes required for TCP biogenesis, either through mutagenesis of the cloned chromosomal region adjacent to *tcpA* or by screening mutations isolated throughout the chromosome for loss of properties associated with TCP expression. Analysis of the active fusions identified nine genes that contribute to TCP biogenesis. Most of these lie adjacent and downstream from *tcpA* and appear to be expressed as an operon (Taylor et al. 1988; Brown and Taylor 1995). DNA sequence analysis has identified five additional open reading frames within the *tcp* locus (Ogierman et al. 1993). These genes appear to encode a series of proteins that share homologous domains with a growing family of proteins involved in type 4 pilus biogenesis and extracellular protein export (for review, see Pugsley 1993). Although all the gene products encoded by genes linked to the *tcpA* pilin gene appear to be specific for TCP biogenesis, a second set of genes that encode products required for a more general export pathway has been identified. These genes, termed *eps* (extracellular protein secretion), are required for secretion of cholera toxin, HA/protease, and chitinase (Sandkvist et al. 1993; Overbye et al. 1993). In addition to these export pathways, a third set of speciality export genes has been identified in *V.*

cholerae that are responsible for the biogenesis of the MSHA pilus and share homologies with the other secretion systems (Hase et al. 1994).

Work on TCP biogenesis has shed light on an additional aspect of protein secretion. Tn*phoA* insertion in a gene unlinked to *tcpA* also exhibited a TCP-negative phenotype, yet produced TCP. This gene product, TcpG, was found to encode a member of a new class of proteins, typified by DsbA of *E. coli*, that are required to accomplish physiological rates of disulfide bond formation in secreted proteins (Peek and Taylor 1992; for review, see Bardwell and Beckwith 1993). Mutants with lesions in TcpG are also affected in the secretion of cholera toxin and HA/protease (Peek and Taylor 1992; Yu et al. 1992). Thus, this new class of proteins seems to integrate with the other secretion pathways.

Regulation of Virulence Gene Expression

Toxin, TCP, ACF, and several other potential virulence factors are not expressed constitutively by *V. cholerae* but vary with environmental conditions such as pH, aeration, osmolarity, and amino acids present in the growth medium, as well as the growth phase (Miller and Mekalanos 1988). The first toxin-negative mutants of *V. cholerae* carried lesions not in the structural genes, but in a regulatory gene called *tox*. It is thought that such mutations were isolated preferentially because of the duplication of the CTX genetic element (cholera toxin genes) in most strains. The *tox* locus has been studied in detail and is responsible for most of the regulation of toxin expression in response to environmental conditions (DiRita and Mekalanos 1991; Miller et al. 1987).

When the *ctx* genes were cloned in *E. coli*, it was observed that toxin production was very low, as expected if a positive regulatory element of *V. cholerae* was required for maximal expression. Utilizing a *ctx-lacZ* transcriptional fusion, additional clones were isolated that overcame the expression defect, indicating that the clones encoded an activator of *ctx* transcription (Miller and Mekalanos 1985). When mobilized back into *V. cholerae tox* mutants, the clones restored toxin production. This technique of using a transcriptional fusion in a regulated gene to identify the regulator itself, through both mutation and cloning, is a powerful and widely used strategy. The gene responsible for activating toxin expression was designated *toxR*. Further mapping studies, as well as Northern analysis of *ctx* RNA levels, support the idea that *tox* and *toxR* are the same and that their regulation of toxin expression is at the level of transcription. The *toxR* gene encodes a protein of 32.5 kD that shares an extensive region of homology with the DNA-binding portion of several transcriptional activators involved in bacterial transmembrane regulatory systems (so-called two-component systems) (Miller et al. 1987).

The protein profiles from strains carrying *toxR* null mutations show dramatic effects on the levels of several proteins in addition to the lack of toxin production. Most strikingly, TcpA synthesis is shut off, and there is a flip in the expression of two major outer membrane proteins, OmpT and OmpU; OmpU disappears and OmpT is expressed instead. Additional genes regulated by ToxR have been identified by isolating strains carrying random ToxR-regulated gene fusions. Such a strategy is particularly useful for identifying genes under control of a given gene product or physiological condition. The identification of the *acf* genes which encode products important for *V. cholerae* colonization was accomplished in this manner (Peterson and Mekalanos 1980). It therefore seems that ToxR in *V. cholerae*, like the *vir* gene product of *Bordetella pertussis*, controls a regulon of virulence-determining genes (Miller and Mekalanos 1988).

Perhaps the most striking feature of the ToxR sequence is the presence of a hydrophobic stretch of amino acids indicative of a membrane-spanning domain in the carboxy-terminal third of the molecule. The use of *phoA* fusions helped to demonstrate that ToxR spans the inner membrane of the bacterial cell, with the amino terminus in the cytoplasm and the carboxyl terminus in the periplasm (Miller et al. 1987). This location provides a mechanism for the regulation of gene expression in response to environmental stimuli. The ToxR-PhoA hybrid proteins were still capable of activating expression of ToxR-regulated genes, although this regulation is altered under some conditions, indicating a role of the periplasmic domain in this regulation.

While the membrane location of ToxR is consistent with its role in modulating gene expression in response to environmental stimuli, ToxR does not act independently. The product of a second gene, *toxS*, which is located in the same transcriptional unit and only four base pairs downstream from *toxR*, modulates its action (DiRita and Mekalanos 1991). The product of this gene is an inner membrane protein as well, but with the bulk of its structure on the periplasmic side. The phenotype exhibited by *toxS* insertion mutations is a decrease in the expression of ToxR-regulated genes under some growth conditions. The role of the ToxS protein may be to stabilize ToxR in a dimerized conformation since *toxS* mutants that express ToxR-PhoA hybrid proteins exhibit a ToxS-independent phenotype (PhoA assumes an active conformation as a dimer).

Not all genes within the ToxR virulence regulon are directly activated by ToxR. It turns out that ToxR action on many genes, including the *tcp* genes, occurs indirectly through another regulator termed ToxT (DiRita et al. 1991). The *toxT* gene was isolated using the same screen for activation of *ctx-lacZ* that was used for isolating *toxR*. Thus, it would appear that both ToxR and ToxT directly contribute to activation of *ctx* expression. ToxR has been demonstrated to bind to the *ctx* and *toxT* promoter regions by gel-shift analysis using membrane fractions containing either ToxR or ToxR-PhoA hybrid proteins (Miller et al. 1987; Higgins and DiRita 1994). However, many of the genes within the ToxR virulence regulon appear to be directly activated by ToxT and not by ToxR, leading to a cascade model for regulation of these genes. ToxT is a cytoplasmic protein and belongs to the AraC family of regulators (Higgins et al. 1992). The gene for *toxT* actually lies within the *tcp* gene cluster (Kaufman et al. 1993). Recent studies have determined that *toxT* is in the same transcriptional unit as the rest of the *tcp* genes, suggesting an autoregulatory role for this protein since ToxT, but not ToxR, activates *tcp* gene expression (Brown and Taylor 1995).

Regulation of virulence gene expression outside of the ToxR regulon is responsive to other host signals. One additional regulon responds to iron concentration. A common signal for inducing virulence gene expression in bacteria is lack of iron availability. Identification of iron-regulated genes was accomplished by identifying *phoA* fusions induced under low-iron conditions (Goldberg et al. 1990). These studies identified genes termed *irg* (for iron-regulated genes) and *viu* (for vibriobactin uptake). The common repressor for these genes is the Fur protein, which is homologous to that of *E. coli* (Litwin and Calderwood 1994). It appears that some of these genes, for example, *viuA*, have a role in a siderophore iron uptake system that is crucial for survival of the bacteria under the iron-limiting conditions of the host (Butterton et al. 1992). Other iron-regulated genes are required for full virulence, but their exact roles have yet to be defined.

SECTION 2
Practical Aspects of Microbial Genetics

"In 1971, all of my research was focused on the temperate *Salmonella* phage P22. I identified completely with the phage community. . .I had some of the best scientific interactions in my career with this community and I learned from them. My work was interesting, it was challenging, it was productive, and it was great fun.

"Why, then, did I even think about doing something different? The reasons derived from a general perception that the end of the road was near for phage molecular genetics. . .Those of us without tenure were earnestly warned that we might not be perceived as having a future were we to stick with our prokaryotic intellectual game for too long. It was time to get eukaryotic, to look to the future.

". . .I believe now as I did then that the intellectual case against phage and bacterial genetics was entirely specious. However, the fact remained that most of my peers and betters thought the end was near, and, unfortunately, it was clear that their thinking it being so made it so. Tenure committees, deans, and study sections, then as now, were impressed mainly by conventional wisdom. It is now clear that the whole prokaryotic/eukaryotic argument was silly. . .The impressive progress that has changed the face of prokaryotic biology since was accomplished by what can only be called a supremely talented, courageous, and tenacious skeleton crew. Bacterial biology remains underfunded and underpopulated even today, showing the power of conventional wisdom even in the teeth of contravening facts."

David Botstein
Reprinted from A Phage Geneticist Turns to Yeast.
In *The Early Days of Yeast Genetics* (ed. M.N. Hall and P. Linder), pp. 361–373.
Cold Spring Harbor Laboratory Press, Cold Spring Harbor, NY. (1993)

Genetic Nomenclature

The general conventions for genetic nomenclature were established in 1966 (Demerec et al. 1966) and have been modified over the years as needed, for example, to designate insertions of transposons. Adherence to standards of nomenclature is essential for the bacterial genetics community; our language is arcane enough that communication and understanding depend on a common vocabulary that everybody understands.

GENOTYPE AND PHENOTYPE

Geneticists draw a sharp distinction between genotype, the particular array of genes in a given organism, and phenotype, the physiological manifestation of gene action. For example, a particular strain of *Salmonella* might have a mutational alteration in its *hisG* gene, encoding ATP-phosphoribosyl transferase. This strain has a *hisG* genotype. The physiological consequence of this genotypic defect is that the strain cannot synthesize its own histidine and thus requires exogenous histidine in the culture medium for growth. Therefore, this strain has a His$^-$ phenotype. A strain carrying a fully functional complement of *his* genes (*his*$^+$ genotype) can synthesize its own histidine for growth (His$^+$ phenotype).

Not all phenotypes are designated with a superscript plus (+) or minus (−). For example, sensitivity and resistance to streptomycin are designated by Sm^s and Sm^r, respectively, corresponding to the *rpsL*$^+$ and *rpsL* genotypes.

Enzymes or gene products are often abbreviated on the basis of gene designations. For example, ATP-phosphoribosyl transferase might be abbreviated as "HisG enzyme." However, this could potentially be confused with "HisG phenotype." Thus, some authors denote gene products in capital letters: "HISG enzyme."

GENES AND OPERONS

Genes are designated with three letters that form a mnemonic descriptive of gene function. For example, *his* genes encode enzymes for biosynthesis or uptake of *his*tidine, *put* genes encode enzymes for *p*roline *ut*ilization (as a carbon and nitrogen source), *nar* genes encode enzymes for *n*itr*a*te *r*eductase, and *tox* genes encode regulatory and structural proteins associated with cholera *tox*in production.

Sometimes, a given gene is the only one required for a particular function. For example, *crp* encodes *c*yclic AMP *r*eceptor *p*rotein, and *rho* encodes the transcription termination factor ρ. More often, several genes are required for a given function, as in the case of histidine biosynthesis. In this case, individual genes are designated with capital letters to distinguish them: *hisG*, *hisD*, etc. Note that all letters in a genotype designation are italicized; commonly seen mistakes include *his*G and *his* G. To designate an operon composed of genes with a common role, the capital letters are strung together; for example, the genes for histidine biosynthesis are arranged in the *hisGDCBHAFI* operon.

By convention, the wild-type (functional) version of a gene is denoted with a superscript plus (+), whereas a mutant version of a gene is simply denoted as written, without a superscript minus (–) designation; *hisG* is assumed to mean "mutant *hisG* gene." However, for clarity, it is occasionally useful to include a minus (–) to distinguish the mutant version of a gene from the wild-type.

ALLELE NUMBERS

A challenging problem in bacterial genetics is designating and cataloging multiple mutant versions of the same gene. Different mutations of a given gene are termed alleles, and different alleles are designated with unique allele numbers. Usually, laboratories are assigned blocks of allele numbers by the appropriate genetic stock centers for use with particular mutant genes. Thus, if 50 independent *his* mutations are isolated, and a block of allele numbers from 251 to 300 has been assigned, then mutant no. 1 is designated as *his-251*, mutant no. 2 is designated as *his-252*, and so forth.

A given allele number remains associated with its particular mutation in perpetuity. Thus, *his-251* might initially be designated as a *hisD* mutation (*hisD251*), but subsequent work might show that this designation is incorrect. Although the mutation is reclassified as a *hisG* mutation, it still retains its allele number (*hisG251*) and can be unambiguously traced through the literature and laboratory records. For the Advanced Bacterial

Genetics Course (1991–1995), the *Salmonella* Genetic Stock Center assigned us the following blocks of *S. typhimurium* allele numbers:

srl-251 through *srl-350*
zxx-7201 through *zxx-7300*

STRAIN DESIGNATIONS

Tracking strains through the literature can be frustrating and confusing. For this reason, each laboratory may request a unique strain designation from the appropriate genetic stock center. Thus, *Escherichia coli* strains in the collection of Valley Stewart are designated "VJS," a designation registered with the *E. coli* Genetic Stock Center, Yale University. Likewise, *Salmonella typhimurium* strains in the collection of Stanley Maloy are designated "MST," a designation registered with the *Salmonella* Genetic Stock Center, University of Calgary. For the Advanced Bacterial Genetics Course (1991–1995), the *Salmonella* Genetic Stock Center assigned us the designation "TSM."

Many culture collections assign their own designations to strains obtained from others in order to maintain a consistent usage and nomenclature. For example, all strains used in this manual are denoted with TSM designations, irrespective of their source. However, when publishing work based on a strain obtained from another laboratory, the original strain designation should be used, both to aid the reader in tracking strains through the literature and to give due credit to the person who developed that strain. For example, strain TSM101 (see Appendix A, Bacterial Strains, Phage, and Plasmids) should be designated TSMTT1704 in publications.

DELETIONS

Deletions are indicated with the Greek capital letter Delta (Δ). A deletion within a single gene is written, for example, as $\Delta hisG251$, and a deletion encompassing more than one gene is written as $\Delta(hisGD)253$. A deletion spanning more than one operon is written to show the endpoints of the deletion; for example, a deletion from *hisD* extending through *gnd* and into *rfb* is designated as $\Delta(hisD\text{-}rfb)$ or $\Delta(hisD\ gnd\ rfb)$.

INSERTIONS

Insertions of transposons, insertion sequences, or other DNA elements are designated with a double colon (::). Thus, a particular Tn*10* insertion in the *hisG* gene is designated as *hisG9425*::Tn*10*; note the placement of the allele number. Likewise, a particular Tn*5* insertion is designated as *hisG9647*::Tn5, whereas a MudJ insertion in *hisD* is designated as *hisD9953*::MudJ.

A problem is encountered when designating insertions that are not known to disrupt a gene. Such adjacent insertions are commonly used in bacterial genetics. Designations for these insertions denote the genetic map position (in minutes) by a three-letter code. The first letter of the code is "z," and the second and third letters denote the map position in tens of minutes and minutes, respectively, where a = 0, b = 1, c = 2, etc. Thus, *zaa-1451*::Tn*10* denotes a Tn*10* insertion at minute 0 on the genetic map, *zab-1452*::Tn*10* denotes an insertion at minute 1 on the map, *zcg-1453*::Tn*10* denotes an insertion at minute 26, etc. Again, note the placement of the allele number. Each insertion is assigned to a unique allele number, which does not change even if subsequent experiments show that the insertion is at a different location on the genetic map. An insertion in an unmapped location is designated as *zxx*. Likewise, insertions in the F factor are designated as *zzf*, without regard to the specific location within the element.

FUSIONS

Fusions of genetic elements come in two flavors: operon fusions, in which only the transcription of the reporter gene is controlled by the upstream foreign elements, and gene fusions, in which both transcription and translation initiation are controlled by the upstream elements. Both types of fusions are designated by the Greek capital letter Phi (Φ). The fused elements are enclosed with parentheses. Thus, an operon fusion of *his* to the *lacZ* gene, where *lacZ* is fused within the *hisD* gene, is written as Φ(*hisD-lacZ*)*9953*. Note the placement of the allele number. A gene fusion, in which the amino-terminal coding region of *hisD* is fused in-frame to the amino-terminal portion of *lacZ*, is written as Φ(*hisD' -lacZ$^+$*)*9960* (Hyb), where (Hyb) denotes production of a hybrid protein (HisD-LacZ protein).

SUPPRESSIBLE AND CONDITIONAL MUTATIONS

A particular property of a given allele known to be suppressible or conditional may also be noted: Amber (UGA), ochre (UAA) and opal (UGA) nonsense mutants are designated as Am, Oc, and Op, respectively; for example, *hisG2148*(Oc) is an ochre mutation in *hisG*. Likewise, temperature (heat)-sensitive and cold-sensitive alleles are designated as Ts and Cs, respectively.

DIRECTED DELETIONS AND DUPLICATIONS

Methods to construct deletions or tandem duplications with defined end-points have been developed by Hughes and Roth (1985). In their nomen-clature, each deletion or duplication is assigned a rearrangement number, analogous to an allele number. Directed deletions constructed with Mud insertions are described in brackets adjacent to the rearrangement num-ber. For example, the designation DEL983[(*hisD*)*MudA*(*hisF*)] de-scribes a strain in which recombination between a *hisD*::MudA insertion and a *hisF*::MudA insertion has resulted in a deletion of the chromosomal material between the two insertions. The MudA element resides at the deletion join-point. (This nomenclature represents a special-case excep-tion to the use of Δ to denote deletions and is reserved for directed dele-tions.) Likewise, the designation DUP984[(*hisD*)*MudA*(*hisF*)] de-scribes a strain in which recombination between a *hisD*::MudA insertion and a *hisF*::MudA insertion has resulted in a duplication of the chromo-somal material between the two insertions. The MudA element resides at the duplication join-point. Nomenclature for more complex rearrange-ments is described by Mahan and Roth (1991).

CORRELATING GENETIC AND PHYSICAL MAPS

The genetic maps of enterobacteria are calibrated in 100 intervals termed minutes because it takes about 100 minutes for an *E. coli* Hfr strain to transfer its entire chromosome to a recipient cell. The advent of molecular mapping methods has led to detailed physical maps for many species. These maps are calibrated in 100 intervals termed centisomes. Thus, for a given species, one centisome is equivalent to 1% of the genome length. Centisomes (determined by physical methods) are not strictly equivalent to minutes (determined by recombination), because recombination frequencies vary at different positions. However, the latest version of the combined genetic and physical map of *S. typhimurium* is calibrated in centisomes (Sanderson et al. 1995).

Genome sequencing projects reveal numerous open reading frames of unknown function. Designations for these genes denote the genetic map position (in centisomes) by a three-letter code. In this case, the first letter of the code is "y," and the second and third letters denote the map posi-tion in tens of centisomes and centisomes, respectively, where a and k = 0, b and l = 1, c and m = 2, etc. Sequential open reading frames at a given map position are denoted by consecutive capital letters. Thus, *yejY*, *yejZ*, *yojA*, and *yojB* denote four contiguous open reading frames at centisome 49 on the genetic map.

Microbiological Procedures

Bacterial geneticists must perform a variety of microbiological procedures, including isolation of single colonies and plaques, inoculation and growth of cultures, preparation of culture media, maintenance of sterile media and solutions, and maintenance and cataloging of stock culture collections. However, the actual techniques involved are often vastly different from those employed by classically trained microbiologists. Bacterial geneticists work with large numbers of isolates, and it is not unusual for a single experiment to involve inoculation of dozens of cultures or single-colony isolation of hundreds of individual strains. In this context, the "classical" methods of microbiology are cumbersome and time-consuming.

The following are three keys to successful bacteriological manipulations in the genetics laboratory:

1. *Work quickly.* Have supplies available and organized. Remove caps from bottles and culture tubes for as brief a time as possible. Classical microbiological technique, with its extensive use of flames, necessarily slows down manual operations. In the time it takes to remove the cap from a bottle and pass the bottle mouth several times through a flame, one can remove the cap, insert a pipette, withdraw the desired amount of liquid, and replace the cap. Classical technique may actually promote contamination in some cases.

2. *Be alert.* If isolating mutants from a plate, do not pick a large green fuzzy colony as a new cell-surface mutant. Visually inspect liquid cultures to help determine if they are contaminated. Visually inspect bottles of culture medium, soft agar, etc. before use to determine if the contents are contaminated.

3. *Maintain a clean working environment.* Clean up spills, especially those involving culture media, thoroughly and promptly. Carefully discard and sterilize petri plates that are no longer wanted; old moldy plates are a good source of contamination.

Finally, remember that the organisms used in bacterial genetics grow very quickly and is one of the main reasons they are used. A single *Bacillus* cell that happens to fall into the tube will not take over a culture inoculated with 10^5 *Salmonella* cells. Few airborne contaminants can cause a serious problem in petri dishes that are incubated for only a day or two.

BASIC BACTERIOLOGICAL TECHNIQUES

Be cautious: Although the following discussion is in the context of the *nonpathogenic Salmonella* and *Vibrio* strains described in this manual, additional precautions or alternate procedures *are required* for other strains, particularly those which are *pathogenic.*

Inoculating Liquid Cultures

Touch the end of a sterile applicator stick to a well-isolated colony and then immerse the end of the stick in the liquid medium and agitate briefly. Applicator sticks can be reautoclaved and reused indefinitely.

Single-colony Purification

Touch the broad, flat end of a sterile toothpick to a colony. Patch the colony on a marked petri plate, flip the toothpick over to the sterile side, and "cross the T." Use a fresh sterile toothpick to streak across the bottom of the T about halfway down the plate, leaving a "tail." Flip the toothpick over to the sterile side and finish the streak (Fig. 2.1). Some workers prefer to streak from the center of the plate to the periphery, where there is more room for isolated colonies to grow. With practice, it is possible to purify up to 12 colonies per petri dish—a valuable skill, when dealing with dozens or hundreds of colonies! Toothpicks can be reautoclaved and reused indefinitely.

Spreading Plates

Pipette the solution to be spread into the center of a petri plate. Immerse a glass spreader in a beaker of alcohol and pass the spreader through a flame to burn off the alcohol. (Be extremely careful not to ignite the ethanol in the beaker.) Do not leave the spreader in the flame—the alcohol does the sterilizing, not the flame. Use the spreader to gently spread the liquid over the surface of the plate, being careful not to force all the liquid to the extreme edge of the dish. Allow the plate to stand for a few minutes while the liquid soaks into the agar and then briefly spread again if necessary to smooth out any wet spots.

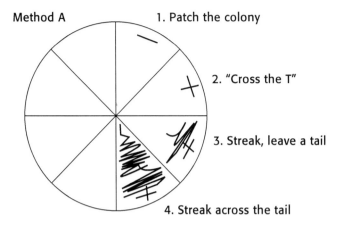

Method A
1. Patch the colony
2. "Cross the T"
3. Streak, leave a tail
4. Streak across the tail

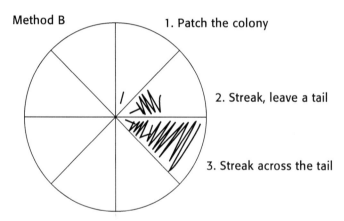

Method B
1. Patch the colony
2. Streak, leave a tail
3. Streak across the tail

Figure 2.1 Single-colony isolation with sterile toothpicks or sticks.

Large-scale Liquid Cultures

Cultures of 10–50 ml often need to be grown for various purposes (e.g., preparation of competent cells and large-scale DNA isolation). It is most convenient to grow a freshly saturated culture of the desired strain, typically overnight. Dilute the culture 1:100 in fresh medium contained in a conical (Erlenmeyer) flask. Flasks fitted with side-arms can be purchased that allow culture density to be monitored in a Klett-Summerson colorimeter or a Spectronic spectrophotometer. The ratio of medium volume to flask volume should be 1:5 or less (e.g., no more than 50 ml in a 250-ml flask) to ensure adequate aeration. Stopper the flask with a foam or cotton plug to allow exchange of gas with the atmosphere. Incubate the culture in a water bath shaker or on a shaker platform at 200–250 rpm.

Serial Dilutions

Many procedures require dilution of a bacterial culture, phage stock, or solution. The most convenient dilutions to make are 1:10 (10^{-1}) and 1:100 (10^{-2}). For 10^{-1} dilutions, pipette 0.1 ml into 0.9 ml of diluent and mix well. For 10^{-2} dilutions, pipette 0.05 ml into 5.0 ml of diluent and mix well. Smaller-scale dilution series can employ microcentrifuge tubes: 10 µl diluted into 90 µl constitutes a 1:10 dilution; 10 µl diluted into 990 µl constitutes a 1:100 dilution. Continue the series until the desired dilution factor is reached.

Adopt a standard description for dilutions for use in keeping notebook records and communicating with lab partners. Remember that a dilution series consists of two parts: the actual dilution itself and the additional dilution made when an aliquot is removed for use. For example, if using 0.1 ml of a 10^{-6} dilution, then the overall dilution factor is 10^{-7}. A useful convention designates such a dilution as 6/.1, indicating that 0.1 ml of a 10^{-6} dilution was used. Likewise, 0.05 ml from a 10^{-4} dilution is 4/.05, and 0.2 ml of a 10^{-1} dilution is 1/.2. What was actually done can thus be recorded and communicated rather than the calculated end result of what was intended.

Replica Plating

This technique was invented by the Lederbergs (Lederberg and Lederberg 1952) to allow testing of many individual colonies for several different phenotypes. Prepare master plates by spreading an appropriate number of cells to yield no more than 300 colonies per plate or by "patching" colonies in a grid pattern to a master plate. Place a sterile (autoclaved and dried) velveteen square face up on a round block whose diameter is slightly smaller than that of the inside of a petri plate (pistons from automobile engines can be used; see Adams 1965). Press the agar face of the master plate gently but firmly into the velveteen, and then follow suit with test plates. Transfer sufficient cells to the velveteen to allow inoculation of several different test plates from a single imprint. This method works best when the master and test plates are somewhat dry, as moisture released from the agar results in a smeared pattern.

Culture Media

Many types of culture media are used in bacterial genetics. Some understanding of the variety of media available, along with their preparation and use, is necessary for full success in designing and completing genetic experiments.

DEFINED MEDIA

Defined media, also known as minimal media or synthetic media, are those in which every component is known in terms of chemical composition and amount. These media contain sources of carbon, nitrogen, phosphorous, sulfur, and magnesium and often include additional trace elements as well. Usually, phosphate serves as the buffer.

A convenient defined medium is medium E of Vogel and Bonner (1956), which is prepared as a 50-fold concentrated stock solution (50x E). This medium contains citrate in addition to the basic elements listed above; citrate is a iron-chelating agent that is efficiently used by enterobacteria. Growth medium is prepared by mixing together sterile 50x E, sterile H_2O and agar, and sterile carbon source (usually glucose). *Salmonella* and many other species can use citrate as a carbon source, however, so E medium is sometimes not useful for selecting for growth on alternate carbon sources. A variation of E medium, termed NCE (no-citrate E), was designed to ameliorate this problem (Berkowitz et al. 1968).

An excellent defined medium for physiology and growth studies was developed by Neidhardt et al. (1974). This medium is buffered with 3-(*N*-morpholino) propane sulfonic acid (MOPS), a good buffer with a pK_a of 7.2. Superior to phosphate buffer in many respects, MOPS allows for the design of a carefully formulated defined medium. This medium provides superior and extremely reproducible growth rates and yields.

COMPLEX MEDIA

Complex media, also known as rich media or undefined media, contain ingredients such as protein digests, infusions, and extracts whose exact chemical composition and amount are not precisely known. Typically, these media provide most growth nutrients in excess and support faster growth rates and higher yields than minimal media. They are suitable for growing cultures prior to genetic manipulations and also as the basis for differential or indicator media. Generally, modern-day complex media are assembled from one or more dehydrated components commercially available from a variety of suppliers. The following are some commonly used components.

Peptone, available in many variations (proteose-peptone, tryptone, etc.), is simply an enzymatic digest of protein. For example, tryptone is a pancreatic digest of casein (milk protein), with conditions controlled to maximize the yield of tryptophan. (Tryptone was developed for use in the indole test, which measures bacterial catabolism of tryptophan.) Peptones consist largely of oligopeptides, which are transported and catabolized efficiently by many bacterial species. Peptones do not contain high levels of vitamins, nucleotides, or minerals.

Casamino acids, also termed acid-hydrolyzed casein, consists of highly purified casein that has been incubated at low pH and high temperature for several hours. This treatment breaks peptide bonds and releases free amino acids, except for tryptophan, which is destroyed by the treatment. (Casamino acids also contains low levels of asparagine, glutamine, and cysteine.) It is possible to obtain highly purified casamino acids, termed vitamin-free, for use as a nutritional supplement in preparing so-called semidefined media.

Yeast extract is the water-soluble portion of autolyzed yeast. This additive is rich in water-soluble vitamins, amino acids, nucleotides, and other organic nutrients. It is often used in conjunction with a peptone to formulate an enriched complex medium.

Infusion is a synonym for extract. Thus, brain-heart infusion is an extract of beef brain and heart and is a particularly rich medium used for fastidious microorganisms. Beef extract is an extract of beef muscle; it is similar to peptone in composition and growth properties.

SPECIFIC COMPLEX MEDIA

Two complex media are used in this manual: Luria-Bertani broth (LB) and nutrient broth (NB). In general, these media are interchangeable for most purposes, although each may have certain advantages for a particular application. LB is marginally more difficult to prepare than nutrient broth, but it promotes slightly faster bacterial growth to slightly higher yields.

Luria-Bertani Broth

This medium, developed by Luria and Bertani (Bertani 1951) for their work on bacteriophage, consists of tryptone, yeast extract, and NaCl. LB promotes very rapid growth and high cell yield for a variety of bacteria and is routinely used as the complex liquid medium in bacterial genetics laboratories. The original Luria and Bertani formula contains 0.1% glucose and 1% NaCl, but most laboratories prepare LB with 0.5% NaCl and no glucose, as described in Appendix B. TY medium contains 20% less tryptone and gives comparable growth rates and yields.

Nutrient Broth

This medium is a premixed formula of beef extract and yeast extract. When supplemented with NaCl, it is similar to LB broth in promoting rapid growth to high final yield. Nutrient agar contains agar in addition to the components of nutrient broth, and thus is particularly convenient for preparing solidified complex medium.

DIFFERENTIAL MEDIA

A variety of media, termed indicator media or differential media, have been developed for special purposes. In general, these media permit equal or near-equal growth of individuals with different phenotypes, but they contain additives that differentiate these alternate phenotypes of individuals in the population. Thus, these media are enormously useful in bacterial genetics for a variety of procedures, including mutant isolation and phenotypic characterization. The indicator media used in this manual are described here.

Tetrazolium Media

Triphenyl tetrazolium chloride (TTC) is a redox-responsive dye that is useful in a variety of biological applications (Fig. 2.2). Oxidized TTC is water soluble and nearly colorless; upon reduction, TTC is converted to an insoluble, dark-red formazan. TTC is reduced by actively metabolizing bacteria, probably at a variety of sites in the respiratory chains.

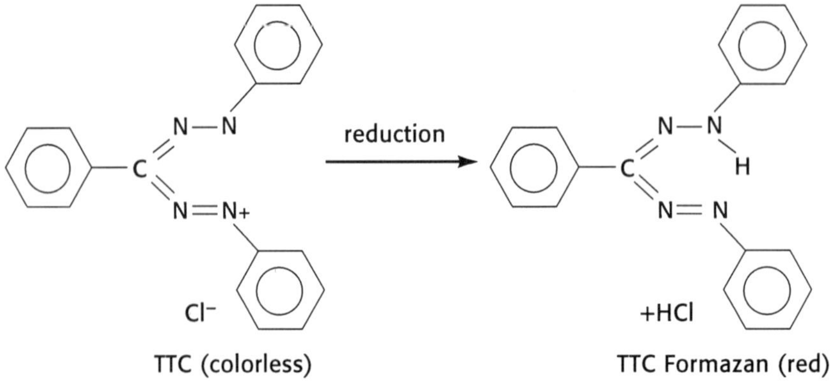

Figure 2.2 Triphenyl tetrazolium chloride (TTC) is colorless, whereas the formazan derivative is brick-red.

Lederberg (1948) was the first to use TTC medium for isolating bacterial mutants. He noted that formazan production is inhibited at low pH. Since carbohydrate utilization by enterobacteria results in strong acidification of the surrounding medium, he reasoned that TTC could discriminate colonies of carbohydrate-utilizing bacteria from those of nonutilizers.

The TTC medium used in this manual is based on nutrient agar, a rich medium that promotes good growth and active metabolism in individual colonies. Supplementation with a high concentration (0.5–1%) of test carbohydrate results in strong acidification in the vicinity of carbohydrate-utilizing colonies. This acidification inhibits TTC reduction (or causes TTC to be reduced to a colorless derivative; B.R. Bochner, pers. comm.) and thus colonies of carbohydrate-utilizing individuals are colorless. Nonutilizers grow well on the nutrient agar, but produce little acid, so the TTC is reduced to red formazan by these colonies. Thus, carbohydrate nonutilizers, the desired mutants in this case, are easily visualized as rare dark-red colonies against a background of white colonies.

A variation of TTC medium was developed by Bochner and Savageau (1977), in which formazan production signals utilization of the test substance. This medium is buffered to maintain neutral pH and contains a limiting amount of nonspecific nutrients (peptone), as well as an excess (0.1–1%) of a test compound. Colonies of individuals that cannot use the test compound are able to grow at the expense of the peptone, but they do not continue active metabolism. Colonies of individuals that can use the test compound continue active metabolism even after the colony has formed; this continued metabolism results in TTC reduction to formazan, so these colonies are colored red. By manipulating medium composition, the indicator (TTC) can thus be used to signal metabolism or lack of metabolism of a particular compound. (This latter medium is not used in this manual.)

MacConkey Agar Medium

A more familiar medium to examine carbohydrate utilization is Mac-Conkey medium, developed by a medical bacteriologist in the early 1900s (MacConkey 1905). This medium is one of several that were developed to differentiate enteric bacteria in stool samples. *Escherichia coli*, a normal inhabitant of the intestine, is Lac⁺, whereas *Salmonella* and many other pathogenic enterics are Lac⁻. Thus, a medium that allows the detection of Lac⁻ enteric bacteria is useful in the presumptive diagnosis of intestinal tract infections.

MacConkey agar contains several components: peptone, which allows growth of both Lac⁺ and Lac⁻ organisms; neutral red and crystal violet, which serve as pH indicators; bile salts, which select against the growth of gram-positive organisms and also participate in the pH-dependent color change; and lactose or other test carbohydrate. Carbohydrate-utilizing organisms form red colonies on this medium, with a precipitated zone of bile salts surrounding the colonies of strongly fermenting organisms. Carbohydrate nonutilizing organisms form white colonies.

The original MacConkey agar recipe includes lactose, but carbohydrate-free MacConkey agar base is commercially available. It is a simple matter to add the carbohydrate of choice to MacConkey agar base to develop a useful indicator for virtually any fermentable carbon source.

XGal

The wide use of *lacZ* gene and operon fusions has made indicators of β-galactosidase activity increasingly useful. MacConkey lactose agar can often be used profitably to examine expression of a particular fusion. Alternatively, a chromogenic substrate of β-galactosidase is useful: 5-bromo-4-chloro-3-indolyl-β-D-galactopyranoside (XGal; Fig. 2.3). This β-galactoside is an efficient substrate for β-galactosidase, which cleaves it to "X" and galactose. Two molecules of "X" then condense, in the presence of molecular oxygen, to form indigo, which of course is dark blue. Thus, Lac⁺ colonies turn blue on XGal-containing medium, and the level of β-galactosidase activity is generally correlated with the intensity of the blue color.

XGal can be used with a variety of complex and defined agar media and is thus the most versatile β-galactosidase indicator available. It is not water soluble, but it is easily dissolved in dimethylformamide prior to its addition to media. One practical drawback of XGal is that it is relatively expensive. In addition, XGal is extremely sensitive, so that it is sometimes difficult to distinguish differences in regulation for highly expressed fusion constructs. Thus, it is advisable to consider alternate indicators (TTC or MacConkey lactose) when possible.

XP

An analogous indicator substrate has been developed for detecting alkaline phosphatase (*phoA* gene product) activity: XP, or 5-bromo-4-chloro-3-indolyl-phosphate. Again, alkaline phosphatase cleaves this substrate to "X" plus phosphate, yielding indigo as an indicator of alkaline phosphatase activity.

COMPOUNDS USEFUL IN *lac* GENETICS

Lactose is the normal substrate for β-galactosidase and is cleaved to form glucose and galactose. XGal is described above. ONPG is the colorimetric substrate for β-galactosidase enzyme assay, cleavage of which forms the dark yellow *o*-nitrophenol. IPTG is a gratuitous inducer of the *lac* operon. The thiogalactoside bond is not cleaved by β-galactosidase, but IPTG efficiently inactivates the LacI repressor, leading to efficient *lac* operon induction (Fig. 2.3).

INDICATOR MEDIA FOR P22 LYSOGENS

Bacteriophage P22 is used for generalized transduction of *S. typhimurium* (see Section 3, Generalized Transduction). One complication of using P22 is that transductants must be tested for the presence of phage, and established to be phage-free, before further work. Strains that are lysogenic for or resistant to phage P22 are useless for subsequent genetic manipulations. Furthermore, colonies arising from transduction experiments often carry phage DNA in an unintegrated state; these pseudolysogens continuously release phage. Indicator media have been developed to differentiate pseudolysogens from nonlysogens and stable lysogens; the latter two classes can be subsequently distinguished by testing for sensitivity to a P22 clear-plaque mutant (see Section 3, P22 Lysogeny and Superinfection Exclusion).

Historically, so-called Green plates have been used by *Salmonella* geneticists to distinguish nonlysogenic colonies and also for plaque assays of P22 itself. More recently, another medium, EBU agar, has been developed for differentiating nonlysogens. The two media probably operate by the same general mechanism. A blue dye present in the medium is internal-

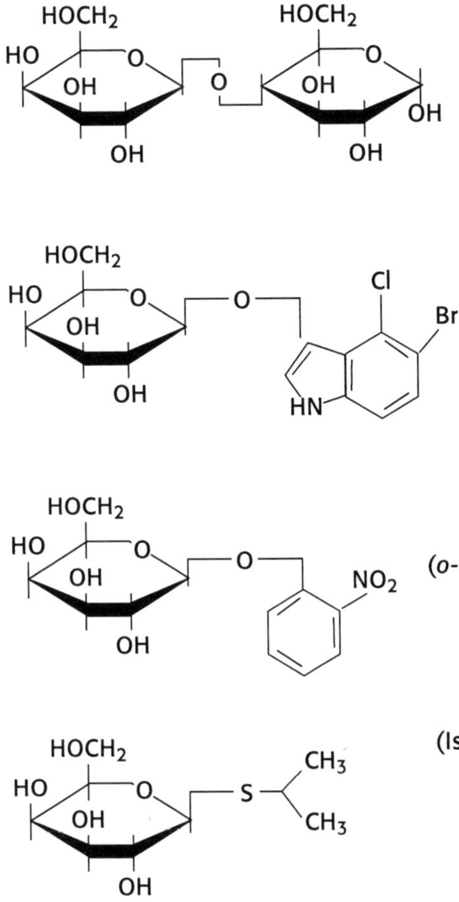

Lactose
(4-*O*-β-D-galactopyranosyl-α-D-glucose)

X-Gal
(5-Bromo-4-chloro-3-indolyl-β-D-galactopyranoside)

ONPG
(*o*-Nitrophenyl-β-D-galactopyranoside)

IPTG
(Isopropyl-β-D-thiogalactopyranoside)

Figure 2.3 Structures of compounds useful in *lac* genetics.

ized by pseudolysogenic cells (phage-infected cells), probably due to changes in cell membrane structure and permeability resulting from the infection (Bochner 1984). Thus, pseudolysogens form dark-colored colonies on these media, and nonlysogens and lysogens form light-colored colonies. Phage-free segregants of pseudolysogens are detected as light-colored colonies in a streak made from the colony of the pseudolysogen.

Green Medium

Many *Salmonella* geneticists use Green plates according to the recipes of Levine (Smith and Levine 1967) or Botstein (Chan et al. 1972), or slight modifications thereof. This medium is based on LB agar and in addition contains a high concentration of glucose and the dyes alizarin yellow GG and methyl blue (aniline blue). Staining of pseudolysogenic colonies apparently results from a complex interaction between the two dyes (Bochner 1984). One drawback of these plates is that they are quite toxic to *Salmonella*; colonies grow poorly and exhibit low viability after extended residence on this medium.

EBU Medium

EBU plates were developed by Bochner (1984) in an attempt to understand the basis for the differential staining of pseudolysogenic colonies and to overcome the toxicity of Green plates. This medium is also based on LB agar and contains a relatively low amount of glucose, phosphate buffer, and the dyes Evans blue and uranine (sodium fluorescein). Pseudolysogens are stained by Evans blue; the uranine serves to increase the colony color distinction. This medium is nontoxic for *Salmonella* and offers superior differentiation of pseudolysogens in comparison with Green plates.

SELECTIVE MEDIA

Distinction is drawn between selective media, upon which only strains of a given phenotype can form colonies, and differential media, upon which strains of different genotypes can form colonies but are differentiated by visual inspection. Defined medium can be a selective medium; for example, His$^+$ strains will form colonies on minimal medium, but His$^-$ strains will not unless the medium is supplemented with histidine. Other types of selective media include antibiotic-containing media, upon which only antibiotic-resistant strains will form colonies.

Fusaric Acid Medium (Bochner-Maloy Medium)

One unusual type of selective medium deserves special attention here, because it has further expanded the usefulness of transposon Tn*10* in genetic experiments. This medium, termed fusaric acid medium or Bochner-Maloy medium, allows for the selection of tetracycline-*sensitive* mutants. Thus, positive selections can be obtained both *for* the presence of Tn*10* (tetracycline resistance) and *against* the presence of Tn*10* (growth on Bochner-Maloy medium).

This medium was originally developed by Barry Bochner (Bochner et al. 1980) and was subsequently modified by Stanley Maloy (Maloy and Nunn 1981). It is based on LB agar and contains phosphate buffer, chlortetracycline, fusaric acid, and $ZnCl_2$. Although the exact mechanism by which these components inhibit the growth of tetracycline-resistant cells is unknown, a specific rationale was used in the original formulation (Bochner et al. 1980). It is known that the tetracycline resistance determinant encoded by Tn*10* results in decreased permeability to the drug (due to active efflux). Thus, Tn*10*-containing cells have a membrane physiology different from that of wild-type cells. Bochner screened a variety of compounds for preferential inhibition of Tn*10*-containing cells. One class consists of lipophilic chelators, such as fusaric acid; this effect is potentiated by Zn^{++}. A noninhibitory inducer of Tn*10*-encoded tetracycline resistance is provided by chlortetracycline, which when autoclaved loses its antibiotic activity but remains effective as an inducer.

Bochner-Maloy medium must be prepared carefully; for example, the time of autoclaving (20 minutes) should be controlled, and the fusaric acid must be well-dissolved in the agar just prior to pouring the plates. However, fresh plates strongly inhibit the growth of tetracycline-resistant cells, particularly at 42°C, and this medium is suitable for most genetic procedures, including replica plating.

Antibiotics, Antibiotic Resistance, and Positive and Negative Selections

Because of the widespread use of transposons and plasmid cloning vectors, bacterial geneticists must use a wide array of antibiotics in various selections. Each antibiotic has its own unique features, and thus it is useful to understand antibiotic action and mechanisms of resistance. The following discussion focuses on the antibiotics used in this manual, and much is drawn from a comprehensive review article (Foster 1983). A series of recent review articles may be consulted for additional perspective (Davies 1994; Nikaido 1994; Spratt 1994).

POSITIVE AND NEGATIVE SELECTION

The ability to select for (positive selection) or against (negative selection) the activity of a specific gene is invaluable in genetics. The term selection is reserved for those situations in which only individuals with the desired phenotype can form colonies (e.g., antibiotic resistance). The term screen refers to situations where colonies expressing the desired phenotype can be visually distinguished from those which do not (e.g., sugar fermentation on a MacConkey plate). Selections are generally preferable to screens, since many more individuals can be examined with many fewer plates. For example, with a strong positive selection, only one plate will be required to recover one resistant mutant from among 10^8 sensitive individuals. In contrast, a screen involving visual inspection of 10^3 colonies per plate would require about 10^5 plates to find that same mutant.

Although selections for antibiotic resistance are perhaps the most common, many other useful positive selections are available, including those for prototrophy and those for carbon-source utilization. The thoughtful geneticist will use any available selection when appropriate. Useful negative selections are less common; some specific examples are described below.

Geneticists use the term counterselection to describe the selective killing of donors and recipients in a genetic cross. For example, mobilization of a Tc^r plasmid might involve the use of an Sm^s *Escherichia coli* donor and an Sm^r *Vibrio cholerae* recipient. The selective medium would contain streptomycin to counterselect the donor *E. coli* cells and tetracycline to select for plasmid-bearing recipient cells. However, the term counterselection is often used as a synonym for the term negative selection.

PHENOTYPIC EXPRESSION

Antibiotic resistance acquired through introduction of a resistance gene is due to the action of the encoded enzyme, which degrades, modifies, or exports the antibiotic in question. In many cases, immediate exposure to a given antibiotic kills the cell before it has time to transcribe, translate, and assemble enough active enzyme to be resistant. In such instances, cultures must be allowed to conduct protein synthesis for a period of time between introduction of the resistance gene and exposure to the antibiotic. This is accomplished by incubating the transduced or transformed culture for about one-half hour in broth medium before plating on antibiotic-containing medium. This delay allows for phenotypic expression of resistance. Most organisms require phenotypic expression before acquiring sufficient resistance to aminoglycoside and β-lactam antibiotics. Enteric bacteria do not require a phenotypic expression period before plating on chloramphenicol and tetracycline, although a phenotypic expression period can enhance recovery of resistant colonies in some cases. When approaching an uncharacterized organism, it is important to determine the necessity of phenotypic expression for each antibiotic used.

AMINOGLYCOSIDE ANTIBIOTICS

Antibiotics of the aminoglycoside-aminocyclitol class are produced by bacteria of the genus *Streptomyces* and related organisms. Specific compounds of this broad group exhibit many structural types; most consist of three or four glycoside residues in a variety of linkages. Three general subgroups are recognized: aminoglycosides containing streptidine (such as streptomycin, Sm), aminoglycosides containing 2-deoxystreptamine (including gentamicin, Gm, and kanamycin, Km), and aminocyclitols (such as spectinomycin, Sp).

Aminoglycoside antibiotics bind to the ribosomes of susceptible cells. Although lethality is due largely to inhibition of protein synthesis, these agents have a broad range of pleiotropic effects that have led to confusion and controversy regarding their mode of action (Davis 1987).

Resistance to aminoglycosides can occur by two types of mutational alterations: those that alter the target site on the ribosome and those that result in reduced antibiotic uptake. Ribosomal mutations can result in resistance to spectinomycin (*rpsE*; ribosomal protein S5) and streptomycin (*rpsL*; ribosomal protein S12). However, single mutations usually do not result in resistance to the 2-deoxystreptamine-containing antibiotics, probably because these compounds bind to multiple sites in their targets. Aminoglycoside uptake requires an energized membrane, so reduced uptake is usually due to mutations in the respiratory chain or in the proton-translocating ATP synthase. Indeed, aminoglycoside antibiotics such as neomycin have been used to select for respiration-deficient mutants.

Transposon- and plasmid-encoded resistance to aminoglycosides is due to enzymes that covalently modify, and thereby inactivate, the antibiotic. Three types of enzymes have been characterized: aminoglycoside-*O*-phosphotransferase (Aph enzyme, encoded by *aph*); aminoglycoside-*O*-nucleotidyl transferase (Aad enzyme, encoded by *aad*); and aminoglycoside-*N*-acetyltransferase (Aac enzyme, encoded by *aac*). The Aph and Aad enzymes use ATP as a cofactor, whereas the Aac enzyme uses acetyl-CoA. Many different subtypes of these three enzymes have been recognized, each of which modifies its substrates at different sites. Thus, although most of these enzymes inactivate a reasonably broad range of different aminoglycosides, they do show considerable specificity for distinct antibiotics.

A Negative Selection

A useful negative selection/counterselection employs the antibiotic streptomycin. Ribosome-encoded Sm^r (*rpsL*) is recessive to Sm^s, so an *rpsL/rpsL$^+$* merodiploid strain is phenotypically Sm^s (Lederberg 1951). Thus, *rpsL* strains carrying a plasmid with the Sm^s *rpsL$^+$* allele fail to grow on streptomycin, whereas plasmid-free segregants regain the Sm^r phenotype and grow. This principle has been used in the design of cloning vectors (Dean 1981) and allele-exchange suicide vectors (Stibitz et al. 1986).

Two features make Sm^s a favorable negative selection. First, spontaneous *rpsL* mutations that confer the Sm^r phenotype are very rare ($\sim$10^{-9} per cell per generation), so there is little problem with background due to spontaneous acquisition of the Sm^r phenotype. Second, the conservation of ribosomal protein structure and function means that the *E. coli rpsL$^+$* gene can be used in a variety of bacterial species.

AMPICILLIN

Antibiotics of the β-lactam class are produced by fungi of the genus *Penicillium*. A variety of β-lactams are produced by this organism, and manipulation of the culture medium can influence the specific type produced by a given culture. In addition, β-lactams can be chemically modified to produce semisynthetic derivatives; ampicillin is one such compound.

β-Lactam antibiotics inhibit penicillin-binding proteins, which are involved in a variety of steps in peptidoglycan biosynthesis. Thus, only growing cells are sensitive to β-lactam action, and penicillin treatment can be used to enrich for auxotrophic and other mutants (Davis 1948; Lederberg and Zinder 1948).

Resistance to β-lactams can occur by mutational alterations in penicillin-binding proteins in some species. Many bacteria have a chromosomally encoded β-lactamase that contributes varying amounts of resistance depending on the species. Thus, for example, *Pseudomonas aeruginosa* and *Klebsiella pneumoniae* are quite resistant to some β-lactams due to high expression of chromosomal β-lactamase. Lack of permeability to β-lactams also contributes to resistance in these species. *Salmonella*, *Escherichia*, and *Vibrio* are relatively sensitive to most β-lactam antibiotics.

Transposon- and plasmid-encoded resistance to β-lactams is due to the action of β-lactamase, which cleaves the β-lactam ring characteristic of this class of compounds. A bewildering variety of β-lactamases have been discovered and characterized, representing a broad range of substrate specificity with respect to natural and semisynthetic β-lactams. Because β-lactamase is a periplasmic enzyme, "leakage" of the enzyme can degrade β-lactams in the medium immediately surrounding a growing colony. Thus, "satellite colonies" are often seen forming around a truly resistant colony, especially on old plates or plates made with relatively low levels of antibiotic.

CHLORAMPHENICOL

The antibiotic chloramphenicol is produced by *Streptomyces venezuelae*, although commercially available chloramphenicol is chemically synthesized. Chloramphenicol diffuses into bacterial cells and inhibits the peptidyl transferase step of polypeptide chain elongation.

Low-level resistance to chloramphenicol can occur by mutational alterations that affect the structure of the outer membrane. Transposon- and plasmid-encoded resistance to chloramphenicol is due to production of chloramphenicol acetyl transferase (Cat enzyme, encoded by *cat*), which covalently modifies and inactivates the drug.

TETRACYCLINE

Antibiotics of the tetracycline class are produced by *Streptomyces* spp. Tetracycline diffuses into bacterial cells and inhibits binding of aminoacyl-tRNA to the ribosome A site. Tetracycline is bacteriostatic rather than bactericidal, a fact to be remembered when using tetracycline in the genetics laboratory. Resistance to tetracycline does not usually occur by mutational alteration. Most transposon- and plasmid-encoded resistance to tetracycline is due to production of a membrane-bound carrier protein (TetA permease, encoded by *tetA*) that catalyzes efflux of tetracycline from the cell. Five distinct classes of tetracycline-efflux systems have been recognized (McMurry et al. 1980; Rubin and Levy 1990).

A Negative Selection

Bochner et al. (1980) determined that cells expressing the Tc^r phenotype are sensitive to lipophilic cation chelators (such as fusaric acid) in the presence of Zn^{++} ion. This provides the basis for a negative selection against Tc^r strains (Bochner et al. 1980; Maloy and Nunn 1981; see Section 2, Culture Media). The ability to select for and against the function of the same gene is both rare and practically useful (see, e.g., Ratzkin and Roth 1978) and has significantly increased the utility of Tc^r elements such as transposon Tn*10*. An important technical issue is that fusaric acid medium is bacteriostatic, and rather heavy background growth can occur on old or crowded plates or on plates that are subject to prolonged incubation. However, carefully prepared fresh medium works well, and this method is widely used.

SUCROSE RESISTANCE: A NEGATIVE SELECTION

The *Bacillus subtilis sacB* gene encodes levansucrase (sucrose:2,6-β-D-fructan 6-β-D-fructosyltransferase). Gram-negative bacteria expressing the *sacB* gene do not grow in culture medium containing 5% sucrose but exhibit no phenotype in the absence of sucrose. Thus, sucrose resistance (Suc^r) provides a useful negative selection (Steinmetz et al. 1983, 1985). This property has been widely used for a variety of counterselection schemes, including allelic replacement. However, the *sacB* gene confers no positive selection in gram-negative bacteria, so useful constructs contain a positive selectable marker as well. Thus, in a two-step allelic exchange, for example, select first for integration of the exchange vector through antibiotic resistance and then for vector-free segregants through sucrose resistance (Ried and Collmer 1987; Quandt and Hynes 1993). The Suc^r phenotype has found utility in a variety of other contexts as well (Gay et al. 1985; Lawes and Maloy 1995).

Bacterial Physiology

The ultimate goal of genetics is to understand the function of the organism in question. Thus, the ability of mutant organisms to carry out a particular process ultimately needs to be examined, whether it be growth under a given set of conditions, regulation of gene expression in response to appropriate signals, or cell-cell interactions involved in virulence. Careful attention to various aspects of bacterial physiology both aid genetic analysis and make it more meaningful. Indeed, contextual understanding of results from genetic and biochemical approaches is achieved only when hypotheses derived from such studies are tested in vivo (see, e.g., Yanfosky et al. 1984; Yanofsky and Horn 1994).

PHYSIOLOGY AND METABOLISM

Detailed knowledge of the physiological process under study is essential for designing meaningful genetic selections and screens. For example, selection of regulatory mutants in the *trp* operon of *Escherichia coli* depended on sophisticated understanding of the Trp biosynthetic pathway and enzymes. Mutations that were key in defining the attenuation mechanism were isolated in selections for decreased *trp* operon expression. This selection took advantage of the knowledge that 5-methyltryptophan inhibits growth. The Trp enzymes convert the precursor 5-methylanthranilate into 5-methyltryptophan. Under appropriate conditions, mutants that are resistant to 5-methylanthranilate have decreased *trp* operon expression; by making less of the Trp biosynthetic enzymes, these mutants accumulate less 5-methyltryptophan and therefore form colonies (Zurawski et al. 1978). This selection would not have been possible without understanding the Trp biosynthetic pathway. Somewhat analogous logic, using the inhibitor aminotriazole, allowed the isolation of mutations in the *Salmonella typhimurium his* operon attenuator (Johnston and Roth 1981). As a final example, consideration of central metabolism allowed the selection of mutants that form disulfide bonds in the cytoplasm. This selection was based on the observation that alkaline phosphatase, a periplasmic enzyme, requires the formation of two intrachain disulfide bonds for activity. Selections were made whereby alkaline phosphatase, acting as a phosphomonoesterase, was able to substitute for the function of fructose-1,6-bisphosphatase in the glycolytic pathway (Derman et al. 1993).

Even seemingly simple aspects of bacterial physiology can profoundly affect a process of interest. For example, the ability of *Salmonella* to invade cultured epithelial cells is maximized by anaerobic growth of the bacterial culture and is enhanced in exponentially growing cells (Ernst et al. 1990; Lee and Falkow 1990). This observation led to an enrichment scheme that identified a previously unknown locus, *hil*, that is involved in the invasion process (Lee et al. 1992). Thus, careful attention to growth conditions and media can greatly facilitate the study of diverse bacterial processes.

Finally, although physiology can aid genetics, the converse is also true. For example, careful physiological studies with well-characterized mutants have revealed the relative influence that repression, attenuation, feedback inhibition, and uptake have not only on *trp* operon expression, but also on the organism's ability to adapt to shifts in the nutritional status of the culture medium (Yanofsky et al. 1984, 1991; Yanofsky and Horn 1994). Such integrated studies provide a more comprehensive understanding of the process under investigation.

GROWTH OF CULTURES

A bacterial culture will grow at a characteristic rate in a given growth medium and temperature. In the laboratory, growth of a culture is typically initiated by inoculating a small volume of a saturated (overnight) culture into a large volume of sterile medium, the dilution factor typically being 1:50 or greater. After a lag phase, ensuing growth of the culture will be unrestricted as long as nutrients are present in excess and toxic products have not accumulated. After an adaptation period of such unrestricted growth, cultures enter a state of balanced growth during which all cell components (DNA, RNA, protein, lipid, etc.) increase at the same rate. The doubling time of each cell component is the same as the doubling time of the population during this balanced growth (Neidhardt et al. 1990; Fishov et al. 1995). Studies of induced enzyme synthesis and other physiological parameters should be made during balanced growth to ensure that other perturbations or influences do not affect the observed phenomena. Additionally, studies with cultures in balanced growth yield the most reproducible results when comparing different experiments. It should be recognized that balanced growth is largely a laboratory phenomenon and rarely occurs for long periods of time in natural environments (see, e.g., Abshire and Neidhardt 1993). In batch cultures, cells eventually leave balanced growth as the composition of the culture medium changes and enter into stationary phase. Therefore, growth conditions must be closely monitored to ensure that measurements are made on balanced cultures.

A critical factor in evaluating growth curves is the relationship between cell number and cell mass. Slow-growing cells are smaller than fast-growing cells, and most measurements of cell growth (e.g., optical density; see below) actually measure the total mass per milliliter of culture. Small stationary phase cells undergo a period of mass increase before initiating cell division, and late exponential phase cultures continue cell division for a period of time after mass has ceased to increase (Maaløe and Kjeldgaard 1966; Neidhardt et al. 1990). Thus, sampling cultures for both mass and cell number yields nonsuperimposable growth curves, and the rates of increase in cell number and mass are equivalent only in balanced cultures.

Choice of Culture Medium

Cultures for physiological experiments are best grown in a defined medium rather than broth medium, so that the influences of specific medium components can be monitored. A superior medium developed by Neidhardt et al. (1974) for such purposes can be modified in a variety of ways to accommodate specific growth parameters (Wanner et al. 1977; Stewart and Parales 1988). Important features to be considered include buffering capacity and initial pH, nature of the carbon, nitrogen, and phosphorus sources, inclusion of nutritional supplements, and nature of the respiratory electron acceptor. These parameters are considered in detail by Neidhardt et al. (1974, 1990).

Aeration

This parameter is critical and often unappreciated, as shake flasks are generally considered to be "aerobic." However, the rate of oxygen diffusion into liquid medium is considerably lower than the rate of oxygen consumption by many bacterial species, including enterobacteria. Sufficient aeration of batch cultures requires the use of small volumes of medium combined with indentations (baffles) on the bottom of the flask, both of which increase the surface area and thus the rate of oxygen transfer. Alternatively, test tube cultures grown on rollers are reasonably well-aerated if the volume of medium is kept low. Even so, aerated cultures are aerobic only during the initial phases of growth, when cell density is relatively low. The influence of culture aeration can have surprising effects on diverse phenotypic properties (e.g., epithelial cell invasion; Ernst et al. 1990; Lee and Falkow 1990).

It is worth bearing in mind that many bacterial species are adapted to anaerobic environments. For example, the mammalian intestine is anaerobic and supports the growth of high densities of strictly anaerobic bacteria; enterobacteria represent a small proportion of the total bacterial mass in feces. Thus, studies on anaerobic cultures in many cases may in fact be more relevant with respect to the natural habitat.

Monitoring Culture Density

The most convenient means to determine the concentration of bacterial cells in a culture is to measure the optical density (OD) with a spectrophotometer or similar device. OD and biomass are directly correlated over the linear range of the instrument. Several considerations are critical to ensure accurate measurements (Fewson et al. 1984). Cells of most species actually absorb little of the incident light (photosynthetic bacteria are obvious exceptions), so the greatest contribution to the measured OD value is actually the amount of light that is scattered by the suspension. Thus, the geometry of the spectrophotometer has a critical influence on the measurements made. An instrument whose cuvette is relatively close to the photodetector will record much more of the scattered light than one whose cuvette is relatively distant. Thus, maximum sensitivity is achieved when the cuvette is as close as possible to the light source, and therefore correspondingly distant from the detector (Fewson et al. 1984).

It is therefore not sufficient to report culture densities solely in terms of OD without also specifying the make and model of the spectrophotometer used. Using another laboratory's value for OD may actually result in a culture of quite different density! Even supposedly identical instruments can give significantly different values, due to differences in light source (the age of the bulb) and calibration. For this reason, a calibration curve must be constructed when using a new spectrophotometer. This is most conveniently done by comparing OD values to colony-forming units per milliliter for cultures in a variety of stages of growth, but measurements of dry weight or other parameters can also be employed. The essentially arbitrary OD values may then be converted into meaningful (and comparable) measures of cell mass or number. Calibration curves should be updated periodically to account for changes in the spectrophotometer, strains, or culture media. In addition, calibration curves based on colony-forming units should be constructed for the range of culture media employed, as the mass per cell changes as a function of growth rate.

The wavelength at which culture density is measured is another important parameter to consider. A shorter wavelength (e.g., 420 nm) provides greater sensitivity, but linear range is rather narrow. A longer wavelength (e.g., 650 nm) increases the linear range over which measurements can be made but is correspondingly less sensitive at low culture densities (Neidhardt et al. 1990). In any case, measurements of relatively high cell concentrations require dilution of the sample prior to determination to ensure that the measurement falls within the linear range of the instrument (Neidhardt et al. 1974).

An effective alternative to a spectrophotometer is provided by the Klett-Summerson photoelectric colorimeter. This simple, sturdy, and inexpensive instrument provides reliable day-to-day performance and accurate measures of culture densities over a wide linear range. Interchangeable filters allow measurements to be made at short or long wavelengths; the red (~660 nm) filter is recommended for routine use. Klett meters are used in conjunction with sidearm flasks; a portion of the culture is tipped into the sidearm, and the sidearm is placed directly into the instrument for measurement. This eliminates the need for withdrawing a portion of culture into a cuvette each time a measurement is made. "Klett units" provide an arbitrary measure of concentration, but certainly no more arbitrary than OD units do, as outlined above. In either case, a calibration curve is essential.

MEASURING ENZYME ACTIVITY

Often, the ultimate goal of growing bacterial cultures is to measure the activity of a specific enzyme, which provides a measure of gene expression in that particular culture. Studies of gene regulation routinely employ *lacZ* operon fusion strains; these strains contain a variety of regulatory mutations and are grown in a variety of media. Attention to the details of the enzyme assay is just as important as attention to the culture conditions to ensure meaningful and reproducible results.

With the wide use of operon and gene fusions, the most commonly performed enzyme assays are those for β-galactosidase and alkaline phosphatase. Both assays share several features in common, and both employ a chromogenic substrate, enzymatic cleavage of which yields the yellow-colored *o*-nitrophenol (ONP). The extent of the reaction can be easily monitored by simply observing the appearance of yellow color, and thus samples of unknown activity can be conveniently assayed. Both assays can be performed in permeabilized cells, obviating the need to prepare cell-free extracts. Other enzyme assay protocols can be modified for use with permeabilized cells. Finally, both assays are sensitive and highly reproducible.

The standard measure of enzyme activity is specific activity, which relates the amount of reaction product formed per minute to the total amount of extract added to the reaction. Specific activity is often reported in terms of product formed per minute per milligram of protein; however, with the permeabilized cell assays, specific activity is reported in terms of product formed per minute per OD unit of the culture assayed. Since OD is proportional to biomass (over the linear range of measurement), this is equivalent to specific activity, although the units are arbitrary.

Enzyme assays must demonstrate linearity with time and with amount of extract added over the range of times and amounts that will be routinely employed. A major cause of nonlinearity comes from allowing the reaction to proceed for too long or from using too much extract; this results in formation of so much product that its concentration exceeds the linear range that the spectrophotometer can measure. Some assays are nonlinear with time due to instability of the enzyme or inhibition of the reaction by its product, which of course accumulates with time. Nonlinearity with amount of extract is sometimes due to inhibitors present in the extract. Fortunately, the β-galactosidase and alkaline phosphatase assays are extremely linear with respect to both time and amount of extract. However, when first setting up these assays, it is good practice to ensure that this is true by assaying a variety of extract volumes for varying periods of time. This simple experiment will provide a good feel for the range of variables tolerated by the assay.

Each sample should be assayed in duplicate (or triplicate), and the values averaged to generate the reported specific activity. Assays in which the duplicate specific activities vary by more than 10% should be repeated. It is also good practice to set up the duplicates with two different volumes of extract, so that the linearity of the assay can be directly monitored. For example, assay 0.1 ml and 0.2 ml of the same extract for the same amount of time. The assay with 0.1 ml of extract should yield half as much product formed per minute as the assay with 0.2 ml of extract. The final specific activities should be very similar, however, since the calculations correct for the amount of extract added. Finally, all assays should be repeated at least once, with different cultures on a different day, to ensure that the observed values are reproducible.

MEASURING THE DIFFERENTIAL RATE OF ENZYME SYNTHESIS

The above section describes endpoint measurements of enzyme activities, whereby cultures are grown to a particular density and the specific activity in that culture is then measured. A more sensitive way to monitor gene expression is to measure the differential rate of enzyme synthesis (Monod et al. 1952; Cohn 1957). This procedure is particularly useful in comparing strains that exhibit only slight differences in gene expression (Stewart and Parales 1988; Egan and Stewart 1990; Mecsas et al. 1993) or in monitoring induction of weakly expressed or regulated genes (Stewart and Yanofsky 1986).

A typical differential rate experiment begins with establishing early exponential phase cultures of the strains to be assayed. At predetermined time points (generally 5–10 minutes apart), the culture density is measured and a sample (0.5–1 ml) is withdrawn into an equal volume of prewarmed medium containing chloramphenicol. After a few minutes' incubation, the sample is placed on ice. This incubation period allows time for assembly of monomeric polypeptide chains into the active oligomeric species (Kepes 1969); β-galactosidase, for example, is active as a tetramer. Often during such an experiment, the first two or three time points are taken from an uninduced culture to establish the basal rate of enzyme synthesis. Subsequent addition of inducer then allows measurement of the induced rate of enzyme synthesis. Each sample is then permeabilized and assayed for enzyme activity as described above. However, specific activity calculations are not made. Rather, the values for amount of product formed per minute per milliliter of culture (dZ) are plotted on the ordinate (y axis), and the measured culture density for each time point (dB) is plotted on the abscissa (x axis). The slope of the resulting line, dZ/dB, gives the differential rate of synthesis in total enzyme activity per milliliter of culture as a function of the total culture mass per milliliter, i.e., the culture density (Monod et al. 1952).

This method has several advantages over the endpoint assay described above. First, measuring the rate of enzyme synthesis helps to average out fluctuations in measurements, so a more precise value is obtained. Of course, differential rate measurements need to be performed on cultures in balanced growth. Second, the plotted rates are independent of the doubling time with which a culture grows, because activity is plotted against the increase in culture mass rather than the time of growth. Thus, strains or culture conditions that have substantially different doubling times can be directly compared. Finally, as mentioned, subtle differences in rates of gene expression can be convincingly documented. The disadvantage of this method is that it is more time-consuming than the endpoint assay, and fewer cultures can be examined in a given day. Thus, this method is generally used only to document specific phenomena selectively, whereas endpoint assays are used for more routine situations.

SECTION 3
Genetic Mapping

"Transduction was discovered by Norton Zinder and Joshua Lederberg as a result of a systematic search for genetic recombination in the genus *Salmonella*. Lederberg began studying *Salmonella* genetics soon after assuming his first faculty position at the University of Wisconsin. Work with *Salmonella* genetics seemed to Lederberg a logical extension of his work with *Escherichia coli*, since the *Salmonella* group is a close taxonomic neighbor of *Escherichia* but has much greater medical significance. . .Lederberg was only 23 years old, and Zinder was 19.

". . .K. Lilleengen in Scandanavia had used various bacteriophages to 'type' numerous strains of the species *Salmonella typhimurium*. . .in the summer of 1950 Zinder began making crosses in all possible combinations between the various Lilleengen serotypes. In all, one hundred pairwise combinations of the 20 types were made. . .Fifteen combinations yielded prototrophs, most of which involved strain LT-22, especially when paired with strain LT-2. Subsequently, it would be shown that strain LT-22 produced a phage (initially called PLT-22 and later P22) active against LT-2. Bacterial strain LT-2 and phage strain P22 would become the foundation for the extensive research on *Salmonella* genetics that this initial work would stimulate."

Thomas D. Brock
Reprinted from The Emergence of Bacterial Genetics, pp. 189–191.
Cold Spring Harbor Laboratory Press, Cold Spring Harbor, NY. (1990)

Genetic Mapping of Chromosomal Genes

One of the first objectives when a previously unknown gene is discovered is to locate it on the genetic map. This information is important for a variety of reasons. First, it determines whether the gene is truly novel or whether it has previously been identified in another context. Second, the location of the gene with respect to its neighbors may provide important clues as to its function and/or regulation. Finally, knowledge of the genetic map position allows the genetic manipulation of a region, for example, to isolate deletions, perform localized mutagenesis, and construct partial diploids for complementation analysis.

MAPPING BY CONJUGATION AND TRANSDUCTION

Traditionally, genetic mapping has been accomplished by Hfr (high frequency of recombination) crosses. The donor Hfr strain carries a conjugative element (e.g., F factor) integrated into the chromosome at a unique position. The donor transfers DNA progressively to the recipient from a fixed location, and with a defined orientation. With an appropriate recipient, recombinants can be selected and scored for inheritance of the donor allele of the gene in question. With appropriate crosses and selections, the newly discovered gene can usually be located to within a few minutes (percent) of the genetic map.

Hfr mapping works well with organisms such as *Escherichia coli* and *Salmonella typhimurium*, for which many Hfr strains are available and for which hundreds of selectable markers have been mapped (Singer et al. 1989; Low 1991). Hfr-type mapping has also been developed for other organisms, such as *Pseudomonas aeruginosa* (O'Hoy and Krishnapillai 1987) and *Rhodobacter sphaeroides* (Suwanto and Kaplan 1992). The development of a Tn5 derivative carrying the *mob* (mobilization) region of the broad host range plasmid RP4 (Simon 1984) has extended Hfr mapping to a variety of other organisms (see, e.g., Klein et al. 1992; Stibitz and Carbonetti 1994).

Once localized by conjugation, the gene of interest is then precisely mapped by generalized transduction with nearby transposon insertions. Useful collections of transposon insertions have been assembled for a variety of organisms, including *E. coli* (Singer et al. 1989), *S. typhimurium* (Kukral et al. 1987; Sanderson and Roth 1988), and *Bacillus subtilis* (Vandeyar and Zahler 1986).

PHYSICAL MAPPING

The high-resolution physical map of *E. coli* (Kohara et al. 1987) was constructed by restriction mapping an ordered set of genomic clones. This laborious procedure has not yet been applied to other species. However, the introduction of pulsed-field gel electrophoresis (PFG) techniques has allowed the construction of low-resolution physical maps for a number of bacterial species, including *E. coli* (Daniels 1990; Heath et al. 1992; Perkins et al. 1992, 1993). In general, a relatively low-resolution physical map is constructed by digesting chromosomal DNA with rare-cutting restriction enzymes and then resolving the fragments by PFG. Specific genes are located on the physical map by comparing restriction digests from a variety of strains carrying defined insertions or by Southern blotting with a variety of cloned DNA probes (Smith and Condemine 1990; Cole and Saint-Girons 1994). Physical maps are divided into 100 physical map units, termed centisomes, that are analogous to the 100 genetic map units, termed minutes.

Correlating Physical and Genetic Maps

A comprehensive physical map, correlated with the genetic map, was constructed for *S. typhimurium* by performing double digests with combinations of the enzymes *Bln*I, *Ceu*I, and *Xba*I (Liu et al. 1993a; Sanderson et al. 1995). The *S. typhimurium* map serves as a basis for constructing maps of other salmonellae, such as *S. enteritidis* strain SSU7998 (Liu et al. 1993b), *S. paratyphi* B (Liu et al. 1994), and *S. typhi* (Liu and Sanderson 1995). The rationale and comprehensive approach used for these maps provide a good model for map construction in general.

The appeal of physical mapping is that sophisticated genetic techniques are not needed to construct at least a crude physical map. However, the precision of the map is greatly enhanced by prior genetic analysis, which also allows exact localization of specific genes. For example, many well-characterized Tn*10* insertions are available in *S. typhimurium*. Tn*10* contains sites for both *Bln*I and *Xba*I, so a genome carrying a Tn*10* insertion has one extra site for each enzyme, these sites being easily localized by PFG mapping. Knowledge of the genetic map position of the insertion thus allows correlation of the genetic and physical maps (Liu and Sanderson 1992; Wong and McClelland 1992a; Liu et al. 1993a). Likewise, a Tn*5* derivative carrying rare sites has also been used (Wong and McClelland 1992b). Similar strategies have been employed in the analysis of *E. coli* (Heath et al. 1992; Perkins et al. 1992, 1993), *R. sphaeroides* (Suwanto and Kaplan 1992), and *P. aeruginosa* (Holloway et al. 1994).

Practical Aspects of Physical Mapping

The choice of restriction enzyme depends on the species. For example, physical maps of the *S. typhimurium* genome have been constructed with the enzymes *Bln*I, which cuts at CCTAGG (Wong and McClelland 1992a); *Xba*I, which cuts at TCTAGA (Liu and Sanderson 1992); and *Ceu*I, which cuts within a 26-bp conserved sequence in each of the seven rRNA operons (Liu et al. 1993a). The utility of the first two enzymes resides in the fact that CTAG sequences are rare in enterobacterial genomes. A different battery of restriction endonucleases was used to construct a physical map of the *R. sphaeroides* genome, which has a high GC content: *Ase*I, which cuts at ATTAAT; *Dra*I, which cuts at TTTAAA; *Sna*BI, which cuts at TACGTA; and *Spe*I, which cuts at ACTAGT (Suwanto and Kaplan 1989).

Intact chromosomal DNA is prepared by lysing bacterial cells within an agarose block, and restriction endonuclease digestion is likewise performed in situ. The block is then loaded into an agarose gel, which is subjected to an electric current of alternating direction for 24 hours or more. High-molecular-weight DNA is not separated by molecular sieving in such a gel; rather, the DNA "snakes" end-on through the gel. The alternating direction of the current causes the DNA to reorient within the gel; larger molecules take longer for reorientation. Thus, size separation is based on the relative time that each fragment spends reorienting versus the time it spends moving toward the cathode; smaller fragments migrate more rapidly. The amount of time that the current is oriented in a particular direction, the pulse time, is the major variable in this method.

Theoretical aspects of PFG have been extensively analyzed (Cantor et al. 1988; Mathew et al. 1988a,b,c; for review, see Cantor et al. 1987), and detailed protocols for a variety of applications have been developed (Smith and Cantor 1987; Smith et al. 1993). Several variations on the PFG principle have been developed and given distinct names (e.g., PFGE, OFAGE, FIGE, and CHEF). Each works on the same basic principle but differs in the geometry and duration of the alternated electric fields (Smith and Condemine 1990). In practice, the choice of a specific method is often dictated by the ready availability of a particular electrophoresis apparatus.

Mud-P22 MAPPING IN *S. Typhimurium*

A general method for genetic mapping in *S. typhimurium* takes advantage of bacteriophage Mud-P22. Although specific for *S. typhimurium*, the

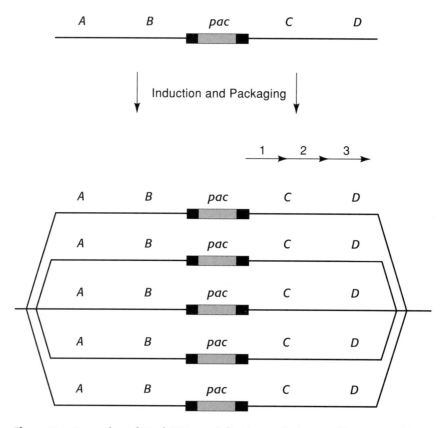

Figure 3.1 Formation of Mud-P22 specialized transducing particles. An excision-deficient Mud-P22 lysogen is induced for DNA synthesis by treatment with mitomycin C. (*Closed boxes*) Mud sequences; (*gray rectangles*) P22 sequences. DNA replication creates "onionskin" amplification in the vicinity of the Mud-P22 insertion. Unidirectional sequential headful encapsulation initiating from the *pac* site results in packaging of markers to one side of the insertion. Thus, phage formed from headful 2 transduce marker *C*, those from headful 3 transduce marker *D*, etc. (Redrawn, with permission, from Benson and Goldman 1992.)

general principle can in theory be applied to other organisms for which appropriate phage have been characterized and engineered. Phage P22 initiates DNA encapsulation unidirectionally from a unique site, *pac*, within the phage genome; once initiated, subsequent encapsulation does not require specific sites (see Section 3, Generalized Transduction). Thus, a P22-defective prophage (Mud-P22), incapable of excision, will encapsulate chromosomal DNA preferentially to one side of its insertion site. Induction of such a defective lysogen (by mitomycin C, for example) causes massive DNA replication initiated from the prophage and results in "onionskin" amplification of DNA in the vicinity of the prophage (Fig. 3.1). Simultaneous phage protein synthesis and encapsulation yields a specialized transducing lysate for the immediately adjacent 3–5-minute region.

The general application of this technique was made possible by the work of Youderian et al. (1988), who constructed a matched pair of hybrid phages, termed Mud-P22s. These elements consist of P22 bacteriophage that are proficient for prophage induction, DNA synthesis, virion protein synthesis, and encapsulation but are deficient in integration and excision. These defective P22 genomes are flanked by the right and left ends of bacteriophage Mu. The MudP element is oriented such that encapsulation proceeds toward the left end of Mu, whereas encapsulation from the MudQ element proceeds toward the right end of Mu. Finally, both MudP and MudQ contain the *cat* gene, allowing selection for chloramphenicol resistance. Donor Mud-P22 elements are carried on F′ elements, which carry no homology with the *S. typhimurium* chromosome. An F′::Mud-P22 insertion can be transduced into a strain carrying a specific MudA or MudJ insertion, with selection for the donor Cm^r marker. Homologous recombination between the Mu ends of Mud-P22 and the Mu ends of the recipient MudA or MudJ insertion allows Mud-P22 to replace the MudA or MudJ insertion, which is monitored by loss of the recipient Ap^r or Km^r marker (Fig. 3.2). Thus, Mud-P22 can be placed at any predetermined position in the genome, with packaging in either orientation. The only general limitation is the availability of specific MudA and MudJ insertions.

A collection of more than 50 MudP and MudQ insertions has been assembled for genetic mapping in *S. typhimurium* (Benson and Goldman 1992). This collection covers the entire genome and greatly simplifies mapping of newly discovered genes in this species. Lysates from each of the strains are placed in individual wells of a microtiter dish, the recipient is spread on selective medium, and portions of each lysate are spotted onto the recipient in a grid pattern. Infection and transduction occur on the agar surface, and wild-type recombinants appear in the positions of lysates that are enriched for DNA corresponding to the site of the recipi-

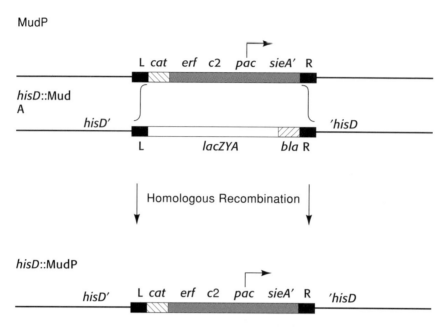

Figure 3.2 Conversion of a *hisD*::MudA insertion into a *hisD*::MudQ insertion by homologous recombination. (Redrawn, with permission, from Youderian et al. 1988.)

recipient mutation. This method allows determination of the position of an unknown mutation to within 5 genetic minutes or so with a single plate. Subsequent conventional transductional crosses with markers from the indicated region then serve to locate the previously unmapped mutation precisely.

The Mud-P22 elements lack gene *9*, which encodes the P22 tail fiber. Thus, the induced Mud-P22 lysates must be "tailed" before they can be used for transduction. Fortunately, tailing is a simple procedure: Portions of a crude extract from a strain carrying a gene *9* on an expression plasmid are simply mixed with each of the mitomycin-C-induced Mud-P22 lysates. For methods of preparing and tailing Mud-P22 lysates, along with a list of available strains for mapping, see Appendix D (Rapid Mapping in *S. typhimurium* with Mud-P22 Prophages).

Generalized Transduction

Generalized transduction is one of the most important techniques for genetic analysis of bacteria. It is invaluable for genetic mapping, strain construction, mutagenesis, and other procedures. Sophisticated *genetic* analysis (as distinct from molecular cloning) can begin in earnest once a system for generalized transduction has been established.

Generalized transduction was discovered by Zinder and Lederberg (1952) during their studies on genetic exchange in *Salmonella typhimurium*. The use of transduction in genetic analysis was pioneered by Demerec and colleagues at Cold Spring Harbor Laboratory (Demerec et al. 1955; Demerec and Ozeki 1959). Nearly 40 years later, strain LT2 and bacteriophage P22 remain the most useful and versatile host/phage system for generalized transduction. This utility is due to distinct features of P22 that make it particularly well-suited for transduction (see below). However, quite useful generalized transduction systems have also been developed for *Escherichia coli, Pseudomonas aeruginosa, Bacillus subtilis, Caulobacter crescentus*, and *Rhizobium meliloti*, to name only a few examples. In principle, it should be possible to develop generalized transduction for any bacterial species if diverse bacteriophage strains active on a given species are successfully obtained and tested.

BACTERIOPHAGE P22

P22 is a member of the "lambdoid" group of bacteriophages, temperate phages that share genetic organization and even homology with bacteriophage λ (Lambda). Researchers who are comfortable with the biology and life history of λ have a good head start in understanding P22. However, P22 differs from λ in many respects, two of which are essential for generalized transduction. (λ does not normally mediate generalized transduction.) One feature of P22 that allows its use for generalized transduction is its headful packaging mechanism. A second feature, and the one that makes P22 the most useful of transducing phage, is the availability of high transduction frequency (HT) mutants. These two features are detailed below.

The genetic map of P22 is very analogous to that of λ, although functionally homologous genes usually have different designations in the two phage (Fig. 3.3).

P22 Regulation and Packaging

Lambdoid phage enjoy a dual lifestyle. First, they can grow lytically by infecting a susceptible cell and replicating themselves a hundredfold or so prior to lysing the cell and releasing the newly formed particles. Second, they can integrate their genomes into the chromosome of the host cell, entering a quiescent state termed lysogeny. The lysogenic state (prophage) is stable for an indefinite number of bacterial generations. However, the prophage can escape the lysogenic state in response to environ-

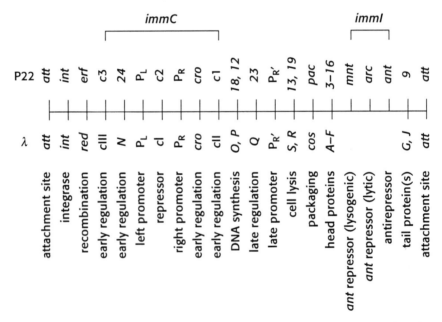

Figure 3.3 Comparison of bacteriophage P22 and λ genetic maps. Not to scale. Functionally homologous genes and sites occupy analogous positions in the two maps. Note that P22 has two immunity regions: *immC*, analogous to the λ immunity region, and *immI*, which encodes antirepressor and accessory regulatory elements.

mental cues (typically DNA damage) by excising the phage genome from that of the host, and entering the lytic cycle.

Maintenance of the prophage state requires the major repressor, termed *c*I in λ and *c*2 in P22. Repressor bound to the left and right operators prevents transcription of genes encoding lytic functions: structural genes for phage head and tail proteins, and genes whose products are involved in DNA synthesis, phage maturation, and so on. (For thorough reviews of the details of phage regulation, see Hendrix et al. 1983; Poteete 1988; Ptashne 1992; Susskind and Botstein 1978.)

Phage DNA replication proceeds through the rolling circle mechanism, which results in the formation of a long iterated series of genomes, or concatemers. An empty phage head, associated with the enzyme terminase, binds to a site within the concatemer (*cos* in λ; *pac* in P22). Terminase makes a double-strand cut at this site, and DNA packaging proceeds (Casjens and Hayden 1988; Schmieger et al. 1990). At this point, an essential difference is found between λ and P22. In λ, the prohead/terminase complex continues along the concatemer until a second *cos* site is encountered. Terminase cuts at this *cos*, completing packaging of the first particle and allowing a second particle to begin packaging. Thus, bacteriophage λ DNA consists of discrete, uniform molecules, each containing precisely one genome equivalent bounded by cut *cos* sites. In P22, on the other hand, terminase initiates the packaging series by cutting at *pac*, but packaging proceeds by a headful mechanism rather than by scanning for a second *pac* site (Fig. 3.4). One P22 headful of DNA corresponds to approximately 102% of the genome size. Thus, P22 phage DNA consists of nonuniform molecules that are circularly permuted and terminally redundant (Tye et al. 1974).

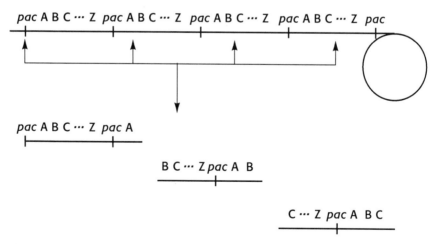

Figure 3.4 Mechanism of sequential headful encapsulation.

P22-mediated Generalized Transduction

Generalized transduction is a consequence of bacterial DNA being packaged into phage particles during lytic growth. In the case of P22, a packaging series occasionally initiates at bacterial sequences that resemble the *pac* site (pseudo-*pac* sites). Once the packaging series has initiated at a pseudo-*pac* site, prohead/terminase complexes continue to bind to DNA and package sequential portions of bacterial DNA until the packaging series spontaneously terminates (after ~5–10 headfuls of DNA have been encapsulated). Thus, transducing particles consist exclusively of bacterial DNA, although they contain all of the phage proteins necessary for infection. Any lysate of P22 will therefore contain a small number of transducing particles, each containing a portion of the bacterial chromosome. One headful of P22 DNA is about 44 kb, so each transducing particle contains approximately 1% (1 minute) of bacterial DNA.

Generalized transduction occurs when a transducing particle infects a susceptible cell and injects its (bacterial) DNA. In approximately 10% of these infections, the injected DNA is incorporated into the recipient chromosome by homologous recombination (Ebel-Tsipis et al. 1972). Thus, if the recipient carries a mutational alteration in, for example, the *his* operon, and the transducing particle is from a lysate that was grown on a *his*+ donor strain, then recombination can replace the recipient *his* lesion with *his*+ material (Fig. 3.5). This *his*+ recombinant can be recovered by plating the transduction mixture on agar medium that lacks histidine, so that only *his*+ colonies will grow. Frequency of transduction is calculated as the proportion of *his*+ colonies per plaque-forming unit (pfu) of phage added.

As mentioned above, this integrative recombination occurs for only

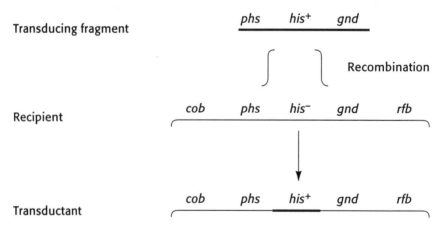

Figure 3.5 Generalized transduction. Two recombination events result in replacement of recipient information with donor information.

about 10% of the transducing particles. The resulting transductants are termed complete transductants. The other 90% are termed abortive transductants (Demerec and Ozeki 1959; Ozeki 1959). In these cells, the transducing DNA is stably maintained, probably in tight association with a P22-encoded protein that protects the DNA ends from exonucleolytic attack. This DNA is not replicated, however, and only one daughter cell will inherit the transducing DNA. Of course, in this example, this his^+ DNA is directing the synthesis of histidine biosynthetic enzymes, and thus the other daughter will be able to divide a few more times before the histidine biosynthetic enzymes are diluted out. Abortive transductants therefore form microcolonies that can be observed with a dissecting microscope.

P22 HT Mutants

Genetic experiments suggest that the *S. typhimurium* LT2 genome contains about seven pseudo-*pac* sites (Schmieger 1982). Thus, the frequency of transduction of any given marker by wild-type P22 depends on the orientation and the position of that marker relative to a pseudo-*pac* site (Chelala and Margolin 1974; Krajewska-Grynkiewicz and Klopotowski 1979; Roth and Hartman 1965). Genetic analysis also shows that, for a given marker, a small number of discrete transducing particles contain the marker, as predicted by the limited number of pseudo-*pac* sites (Roth and Hartman 1965; Mandecki et al. 1986).

Schmieger (1972) isolated mutants of P22 that exhibited increased frequencies of generalized transduction. In the best of these mutants, approximately 50% of the phage in a given lysate are transducing particles! Furthermore, virtually all bacterial markers are transduced at high frequency, in contrast to wild-type P22, where the frequency of marker transduction strongly depends on its location relative to the pseudo-*pac* sites. Thus, it seems that these P22 HT mutants have a substantially altered specificity for DNA packaging and that bacterial and phage DNAs are packaged with approximately equal frequencies.

Genetic analysis of HT mutants has shown that the increased transduction frequencies are due to mutational alterations in gene *3*, which encodes terminase (Raj et al. 1974; Jackson et al. 1982; Casjens et al. 1987). This is fully consistent with the idea that these mutants have altered packaging specificity, since it is terminase that initially recognizes *pac* (or pseudo-*pac*) and makes the cut to initiate the encapsulation series (Casjens and Hayden 1988). Physical analysis of one mutant, HT12/4, indicated that the altered terminase of this mutant recognizes two cryptic *pac* sites within P22 itself (Tye 1976; Casjens et al. 1987). Although this finding explains previous observations concerning the nature of phage DNA in HT12/4 particles, it does not explain the increased frequency of generalized transduction, because the *S. typhimurium* genome is not expected to contain large numbers of these cryptic *pac* sites. (*pac* and cryptic *pac* sites are 8–12 bp long, so their random occurrence by chance is small.) Genetic analysis also suggested that HT mutants have an altered but nonrandom specificity for packaging bacterial DNA (Chelala and Margolin 1976; Schmieger and Backhaus 1976).

Regardless of mechanism, the availability of HT mutants makes P22 unsurpassed as a vector for generalized transduction; similarly efficient mutants are not available for other transducing phages, such as coliphage P1. The very high frequency of transduction for all markers ensures success and allows for sophisticated genetic experiments requiring rare double transduction events (Schmid and Roth 1983; Hughes and Roth 1985). Most *Salmonella* geneticists use the HT105/1 mutant for generalized transduction.

P22 Lysogeny and Superinfection Exclusion

One potential complication of using P22 for generalized transduction is that P22 lysogens are immune to P22 infection. This immunity is due to two superinfection exclusion systems, encoded by the *sieA* and *sieB* regions, respectively (Susskind and Botstein 1978), and to production of c2 repressor by the lysogen. Thus, if a transductant is isolated from a genetic cross, there is a finite chance that that particular cell will also have been lysogenized by P22. Such a P22 lysogen is worthless for further genetic analysis, because it cannot serve either as a donor or as a recipient for further crosses.

This complication can be overcome by using a P22 HT105/1 phage that also carries a second mutational alteration, termed *int-201*. The *int* gene encodes integrase, which is essential for the establishment of stable lysogeny by catalyzing integration of the phage genome into that of the host. However, even *int-201* phage can form unstable pseudolysogens that continuously release phage (Fig. 3.6). P22 HT105/1 *int-201* (hereafter termed P22 HT *int*) was constructed by Gary Roberts in John Roth's laboratory then at the University of California-Berkeley.

It is therefore essential to isolate and characterize phage-free and non-lysogenic clones of each desired transductant. This is done by using an indicator medium (Green plates or EBU plates) that differentiates pseudolysogenic colonies from those that are not releasing phage and by testing transductant colonies for their sensitivity to a c2 mutant of P22 (P22 H5), which differentiates lysogens from nonlysogens.

EBU plates are described more fully in Section 2 (Indicator Media for P22 Lysogens). Briefly, several transductants from a given cross are

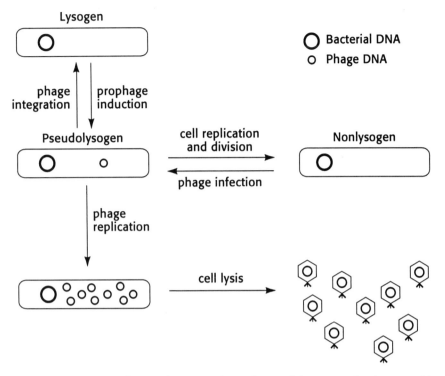

Figure 3.6 Formation and segregation of pseudolysogens. (Redrawn, with permission, from Bochner 1984.)

streaked for single colonies on EBU plates, which are incubated overnight. Pseudolysogens form dark-blue colonies on this medium, whereas nonlysogens and stable lysogens form easily distinguished light-colored colonies. Several light-colored colonies are then streaked across a line of the P22 H5 lysate that has been placed on an EBU plate (Fig. 3.7). Nonlysogens are fully sensitive to P22 H5, which grows lytically due to its defect in *c2*, encoding the major repressor. Thus, the nonlysogenic cells are killed, and little or no growth in the streak occurs once the cells have passed over the P22 H5. In contrast, lysogens are fully immune to P22 H5 infection because lysogens contain c2 protein synthesized by the resident prophage. When a lysogen is infected by phage H5, the c2 repressor already present binds to the H5 operators and prevents lytic growth. Lysogens are also immune to H5 infection because of superinfection exclusion. Therefore, lysogens (and P22-resistant cells) will survive and grow in the streak even after exposure to P22 H5.

After each transductional cross, it is therefore essential to isolate phage-free, phage-sensitive clones for further work not only to allow subsequent genetic manipulations via P22-mediated transduction, but also to

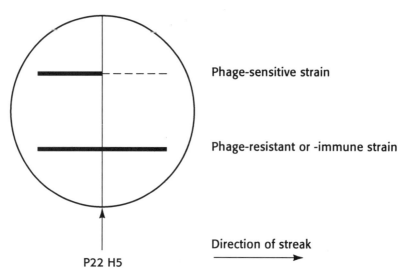

Figure 3.7 Testing colonies for phage sensitivity.

ensure that pseudolysogeny does not affect the observed properties of the strain in question. For example, a pseudolysogenic strain is unlikely to be as virulent as an otherwise identical phage-free strain, so pseudolysogeny can potentially (and unknowingly) complicate data interpretation in studies designed to assess the role of a specific gene product in pathogenesis.

Plasmid Transduction by P22

P22 can also be used to transduce plasmids from one strain of *Salmonella* to another. Upon infection of a plasmid-carrying host, the P22 *abc* gene product (analogous to the *gam* gene product of bacteriophage λ) inhibits the host RecBCD function, allowing the plasmid to undergo rolling circle replication (Poteete et al. 1988). This concatemeric plasmid DNA is efficiently encapsulated. The resulting lysate can be used to infect the desired recipient strain, whereupon the linear concatemeric plasmid DNA recircularizes by homologous recombination. This process requires a cloned *pac* site when using P22 standard, but it is *pac*-independent when using P22 HT (Orbach and Jackson 1982; Schmieger 1982).

EXTENSION OF P22 HOST RANGE TO OTHER SPECIES

The use of P22 generalized transduction has until recently been limited to *S. typhimurium*, largely because of the unique structure of the O antigen component of *S. typhimurium* lipopolysaccharide (LPS), which provides the cell-surface receptor for P22 adsorption. O antigen is a polysaccharide, and its synthesis is directed by the *rfb* gene cluster (which encodes enzymes for synthesis of the oligosaccharide units) and the *rfc* gene (which encodes the enzyme responsible for polymerization of the oligosaccharide units).

The potential for extending P22 host range to other genera comes from recent work by Neal et al. (1993), who have constructed a cosmid that carries the *S. typhimurium rfb* cluster and the *rfc* gene. *Escherichia coli* strains carrying this cosmid are sensitive to P22 infection, and thus chromosomal genes can be transduced by P22. In principle, this cosmid should allow the host range of P22 to be extended to any species in which it is capable of replicating and packaging. One potential complication might arise if the organism's indigenous LPS is of sufficient length or structure as to occlude the *S. typhimurium* LPS and prevent its recognition by P22. In such cases, rough mutants lacking LPS would have to be employed.

P22 has not been useful for transductional analysis of the human pathogen *Salmonella typhi*. Plasmids can be transduced into *S. typhi*, indicating that P22 can recognize *S. typhi* LPS and inject DNA. Recent progress has come with the recognition that *S. typhi* strains that are deficient in methyl-directed mismatch repair (*mutL* or *mutS* mutants) act as efficient recipients for interspecies transduction from *S. typhimurium* (Zahrt et al. 1994). However, P22 does not replicate in *S. typhi*, so additional modifications will be necessary to allow P22 transduction between different strains of *S. typhi*.

BACTERIOPHAGE P1

Phage P1 was isolated from a lysogenic strain of *E. coli* (Bertani 1951), and its use as a generalized transducing phage followed soon after the discovery of P22-mediated transduction (Lennox 1955; Yanofsky and Lennox 1959). Since then, P1 has been the workhorse of *E. coli* genetics. Although generalized transduction by P1 is superficially similar to that by P22, P1 is not a lambdoid phage, and certain of its properties as a transducing phage are substantially different from those of P22. First, the P1 genome is approximately 100 kb, slightly more than twice the size of the P22 genome. Thus, P1 transducing particles carry 2 minutes' worth of bacterial DNA, which allows linkage analysis of more widely spaced markers. Second, the mechanism of P1 encapsulation differs from that of P22 encapsulation, particularly in terms of *pac* site recognition (Sternberg and Coulby 1987a,b, 1990). Two practical consequences of this difference are that transducing particles do not arise from recognition of pseudo-*pac* sites in the host chromosome, but rather seem to be packaged more at random (Masters 1977; Newman and Masters 1980; Masters et al. 1984), and stable HT-type (high-frequency transducing) mutants have not been isolated (Wall and Harriman 1974). Finally, although P1 is a temperate phage, lysogens carry the P1 genome as an autonomous plasmid (Yarmolinsky and Sternberg 1988).

A most useful property of phage P1 is its potentially wide host range. Some strains, such as *Klebsiella oxytoca* (*pneumoniae*) M5al, are naturally sensitive to P1 infection. It is also possible to isolate P1-sensitive mutants of normally resistant enterobacteria by infecting with a drug-resistant version of P1 and directly selecting for lysogens (Goldberg et al. 1974). *Salmonella* represents a special case. Mutant strains lacking galactose epimerase (*galE*) produce a truncated LPS that is effectively recognized by the P1 tail fiber. These strains are resistant to P22. However, growth of a *galE* mutant in the presence of 1% glucose and 1% galactose restores synthesis of the normal LPS. Thus, *S. typhimurium galE* mutants can be infected by P1, or by P22, depending on the composition of the medium in which they are grown (Enomoto and Stocker 1974; Ornellas and Stocker 1974).

Finally, P1 can be useful even for species in which it fails to replicate. P1 will adsorb to cells of *Myxococcus xanthus* and inject its DNA, but it fails to replicate or to establish lysogeny. Even so, P1 serves as a convenient means of introducing plasmids from *E. coli* to *M. xanthus*, by growing P1 on the plasmid-containing *E. coli* strain and using the resultant lysate to transduce recipient *M. xanthus* cells (Shimkets et al. 1983).

ISOLATION OF TRANSDUCING PHAGE

Generalized transducing phage have been isolated for a number of bacterial species. Two general approaches have been taken to find such phage. One method is to identify lysogenic strains of a given species, recover the resident temperate phage, and test it for transduction. This is in essence the method used to identify phages P22 and P1 (Bertani 1951; Zinder and Lederberg 1952). Similarly, useful transducing phage for *P. aeruginosa* were initially recovered from lysogens (Holloway 1969).

Alternatively, environmental samples may be screened for phage that are active on the species of interest; such samples should be characteristic of the environment in which that species is normally found. For example, the *M. xanthus* transducing phage Mx4 and Mx8 were identified in collections of myxophages that had been isolated from soil and dung, two characteristic habitats for myxobacteria (Campos et al. 1978; Martin et al. 1978). Indeed, soil samples from around the world were solicited by Dale Kaiser in the "Dirt for Dale" crusade (Casjens and Manoil 1991). Successful isolation of transducing phage for *C. crescentus*, whose normal habitat is fresh water, came with the realization that tropical fish are shipped in the water from which they were captured, and therefore that pet stores contain water samples from all over the world (Ely and Johnson 1977; Ely and Shapiro 1989). Finally, transducing phage for *R. meliloti*, the symbiant of alfalfa, were isolated from alfalfa field soils or alfalfa inoculants (Finan et al. 1984; Martin and Long 1984), whereas phage for the plant pathogen *Xanthomonas campestris* were isolated from soil samples (Weiss et al. 1994). Even if a transducing phage cannot be developed, phage can often be put to other uses (see, e.g., Hatfull and Sarkis 1993).

Genetic Mapping by Generalized Transduction

One important application of generalized transduction is in constructing genetic maps. The frequency with which two markers are coinherited in a transductional cross depends on the distance between them, so measurements of cotransduction frequency represent the simplest means of expressing genetic distances. This is also the most useful representation for the practicing bacterial geneticist, because frequency of cotransduction is an essential value to know when carrying out transductional crosses for strain construction, backcrossing mutations, etc. Furthermore, analysis of cotransduction frequencies among three or more markers (three-point cross) allows deduction of the relative order of the markers on the genetic map.

DETERMINING GENETIC LINKAGE AND GENE ORDER FROM COTRANSDUCTION DATA

The cotransduction frequency of two markers provides a measure of their linkage (i.e., the genetic distance between them). In any given transductional cross, one (dominant) marker serves for positive selection of recombinants, and the coinheritance of unselected markers is measured. Crosses involving *srl* and adjacent markers serve to illustrate general principles.

Consider a series of P22 HT *int* crosses involving the markers *srl-251*::MudJ (marker Km^r), *zgc-7203*::Tn*10*d(Tc) (marker Tc^r), and *zgc-1623*::Tn*10*d(Cm) (marker Cm^r). In the first cross, P22 HT *int* was grown on a strain carrying Cm^r and used to transduce a strain carrying Km^r and Tc^r. Chloramphenicol-resistant transductants were selected, purified, and scored for inheritance of the donor markers Km^s and Tc^s. The following results were obtained:

Cross No. 1   Donor: Cm^r Tc^s Km^s Recipient: Cm^s Tc^r Km^r

Class	Cm	Tc	Km	Number
1	Cm^r	Tc^r	Km^r	146
2	Cm^r	Tc^s	Km^r	40
3	Cm^r	Tc^s	Km^s	12
4	Cm^r	Tc^r	Km^s	2

$$\text{Cotransduction between Cm}^r \text{ and Tc}^s = \frac{\text{class 2 + class 3}}{\text{total}} \quad \frac{40 + 12}{200} = 26\%$$

$$\text{Cotransduction between Cm}^r \text{ and Km}^s = \frac{\text{class 3 + class 4}}{\text{total}} \quad \frac{12 + 2}{200} = 7\%$$

Thus, 26% of all Cm^r recombinants also inherited the donor Tc^s marker, and 7% of all Cm^r recombinants also inherited the donor Km^s marker. Since map distance = (1 − cotransduction frequency), the linkage between Cm^r and Tc^s is 0.74 and that between Cm^r and Km^s is 0.93.

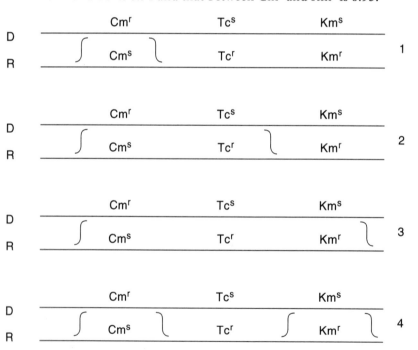

Figure 3.8 Three-point cross, example 1. The Cm^r Tc^s Km^s donor (D) is crossed with the Cm^s Tc^r Km^r recipient (R), selection being made for Cm^r. Recombination events leading to each of the four recombinant classes are diagrammed schematically.

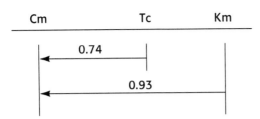

Figure 3.9 Genetic map derived from example 1. By convention, the arrowhead points to the selected marker.

Once the cotransduction frequencies are calculated, the relative order of the markers must be discerned. Simple inheritance of a single marker in a cross requires two crossovers to incorporate the linear donor fragment into the circular recipient chromosome. The least frequent class *often* represents recombinants that required four crossovers to generate, because the probability of four crossover events is much less than that of only two crossover events. In this case, class 4 (Cm^r Tc^r Km^s) is the least frequent recombinant class, implying that locus Tc^r is between loci Cm^r and Km^r. This is best visualized with pencil and paper (Fig. 3.8). In this case, the analysis is straightforward, as the order Cm^r Tc^r Km^r determined from analyzing the least frequent class is also consistent with the calculated map distances between the markers (Fig. 3.9).

In a second reciprocal cross, P22 HT *int* was grown on a strain carrying Tc^r and used to transduce a strain carrying Km^r and Cm^r. Tetracycline-resistant transductants were selected, purified, and scored for inheritance of the donor markers Km^s and Cm^s. The following results were obtained:

Cross No. 2	Donor: Cm^s Tc^r Km^s		Recipient: Cm^r Tc^s Km^r	
Class	Cm	Tc	Km	Number
1	Cm^s	Tc^r	Km^s	117
2	Cm^r	Tc^r	Km^s	27
3	Cm^s	Tc^r	Km^r	50
4	Cm^r	Tc^r	Km^r	6

$$\text{Cotransduction between } Tc^r \text{ and } Km^s = \frac{\text{class 1 + class 2}}{\text{total}} \quad \frac{117 + 27}{200} = 72\%$$

$$\text{Cotransduction between } Tc^r \text{ and } Cm^s = \frac{\text{class 1 + class 3}}{\text{total}} \quad \frac{117 + 50}{200} = 84\%$$

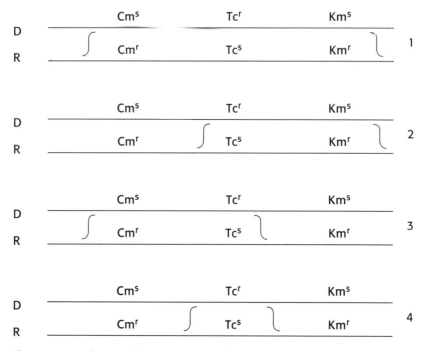

Figure 3.10 Three point cross, example 2. The Cm^s Tc^r Km^s donor (D) is crossed with the Cm^r Tc^s Km^r recipient (R), selection being made for Tc^r. Recombination events leading to each of the four recombinant classes are diagrammed schematically.

Thus, 72% of all Tc^r recombinants also inherited the donor Km^s marker, and 84% of all Tc^r recombinants inherited the donor Cm^s marker. Since map distance = (1 − cotransduction frequency), the linkage between Tc^r and Km^s is 0.28 and that between Tc^r and Cm^s is 0.16.

The least frequent class *often* represents recombinants that required four crossovers to generate. In this case, class 4 (Cm^r Tc^r Km^r) is the least frequent recombinant class, at first implying that locus Tc^r is between loci Cm^r and Km^r. However, this class retains both of the recipient markers, so it cannot be due to a quadruple crossover; there are no quadruple crossover events that lead to inheritance of a central donor marker. This is best visualized with pencil and paper (Fig. 3.10). In this case, the analysis again is straightforward, as the order Cm^r Tc^r Km^r is also consistent with the calculated map distances between the markers (Fig. 3.11).

These two examples illustrate a variety of important points in mapping. First, it is wise to perform reciprocal crosses whenever possible to confirm the deduced order by independent means. Second, marker effects can influence the observed frequency of cotransduction. The distance between Cm^r and Tc^r was found to be 0.74 in cross 1 and 0.16 in cross 2—a substantial difference. Several factors, including nonrandom

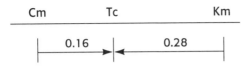

Figure 3.11 Genetic map derived from example 2. Compare the Cm^r–Tc^r map distance with that in example 1.

recombination and nonrandom packaging of transducing fragments, can lead to such discrepancies. Finally, when insertions are used as donor markers, the observed map distances will be different from those observed with wild-type or point alleles, as the insertion reduces the amount of flanking DNA that is packaged in the transducing particle (see below).

Recently, a computer program that analyzes data from transduction crosses has been developed; the description of this program also contains a good theoretical analysis of transductional mapping (Berlyn and Letovsky 1992).

RELATIONSHIP BETWEEN COTRANSDUCTION FREQUENCY AND PHYSICAL DISTANCE

Estimation of the physical distance between two markers may be carried out, for example, when constructing a map of an entire genome. The formula for calculating physical map distance from measurements of cotransduction frequency was derived by Wu (1966). Some investigators argue that Wu's formula is valid only for cases where a large number of different transducing fragments are formed (Mandecki et al. 1986). As described above, genetic evidence suggests that only a small number of different transducing fragments contain a given marker; the mean has been estimated as two (Mandecki et al. 1986). However, the Wu formula is widely used and served as the basis for constructing the genetic linkage maps of *Salmonella typhimurium* (Sanderson and Roth 1988) and *Escherichia coli* (Table 3.1) (Bachmann 1990).

Wu's formula relating transduction frequency to physical distance (in minutes) is

$$f = (1 - \frac{d}{L})^3$$

where *f* is the percent cotransduction; *d* is the physical distance between the markers; and *L* is the length of the transducing DNA (i.e., the amount packaged).

Table 3.1 Sample Values Based on Wu's Equation

d	$f(\%)$
0	100
0.1 L	72
0.2 L	50
0.3 L	35
0.4 L	22
0.5 L	12
0.6 L	6
0.7 L	3
0.8 L	1

For phage P22, $L \approx 1.0$ minute or 44 kb.
For phage P1, $L \approx 2.0$ minutes or 90 kb.

Assumptions made by the Wu formula include:

1. Packaging of chromosomal DNA is random.

2. Multiple crossovers (more than two) between the transducing fragment and the homologous portion of the chromosome are sufficiently rare that they can be ignored.

3. The probability of recombination is proportional to the physical distance of the interval.

4. The entire length of the transduced fragment is homologous to the chromosomal equivalent, and the donor and recipient alleles do not differ significantly in size.

MODIFICATION OF THE TRANSDUCTION MAPPING FUNCTION

As stated above, this formula assumes that donor and recipient are homologous over their entire lengths. With the extensive use of transposons in genetic mapping, this assumption is often incorrect. Inclusion of an insertion in the donor fragment reduces the amount of flanking, homologous DNA that is available for recombination. Sanderson and Roth have derived a modification of the Wu formula that allows these corrections to be made (see Fig. 2 in Sanderson and Roth 1988):

Table 3.2 Approximate Values of *m* and *n* for Different Transposons

Element	Size (kb)
MudJ	10
Tn*10*	10
Tn*10*d(Tc)	3.0
Tn*10*d(Cm)	1.5

Data from Sanderson and Roth (1988).

$$\frac{100}{f} = 1 + \frac{(L\text{-}m)^3 - (L\text{-}m\text{-}d)^3}{(L\text{-}m\text{-}n\text{-}d)^3}$$

where f is the percent cotransduction; d is the physical distance between the markers (in kb); L is the length of the transducing DNA (i.e., the amount packaged, in kb); m is the excess size of the selected donor marker (in kb); and n is the excess size of an unselected donor marker (in kb). Examples of commonly used values of m and n are shown in Table 3.2.

FACTORS THAT INFLUENCE GENETIC LINKAGE

One important consideration for understanding transductional crosses is that a linear fragment recombines with a circular chromosome. A single reciprocal exchange will result in a linearized chromosome. Therefore, at least two reciprocal exchanges are required to recover a viable cell, but the total number of exchanges must be an even number. This means that the closer two markers are on the chromosome, the greater the likelihood that they will be coinherited. Conversely, the size of the packaged transducing fragments limits the distance by which two markers can be separated and still exhibit linkage. Thus, transductional mapping is useful only for analyzing the relationships between closely linked genes. Other methods (e.g., Hfr crosses or physical mapping techniques) must be employed to localize newly discovered genes to a general region of the chromosome before transductional analysis can be used.

Recombinational hot spots, such as Chi sequences, can influence the genetic distance between two markers (assumption 3 above). Likewise, a restricted number of sites for the packaging of transducing DNA leads to a nonrandom distribution of fragments (assumption 1 above). In practice, both of these factors significantly influence linkage values, and therefore, it is emphasized that correlations between genetic and physical distance are crude estimates at best.

Deletion Mapping

Deletion mapping provides a means to localize rapidly the genetic position of a large number of point mutations. This method as applied to microorganisms was pioneered by Seymour Benzer (1955, 1957, 1961), who determined the genetic map position of many hundreds of single-site mutations within the *r*IIA and *r*IIB genes of bacteriophage T4.

A genetic map may be constructed by simply crossing all available mutants with each other. The frequency with which a pair of mutant alleles will recombine to yield wild-type recombinants is a function of the distance by which they are separated. However, as the number of alleles to be mapped increases, so does the number of crosses required to determine their genetic map positions. Deletion mapping provides an efficient means for simplifying map construction. A relatively small number of characterized deletions within the region of interest can be used as recipients in an initial set of crosses, which will localize the general map position for each of the donor alleles tested (see, e.g., Benzer 1957, 1959). Additionally, deletion mapping is more precise than three-point crosses when ordering closely linked mutations under certain circumstances (Beckwith 1978).

For example, consider a point mutation within the *srlB* gene (Fig. 3.12). If this mutant is crossed with a strain carrying a deletion that lies within the *srlA* gene, then Srl⁺ transductants will be recovered: The donor *srlB* strain provides *srlA⁺* information, whereas the recipient Δ*srlA* strain provides *srlB⁺* information. A crossover between the deletion endpoint and the *srlB* mutant allele results in the reconstruction of a wild-type *srl* operon. However, if the recipient deletion extends into the region of the *srlB* gene that includes the site of the donor mutant allele, then wild-type recombinants cannot be recovered; neither donor nor recipient carries wild-type information for that position. Thus, the mutant allele in question maps between the two deletion endpoints and is said to map within that deletion interval. A larger set of recipient deletions provides correspondingly greater resolution for the genetic map. An inspiring example is the comprehensive deletion map that was constructed for the *Salmonella typhimurium hisG* gene (Hoppe et al. 1979).

Why bother to construct a deletion map, particularly when facile methods are available for DNA sequence analysis? Indeed, initial construction of a deletion map can be time-consuming, although methods for construction of directed deletions (see below) have greatly simplified this process. One example illustrating the utility of deletion mapping is found in studies of the *S. typhimurium putP* gene, encoding the catabolic proline permease. Mutational analysis of the *putP* gene has identified a number of alleles that alter the specificity of proline permease for either

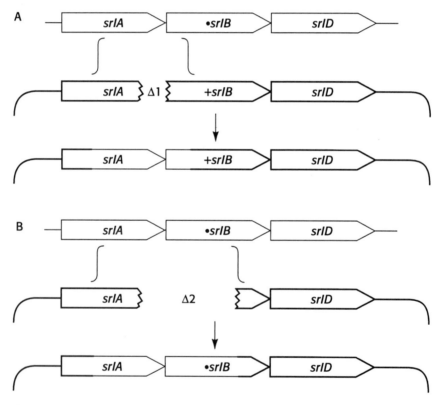

Figure 3.12 Deletion mapping. (*Thin lines*) Donor material; (*heavy lines*) recipient material. In cross *A*, the donor *srlB* mutant allele (indicated by •) is in a region that is present in the recipient deletion strain, Δ1 (indicated by +). Thus, any recombinational event that occurs to the right of the donor mutation and to the left of the deletion endpoint will result in a wild-type recombinant. In cross *B*, the donor *srlB* mutant allele is in a region that is missing in the recipient deletion strain, Δ2. Thus, wild-type recombinants are never recovered.

of its substrates, proline or sodium; 17 proline specificity mutants and 24 sodium specificity mutants were each localized to 1 of 11 deletion intervals. Proline specificity mutations fell into deletion intervals 2, 4, 6, 7, or 8, whereas sodium specificity mutations fell into deletion intervals 1 or 11. Thus, a simple series of crosses localized these mutations to specific regions of the gene and demonstrated that the two types of mutations affect distinct regions of the PutP polypeptide (Dila and Maloy 1986; Myers and Maloy 1988). Additionally, this genetic localization greatly reduced the amount of DNA sequencing necessary to determine the molecular nature of each mutation. The *putP* gene is 1506 nucleotides long, so localization of a mutant allele to 1 of 11 deletion intervals means that only approximately 150 nucleotides need to be sequenced to determine its exact nucleotide change—a considerable savings in time and effort.

DELETION MAPPING IN PRACTICE

The *S. typhimurium*/P22 system allows deletion mapping to be performed to extremely high resolution. As a first step, an individual mutant is crossed against a relatively small battery of widely spaced deletions, and wild-type recombinants are selected. This can often be done by spot transductions. The results provide a rough location for the mutant allele in question. The mutant can then be crossed against a number of more closely spaced deletions that further delimit the region in question. Appearance of even a single recombinant indicates that the mutant allele lies outside of the deletion interval tested (assuming of course that appropriate reversion controls are negative). Finally, to confirm that the mutant allele lies within the identified deletion interval, full-scale crosses are conducted. For example, in mapping *hisG* mutations, full-scale crosses with a wild-type donor and a given deletion usually yield about 10,000 His+ recombinants per plate. A mutant allele is assumed to lie within the deletion interval in question if no recombinants are observed when five plates worth of transduction are performed (i.e., the resolution is approximately 1 in 50,000; Hoppe et al. 1979).

In deletion mapping, the point mutants are used as donors for crosses with deletions as recipients to reduce any potential background due to reversion of the point mutation.

DIRECTED DELETIONS

In *S. typhimurium*, a general method for constructing deletions with defined endpoints was developed by Hughes and Roth (1985). This method involves recombination between Mud transposon insertions. For ex-

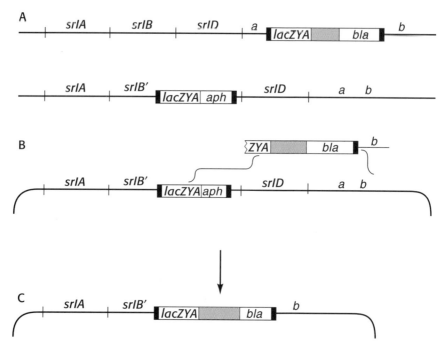

Figure 3.13 Formation of directed deletions. (*A*) Structure of starting strains. The donor strain carries a MudA insertion downstream from the *srl* operon, and the recipient strain carries a MudK insertion in the *srlB* gene. (*B*) Transduction and recombination. Most *bla* (Ap^r) gene transducing fragments do not carry the entire MudA insertion, because of restrictions on the amount of DNA encapsulated by P22. Homologous recombination between the bacterial DNA in region *b* and the Mud DNA within the *lac* operon results in replacement of the MudJ element and downstream DNA with the MudA element. (*C*) Resulting deletion strain (Ap^r Km^s). The left deletion endpoint, within the *srlB* gene, is specified by the position of the MudJ insertion, and the right deletion endpoint, downstream from the *srl* operon, is specified by the position of the MudA insertion.

ample, consider the *srl* operon (Fig. 3.13). A collection of *srl*::MudJ insertions (Km^r) may be isolated in various locations within the *srl* operon (see Experiment 1), and a MudA insertion (Ap^r) may be isolated adjacent to the *srl* operon (see Experiment 2). MudA is 37 kb in size, so most P22 transducing particles package only a portion of the MudA element, along with bacterial DNA flanking one side of the MudA insertion. Thus, transduction for Ap^r into an otherwise wild-type strain will be very inefficient, because few transducing particles will carry the entire MudA element along with chromosomal DNA flanking both sides of the insertion; this flanking bacterial DNA is necessary for homologous recombination to place the MudA insertion into the chromosome. However, if a *srl*::MudJ insertion strain is used as a recipient, the frequency of Ap^r transduction is considerably increased, because the *lac* sequences common to the MudA and MudJ elements provide homology for recombination. Thus, many Ap^r Km^s transductants carry a deletion whose endpoints are determined by the sites of the MudA and MudJ insertions. (Other Ap^r Km^s transductants result from replacement of the recipient MudJ element with the donor MudA element.) Therefore, if a series of MudJ insertions is isolated in the course of characterizing operon structure, it is a relatively simple matter to use these insertions also for the construction of a deletion map. Particularly good examples of using directed deletions for mapping are provided by Roof and Roth (1988) and by Hughes et al. (1991).

The critical feature of this method is the fact that most P22 transducing particles do not carry the entire donor MudA element. Thus, this method is unsuitable for use with transducing phage such as P1, which encapsulate a much larger region of chromosomal DNA. However, the recent extension of P22 host range to *Escherichia coli* (and presumably other species) potentially makes this method even more generally applicable.

SECTION 4
Mutants and Their Analysis

"The utility of genetic analysis does not end with the isolation of a null mutation."

Thomas J. Silhavy
Personal Communication

"In our studies on the tryptophan synthetase of *E. coli*, well over 600 independently isolated *trpA* mutants were examined genetically and biochemically. We were surprised to discover that only a small percentage of *trpA* mutants were of the missense variety and of these, many were identical. Apparently, either there are relatively few positions in the α-chain at which an amino acid substitution will completely inactivate the protein, or there are several missense mutational hot spots and changes at these sites are therefore most commonly seen. To examine these alternatives we have isolated missense mutants in a *trpA* revertant strain which produces a functional modified protein; the α-chain of this strain contains valine instead of glycine at position 211. Although this strain is a prototroph, its α-chain is functionally less effective than the wild-type (glycine) protein. Starting with the valine revertant, we were able to isolate missense mutants which were not detected in analyses with the wild-type (glycine) strain. Studies with one missense mutant revealed that valine must be present at position 211 in the protein if the new missense change is to inactivate the protein. . .This finding shows that functional considerations are of importance in limiting the recovery of different types of missense mutants. Therefore, it would appear that of the many possible amino acid substitutions, only a small fraction can completely inactivate the α-chain. Limitations on permissible amino acid changes imposed by the genetic code and by mutational probability (only one base change per mutation) also are pertinent, especially in view of the structural similarities between many amino acids specified by codons which are identical at two of the three nucleotide positions."

Charles Yanofsky
Reprinted, with permission, from (©1971 by the American Medical Association) Tryptophan Biosynthesis in *Escherichia coli:* Genetic Determination of the Proteins Involved. *J. Am. Med. Assoc.* 218: 1026–1035 (1971)

"In early genetic studies with mutants, Yanofsky and colleagues. . .identified eight sites in the tryptophan synthetase α subunit from *Escherichia coli* at which one or more single amino acid replacements result in inactivation. Recent x-ray crystallographic studies. . .show that seven of these eight sites. . .are located in the active center of the α subunit. . ."

Shinji Nagata, C. Craig Hyde, and Edith Wilson Miles
Reprinted, with permission, from The α Subunit of Tryptophan Synthase. *J. Biol. Chem.* 264: 6288–6296 (1989)

Mutagenesis

Mutations provide genetic tools to construct strains with desired properties, to determine the role of the normal gene product in the cell, and to localize the position of a gene in the genome. Mutations are thus extremely valuable for understanding molecular biology, physiology, and genetic regulation (see Maloy et al. 1994). Since different types of mutations have different uses, it is beneficial to understand the properties of each type of mutation and how to isolate mutants.

WHAT IS A MUTATION?

A mutation is any heritable change in the DNA sequence of an organism. For example, one nucleotide may be substituted for another, nucleotides may be deleted or inserted, or a region of the chromosome may be inverted or duplicated. A strain with a mutation is called a mutant. The mutation may or may not change the observable properties of the mutant compared to the parental strain, so not all changes in the genotype (the DNA sequence of the organism's genome) result in changes in the phenotype (the observable properties of an organism). Furthermore, different mutations in a single gene, i.e., different alleles of the gene, may produce different phenotypes.

Most mutations are random events. Although an organism with the mutation may have a selective advantage in some particular environment, the mutation is not induced by these conditions. Every gene mutates spontaneously at a characteristic rate that depends on the size and nucleotide sequence of the gene. Thus, it is possible to estimate the probability that a given gene will mutate in a particular cell and the corresponding probability that a mutant allele of the gene will occur in a population of a particular size.

TYPES OF MUTATIONS

Mutations may be described by their effects on the DNA itself or their effects on the resulting gene products. In both cases, the description of different types of mutations provides insight into the biology of the resulting mutants (Table 4.1).

One way of describing mutations is based on the changes that have occurred in the DNA. A point mutation only changes a single base pair. Point mutations that substitute a purine for a purine (e.g., A-T→G-C) or a pyrimidine for a pyrimidine (e.g., T-A→C-G) are called transition mutations. Point mutations that substitute a purine for a pyrimidine (e.g., A-T→T-A) or a pyrimidine for a purine (e.g., T-A→A-T) are called transversion mutations. A deletion mutation results from the removal of multiple contiguous base pairs. An insertion mutation is the addition of multiple base pairs without removal of any other material. A duplication mutation is the acquisition of a second copy of a number of base pairs. An inversion mutation changes the orientation of a number of contiguous base pairs.

Another way of describing mutations is based on their phenotypic effects on mutant gene products. The change in DNA sequence caused by a mutation can affect a gene encoding a protein, a gene encoding a structural RNA (such as ribosomal RNA), or a regulatory site. Although mutations in each of these different targets will have different properties, the way in which mutations change the phenotype of a cell can be seen by considering the effects of a mutation on a gene encoding a protein. The chemical and physical properties of proteins are determined by their amino acid sequence. An amino acid substitution can change the structure and hence the biological activity of a protein. Even a single-amino-acid change is capable of altering the activity of, or even completely inactivating, a protein. For example, an enzyme with a histidine residue that is essential for its catalytic function could be inactivated by a single base change of proline for the histidine. Other mutations may not change specific residues at the active site of the protein, but they may inactivate a protein by disrupting its three-dimensional structure. Such mutations that completely eliminate activity of a gene product are called null mutations.

In addition to simple amino acid substitutions, several other types of mutations may eliminate activity of a protein: (1) a deletion mutation that causes one or more amino acids to be absent in the completed protein, (2) a frameshift mutation due to deletion or insertion that shifts the reading frame such that all the codons after the mutation are changed, and (3) a nonsense mutation, in which a base change generates a stop codon, resulting in a shorter polypeptide that lacks the carboxyl terminus of the protein.

Table 4.1 Typical Properties of Different Types of Mutations

Type of mutation	Missense	Nonsense	Frameshift	Insertion	Deletion
Phenotype	silent, weak, or null	null[a]	null[a]	null[a]	null[a]
Effect on protein	amino acid substitution	premature translational termination, truncated protein	shift in reading frame, multiple amino acid substitutions, usually premature translational termination	large block of additional amino acids, often transcriptional and/or translational termination	loss of amino acids, often shifts downstream reading frame also
Revertants	true revertants, missense suppressors, intragenic suppressors	true revertants, nonsense suppressors, intragenic suppressors	true revertants, frameshift suppressors, intragenic suppressors	true revertants, bypass suppressors	**no** true revertants, bypass suppressors
Mapping properties	point	point	point	like a point mutation[b]	deletion

[a]These types of mutations nearly always have a null phenotype, but they may have a silent or weak phenotype if the mutation does not disrupt an essential part of the gene (e.g., if the mutation occurs very close to the 3′ end of the gene).

[b]Although insertion mutations may add a substantial amount of extraneous DNA, since none of the DNA adjacent to the insertion is lost, the gene can be repaired by recombination between any homologous sites within the gene as is true for point mutations.

Protein structure and function are determined by such a variety of interactions that sometimes an amino acid substitution is only partially disruptive or "leaky." This may cause a reduction, rather than a complete loss, of enzyme activity. For example, a bacterium with such a mutation in an enzyme in a gene required for sorbitol utilization might form small colonies on medium with sorbitol as a carbon source.

Conditional mutations only produce a mutant phenotype under certain circumstances. Conditional mutants may be affected by environmental conditions or the presence of other mutations. Temperature-sensitive (Ts) mutants have a wild-type phenotype at one temperature (e.g., 30°C) and a mutant phenotype at another temperature (e.g., 42°C). Temperature-sensitive mutations fall into two classes (Beckwith 1991). (1) Most temperature-sensitive mutations affect the folded conformation of a protein: At the permissive temperature, the protein folds into a functional conformation, but at the nonpermissive temperature, the protein is partially unfolded and typically quickly degraded by cellular proteases. (2) Some temperature-sensitive mutations affect the proper assembly of proteins during synthesis; once synthesized, such temperature-sensitive synthesis (Tss) mutant gene products remain active even at the nonpermissive temperature. In contrast to temperature-sensitive mutations, cold-sensitive mutations result in an active gene product at high temperatures (e.g., 37–42°C) but in an inactive product at lower temperatures (e.g., 20–30°C). Cold-sensitive mutations often affect hydrophobic interactions that mediate protein-protein or protein-membrane associations (Guthrie et al. 1969; Tai et al. 1969; Moir et al. 1982). Another example of conditional mutations is the suppressor-sensitive mutations, which exhibit a wild-type phenotype in some bacterial strains and a mutant phenotype in others (see Suppression in this section).

Because it may be difficult to vary the conditions in a eukaryotic host, most types of conditional mutations are not directly useful for studying host-pathogen interactions. However, they have been used (1) to identify and characterize essential genes, i.e., genes required for growth of the organism, (2) to identify proteins that interact with other proteins, and (3) to inactivate a single gene and follow the effects of the inactivation over time.

Not all amino acid substitutions produce a mutant phenotype. Silent mutations occur when a base change has no detectable effect on phenotype. For example, because the genetic code is redundant, many base changes do not result in an amino acid substitution. Furthermore, many amino acid substitutions do not have much effect on the structure and function of a protein (Yanofsky 1971; Nagata et al. 1989). For example, a protein might be virtually unaffected by a replacement of a leucine by another nonpolar amino acid such as isoleucine.

ISOLATION OF MUTANTS

Genetics requires the isolation and characterization of mutants that affect some process of interest. Typically, spontaneous mutations in a given gene occur at a frequency of 10^{-6} or less per cell per generation. Because mutations are so rare, isolation of the mutants from a large population of wild-type cells requires a selection or screen (Beckwith 1991).

Genetic Selections

A selection is a condition that only allows growth of specific mutants. Thus, genetic selections allow the direct isolation of very rare mutations from a large population of cells. For example, antibiotic-resistant mutants can be selected by simply plating a large number of bacteria on solid medium containing the antibiotic—only resistant mutants can form colonies.

Genetic Screens

If the mutation is relatively common and there is no direct selection for mutants, it is possible to simply look for mutants on media where both the mutant and parental cells grow but where the phenotype of the mutant can be distinguished from the phenotype of the parental cells. For example, mutants unable to use sorbitol could be identified by plating several hundred cells onto tetrazolium medium with sorbitol: The wild-type Srl^+ cells would form white colonies and the rare Srl^- mutants would form red colonies.

The difference between a selection and a screen has important practical consequences. For example, compare the selection for a spontaneous streptomycin-resistant (Sm^r) mutant with a screen for a spontaneous mutant that is unable to use sorbitol. It is possible to select directly for an Sm^r colony by spreading a large number of cells on a plate containing streptomycin; as many as 10^{10} cells can be spread on a plate, allowing mutants as rare as 10^{-10} to be isolated on a few plates. In contrast, thousands of plates would be needed to find a single Srl^- mutant by screening on tetrazolium-sorbitol plates because only a few hundred colonies can be examined on a single plate.

The frequency of mutations can be increased by treatment with chemical or physical mutagens; however, even after heavy mutagenesis, it may be difficult to find a rare mutant by screening. It is sometimes possible to enrich for mutants that are unable to grow under specific conditions from a population of cells that can grow under these conditions. Two common enrichment procedures use the antibiotics ampicillin or cycloserine, which kill bacteria by interfering with the synthesis of the bacterial cell wall. When ampicillin or cycloserine is added to a culture of growing bacteria, cell wall synthesis stops but growth of the cell continues. Without an intact cell wall, the high internal osmotic pressure ultimately causes the cell to burst. Thus, ampicillin and cycloserine only kill growing cells, not a nongrowing mutant. By exposing the cells to the ampicillin or cycloserine under conditions that allow growth of the wild-type cells but not the mutant cells, many of the wild-type cells will die, whereas the nongrowing mutant cells will survive. The cells that survive this enrichment are then grown under conditions that are permissive for the mutant cells before screening for the mutants. This typically results in about a 100-fold enrichment for mutants. However, the problem with this approach is that each enrichment may contain siblings.

MUTAGENESIS

Mutations may be due to spontaneous errors or may be induced by specific mutagens (Friedberg et al. 1995). Examples of some potentially useful mutagens are described below (Table 4.2). Some general considerations about how to choose a mutagen are described in the Overview.

Spontaneous Mutations

Rare mutations arise due to spontaneous errors occurring during DNA replication and repair or by chemical modification of nucleotides in DNA. Spontaneous mutations encompass a wide variety of alterations, including base substitutions, frameshift mutations, and chromosomal rearrangements.

Point mutations may be produced by incorporation errors due to tautomerism of the nucleotide bases. Some of the bases exist in alternative forms with different base-pairing properties. For example, the rare imino form of A can pair with C, and the rare enol form of T can pair with G. Such tautomers are unstable and thus rapidly isomerize to the more stable form. However, if such tautomers occur at the replicating DNA fork, an incorrect base will be correctly hydrogen-bonded to the template strand.

Another major source of spontaneous point mutations is the deamination of C or 5-methylcytosine (MeC), a methylated form of C that also pairs with G. In the DNA of many bacteria and viruses, approximately 5% of the C is methylated. Both C and 5-MeC are occasionally deaminated. Deamination of C yields uracil. Since uracil pairs with A instead of G, replication of a molecule containing a G-U base pair will ultimately lead to substitution of an A-T pair for the original G-C pair (by

Table 4.2 Some Useful Mutagens

Mutagen[a]	Mechanism	Types of mutations produced[b]
Spontaneous	DNA replication and repair errors, spontaneous modification of nucleotides	all types of mutations produced
UV irradiation	pyrimidine dimers induce error-prone repair (SOS)	mainly G-C→A-T transitions; all other types of mutations at somewhat lower frequency
2-Aminopurine	base analog	G-C→A-T transitions
Bromouracil	base analog	G-C→A-T and A-T→G-C transitions
Hydroxylamine	alkylating agent	G-C→A-T transitions
N-methyl-N'-nitro-N-nitrosoguanidine (MNNG)	alkylating agent	G-C→A-T transitions; multiple, closely spaced mutations are common
Ethylmethane sulfonate (EMS)	alkylating agent	G-C→A-T transitions
Ethylethane sulfonate (DES)	alkylating agent, requires SOS response	G-C→T-A transversions; other base substitutions at lower frequency
Nitrous acid	oxidative deamination	G-C→A-T and A-T→G-C transitions
ICR-191	intercalating agent	frameshifts

[a]Three excellent references which give detailed protocols for using these mutagens are Roth (1970), Foster (1991), and Miller (1992).
[b]Modified from Miller (1992).

the process G-U→A-U→A-T in successive rounds of replication). However, cells possess an enzyme (uracil glycosylase) that specifically removes uracil from DNA, so the C-to-U conversion rarely leads to mutation. In contrast, deamination of 5-MeC yields 5-methyluracil, or T. Because T is a normal DNA base, no T glycosylase exists, so the G-MeC pair becomes a G-T pair. The G-T pair is subject to correction by mismatch repair. The direction of correction will be random, sometimes yielding the correct G-C pair and sometimes an incorrect A-T pair. MeC sites are thus mutation hot spots.

Frameshift mutations usually occur in sequences with monotonous repeats of one or a few base pairs. The strand-slippage model proposes that during DNA replication, the strands may separate, slip, then re-pair in such a way that one or two bases are looped out. As the DNA is elongated, extra bases would be added to one strand or deleted from one strand of the DNA. The DNA sequence remains unchanged beyond the point of the insertion or deletion, but because the sequence is translated into an amino acid by reading groups of three bases, the addition or deletion of one or two bases changes the reading frame. Downstream from the frameshift mutation, out-of-frame codons will be read until translation is terminated by a nonsense codon in the improper reading frame.

Table 4.3 Types of Mutations Produced by Different Mutagens

	Base substitutions[a]	Frameshifts[b]	Insertions[c]	Deletions
Spontaneous	+	+	+	+
Base analogs	+	−	−	−
Alkylating agents	+	−/+	−	−
Intercalating agents	−	+	−	−
SOS induction	+	+	+	+

[a]Missense or nonsense mutations.
[b]−/+ indicates that frameshift mutations are rarely produced by this mutagen.
[c]Insertion mutations refer to large insertions, not simply insertion of one or a few nucleotides resulting in a frameshift mutation. In some bacteria, insertion mutations caused by endogenous IS elements and transposons are the major class of spontaneous null mutations. In other bacteria, such mutations are very rare.

Chromosomal rearrangements (such as deletion and duplication mutations) often occur by homologous recombination between DNA sequences with direct repeats. They may also be caused by incomplete transposition of a transposon or insertion sequence. Although these two types of chromosomal rearrangements are relatively common, those due to inversion mutations are extremely rare.

MUTAGENS

Sometimes it is not possible to devise a selection or a screen that is sensitive enough to isolate rare spontaneous mutations. In such cases, it may be necessary to treat the DNA with a mutagen. Mutagens are chemical or physical agents that increase the frequency of physical changes in DNA. Mutagens may directly stimulate mutations by causing mispairing or may indirectly stimulate mutations by stimulating error-prone repair mechanisms. Some examples of the effects of mutagens include loss of a base by depurination, alteration of a base due to alkylation of a nucleotide, a bulge in the double helix due to an intercalating agent, strand breaks due to ionizing radiation, intrastrand cross-linking due to pyrimidine dimers formed by UV irradiation, and intrastrand cross-linking due to treatment with mitomycin C. Different types of mutations are induced by specific mutagens (Table 4.3). The effects of mutagens often depend on a variety of conditions, including growth medium, pH, temperature, and genotype of the cell.

Base-analog Mutagens

A base analog is a compound sufficiently similar to one of the four DNA bases that it can be incorporated into a DNA molecule during normal replication. Base analogs increase the frequency of point mutations in two ways. (1) Because nucleotide synthesis is carefully regulated to maintain each nucleotide in a balanced ratio, nucleoside analogs often inhibit production of the corresponding NTP. This changes the intracellular ratio of the dNTPs, resulting in increased rates of misincorporation. (2) Many base analogs have substitutions that increase the frequency of tautomerization. For example, the substituted base 5-bromouracil (BU) is a structural analog of T (the bromine is about the same size as the methyl group of T). When incorporated in DNA in place of T, BU usually forms base pairs with A. However, the bromine atom shifts the keto-enol equilibrium so that BU has a much greater tendency to tautomerize than T. If this happens during DNA replication, BU may pair with G, converting a T-A base pair to a C-G base pair.

Alkylating Agents

Addition of alkyl groups can occur at various positions on the bases. This chemical modification often causes mispairing. Three commonly used alkylating agents are hydroxylamine, ethylmethane sulfonate (EMS) or ethylethane sulfonate (DES), and N-methyl-N'-nitro-N-nitrosoguanidine (MNNG).

Hydroxylamine (NH_2OH) reacts specifically with C, converting it to a modified base (N_4-hydroxyC) that pairs with A instead of G. Thus, treatment of DNA with hydroxylamine in vitro has two consequences: (1) hydroxylamine only produces GC→AT transitions and (2) mutations induced by hydroxylamine cannot be precisely reverted by hydroxylamine. In contrast, when used in vivo, hydroxylamine may produce free radicals that damage the DNA—the resulting DNA damage may induce the SOS system, resulting in a wide variety mutation types.

Ethylmethane sulfonate (EMS) and MNNG both add an alkyl group to the hydrogen-bonding oxygens of G and T. The alkylation impairs the normal hydrogen bonding of the bases, causing mispairing of G with T. This results in transition mutations, mostly A-T→G-C, but G-C→A-T substitutions also occur. EMS also reacts with A and C, but these modifications are much less mutagenic. MNNG acts on DNA at the growing DNA replication fork and often produces multiple mutations; it is a very powerful mutagen that is often useful for initial studies to determine whether or not it is possible to isolate a particular type of mutation. Because MNNG causes multiple mutations, it should be avoided for subsequent fine-structure genetic analysis when specific, single point mutations in a gene are desired.

Alkylating agents may also stimulate mutations indirectly. For example, alkylation of G can result in depurination, breaking the bond joining the purine nitrogen and deoxyribose. Depurination is not always mutagenic, because the gap left by loss of the purine can be efficiently repaired. However, sometimes the replication fork may reach the apurinic site before repair has occurred. When this happens, replication stops just before the apurinic site, the SOS system is activated, and replication proceeds, often putting an A nucleotide in the daughter strand opposite the apurinic site. Since the original parental base (which was removed) was a purine, the base pair at that site will be a mismatch (Pu-A), resulting in a transversion mutation after DNA replication.

Oxidizing Agents

Nitrous acid primarily converts amino groups to keto groups by oxidative deamination. For example, C and A are converted to U and hypoxanthine (H), which form the base pairs U-A and H-C. This results in G-C→A-T transitions due to deamination of C and A-T→G-C transitions due to deamination of A. Oxidizing agents also damage many other cell components. Thus, the frequency of cell killing is often much greater than the frequency of mutagenesis by these agents.

Intercalating Agents

Acridine orange, proflavine, and ICR-191 are planar, heterocyclic molecules with dimensions roughly the same as those of a purine-pyrimidine pair. In aqueous solution, these substances can insert (or intercalate) into DNA between the adjacent base pairs, resulting in a bulge in the DNA double helix. The intercalated molecules stabilize non-hydrogen-bonded nucleotides that may have looped out of the double helix due to strand-slippage. When DNA containing intercalated agents is replicated, additional bases are often inserted, resulting in frameshift mutations. Usually a single base is added, but sometimes two bases are added. Deletion of one or two bases also occurs, but this is far less common than base addition.

Ultraviolet Light

The major type of damage caused to DNA by irradiation with UV light (200–300 nm) is the formation of cyclobutane dimers between adjacent pyrimidines (T or C). Pyrimidine dimers block DNA replication. Several mechanisms can repair pyrimidine dimers: a light-dependent mechanism called photoreactivation, excision by UvrABC, recombination with ho-

Table 4.4 Mutations in DNA Repair Systems That Have Mutator Phenotypes

Repair system	Mutations[a]	Increase in mutation frequency	Major types of mutations
Mismatch repair	*muthSL*	10^2-10^3	transition, frameshifts
Proofreading	*mutD*[b]	10^3-10^4	transitions, transversions, frameshifts
G-A mismatch repair	*mutT*	10^3-10^4	AT→CG transversions
G-A mismatch repair	*mutY*	$>10^2$	GC→TA transversions

Modified from Foster (1991).
[a]Because these functions play critical roles in all cells, the genes encoding DNA repair enzymes tend to be highly conserved. Thus, expression of cloned, dominant-negative alleles of mutator genes from *E. coli* may be used to produce mutator strains of different species of bacteria (Wu and Marinus 1994; T. Zahrt and S. Maloy, unpubl.).
[b]*mutD* is an alternative name for the *dnaQ* gene. *dnaQ* encodes the ε-subunit of DNA polymerase III required for proofreading. The frequency of mutations in *dnaQ* mutants depend upon the growth medium, temperature, and salt concentration. Faster growth conditions result in 10–100-fold greater mutation frequencies because under these conditions saturation of the mismatch repair system prevents repair of the accumulated mutations.

mologous DNA, and SOS repair. The first three mechanisms precisely repair the original nucleotides, and no mutation results. In contrast, the SOS response is error-prone, often resulting in mutations (see below).

DNA REPAIR

Although mutations are potentially lethal events, errors in nucleotide incorporation during DNA replication occur at a relatively high frequency. Most of the resulting errors are repaired by the cell. More than 100 genes are involved in DNA repair mechanisms. However, three mechanisms are responsible for correcting most potential errors: proofreading by DNA polymerase, methyl-directed mismatch repair, and G-A mismatch repair (Table 4.4). These three repair mechanisms are very efficient, but they are not perfect. If an error is incorporated into the DNA, then a race between replication and repair determines whether the error will become an inherited mutation.

Certain types of DNA damage induce the SOS response, a complex regulon of genes that help the cell survive DNA damage (Peterson et al. 1988; Friedberg et al. 1995). SOS-regulated genes are repressed by the LexA protein. DNA damage activates the RecA protein, which then stimulates the proteolytic cleavage of LexA protein. Inactivation of the LexA repressor results in derepression of genes in the SOS regulon. Three DNA repair systems are induced with the SOS regulon: (1) error-prone repair, (2) recombination repair, and (3) excision repair.

Error-prone Repair

Induction of SOS results in a high frequency of error-prone repair that accounts for most UV-induced mutations. Two mechanisms may be responsible for repair of pyrimidine dimers produced by UV irradiation, depending on whether T-T dimers are formed or C-C/C-T dimers are formed. When DNA polymerase III encounters pyrimidine dimers, it stalls because the 3′ to 5′ proofreading activity repeatedly removes any nucleotide incorporated opposite the lesion. The distorted template prevents proper base pairing, so even the correct nucleotide is removed by the proofreading complex. Error-prone repair is due to translesion synthesis across the pyrimidine dimer. When DNA polymerase stalls at T-T dimers, the proofreading function is inhibited, allowing DNA polymerase to continue but increasing the error frequency (Echols and Goodman 1991). In contrast, when DNA polymerase stalls at C-C or C-T dimers, the replication bypass is delayed, the Cs are deaminated to uracil, and DNA polymerase inserts A on the complementary strand, resulting in C-T transitions (Tessman et al. 1992).

The *umuDC* gene products are required for error-prone repair. Expression of the *umuDC* genes is regulated by the LexA protein. In addition, activated RecA protein causes proteolytic cleavage of the UmuD protein producing UmuD′, the active carboxy-terminal fragment. The biochemical mechanism of the UmuD′ and UmuC proteins is not known, but it seems likely that the UmuD′-UmuC complex directly interacts with RecA protein and the stalled DNA polymerase to promote error-prone replication (Echols and Goodman 1991). Not all bacteria seem to have functional *umuDC* genes, but functional homologs of the *umuDC* genes are often present on plasmids. Bacteria that lack functional *umuDC* genes are poorly mutable by UV irradiation. For example, *Salmonella typhimurium* is poorly mutated by UV unless *umuDC* homologs are provided on a plasmid (such as the *mucAB* genes derived from pKM101) (Walker 1984; Smith et al. 1990; Nohmi et al. 1991; Friedberg et al. 1995).

MUTATOR STRAINS

Mutations in genes involved in DNA repair often have a mutator (Mut⁻) phenotype; i.e., the resulting defect in DNA repair increases the frequency of mutagenesis (Table 4.4). For example, mutations in the *dnaQ* gene (also called *mutD*), which is required for the proofreading function of DNA polymerase III, and mutations in the *mutH*, *mutL*, and *mutS* genes, which eliminate methyl-directed mismatch repair, all confer powerful mutator phenotypes. A number of other mutator genes have been identified (Miller 1992), but some of the mutators are still poorly understood.

Such *mut⁻* strains are often called mutator strains. The mutation frequency of DNA passed through mutator strains (e.g., on a plasmid or phage) is greatly enhanced. An important practical consideration is that efficient mutator strains generally produce "sickly" colonies because of the accumulation of deleterious mutations in the dividing cells. This imposes a strong selection for suppressors of the mutator phenotype, which can be seen as healthy, fast-growing colonies. Therefore, when using mutator strains, avoid these healthy-looking colonies that may have lost the mutator phenotype.

IN VITRO MUTAGENESIS

DNA can also be mutagenized in vitro. Two alternative in vitro mutagenesis approaches are to mutagenize a fragment of DNA randomly and to make mutations in specific sites on a fragment of DNA.

Because it is often difficult to predict which bases of a gene are important, random mutagenesis is useful for the initial characterization of a gene. With appropriate selections or screens, the resulting mutations can identify specific residues involved in the structure and function of the gene product. Once a gene has been characterized and cloned and the DNA sequence determined, it is possible to mutagenize specific base sequences to determine the effect of particular changes on the gene product.

Localized Mutagenesis

A DNA fragment can be randomly mutagenized in vitro by exposing it to a chemical mutagen. If the DNA fragment contains a selectable genetic marker, it can then be moved into an appropriate recipient cell. An advantage of this approach is that it allows heavy general mutagenesis of a small, specific fragment of DNA without producing secondary mutations elsewhere on the chromosome. Thus, localized mutagenesis is especially useful for obtaining rare point mutations in or near a gene of interest, such as cis-dominant regulatory mutants linked to a gene (Hahn and Maloy 1986; Maloy 1989), or mutations that affect the active site of an enzyme (Dila and Maloy 1987; Myers and Maloy 1988).

Hydroxylamine is a very useful mutagen for localized mutagenesis. Unlike many mutagens, hydroxylamine can even modify DNA packaged

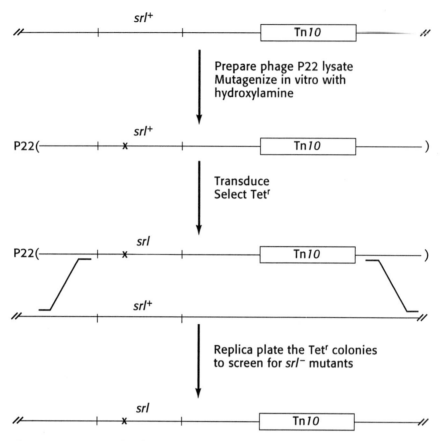

Figure 4.1 An example of localized mutagenesis with hydroxylamine.

inside of phage heads, allowing localized mutagenesis of transducing particles (Hong and Ames 1971). A phage lysate containing transducing particles is mutagenized with hydroxylamine in vitro, and cells are then transduced with this lysate, selecting for a specific marker. When the transducing fragment is recombined onto the chromosome, only the small, localized region carried on that transducing fragment is mutagenized (Fig. 4.1). The extent of mutagenesis of the transducing DNA can be monitored indirectly by following the decrease in phage titer (or "killing") due to mutations in essential phage genes (as well as any nonspecific chemical damage to the phage) or directly by following the increase in clear plaque phage mutants in the lysate. Since each mutant is directly selected from an independently mutagenized transducing fragment, problems with isolation of siblings are avoided.

Localized mutagenesis of P22 transducing lysates is described in Experiment 4. This is also an effective method for isolating mutations in plasmid DNA (see Appendix D, Hydroxylamine Mutagenesis of Plasmid DNA).

Use of Nucleases and PCR to Construct Mutations

Once a gene has been cloned and sequenced, it is straightforward to make deletions or insertions using a variety of nucleases in vitro. These methods are described in detail elsewhere (Sambrook et al. 1989). It is also possible to use polymerase chain reaction to construct mutations in vitro (see Section 7, Polymerase Chain Reaction).

Site-directed Mutagenesis

It is sometimes useful to mutagenize a specific base pair in a DNA sequence to test its role directly—for example, a base pair that encodes an amino acid believed to be at the active site of a protein or a base pair believed to be part of the DNA-binding site of a regulatory protein. In such situations, it is possible to change one (or more) specific base pair directly to a different base pair by site-specific mutagenesis. The basic protocol for site-directed mutagenesis simply involves annealing a mutagenic oligonucleotide to the DNA, followed by chain elongation with DNA polymerase and sealing with DNA ligase. However, there are many approaches to select for the mutant strand of DNA after the heteroduplex DNA is transformed into cells (McPherson 1991). Two of the most common methods involve transformation of the mutated DNA into *dut ung* recipients or *mutS* recipients (Fig. 4.2).

To test specifically the role of an amino acid residue, keep the following considerations in mind when making amino acid substitution mutations. First, to avoid simply decreasing expression of the gene by substitution of rare codons, the codon usage of the substituted codon should be similar to the codon replaced (Table 4.5). Second, to avoid simply disrupting the protein structure by inserting a bulky amino acid or producing unfavorable charge interactions, the shape and charge of the amino acid should be carefully chosen (Fig. 4.3).

RANDOM VS. DIRECTED MUTAGENESIS

Why spend the time and effort isolating random mutations that affect a process when it seems much more straightforward to simply clone a gene, determine the DNA sequence, and make site-directed mutations based on similarity to other genes in the DNA database? First, by isolating a number of mutants using random mutagenesis in vivo, it is often discovered that more genes are involved in the process than predicted—a potentially important finding that would not have been obvious from site-directed mutagenesis experiments. Second, it is often difficult to predict which regions of a gene are important by sequence gazing, but given a suitable

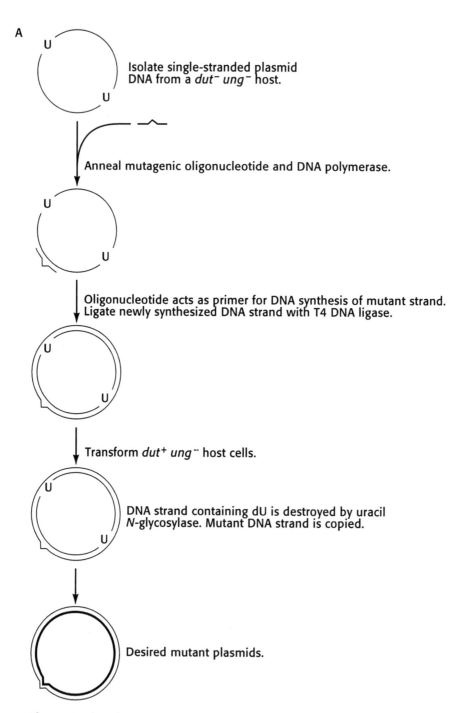

A

Isolate single-stranded plasmid
DNA from a *dut⁻ ung⁻* host.

Anneal mutagenic oligonucleotide and DNA polymerase.

Oligonucleotide acts as primer for DNA synthesis of mutant strand.
Ligate newly synthesized DNA strand with T4 DNA ligase.

Transform *dut⁺ ung⁻* host cells.

DNA strand containing dU is destroyed by uracil
N-glycosylase. Mutant DNA strand is copied.

Desired mutant plasmids.

Figure 4.2 (*See facing page for part B and legend.*)

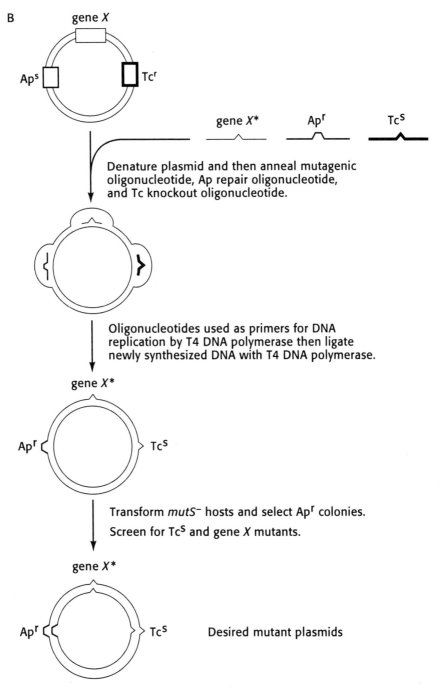

Figure 4.2 (*A*) Site-directed mutagenesis using the *dut ung* selection. (*B*) Site-directed mutagenesis using the *mutS* selection.

Table 4.5 Codon Usage in E. coli Genes

Codon	Amino acid	Frequency[a]	Ratio[b]	Codon	Amino acid	Frequency	Ratio	Codon	Amino acid	Frequency	Ratio	Codon	Amino acid	Frequency	Ratio	
UUU	Phe	19	0.51	UCU	Ser	11	0.19	UAU	Tyr	16	0.53	UGU	Cys	4	0.43	U
UUC	Phe	18	0.49	UCC	Ser	10	0.17	UAC	Tyr	14	0.47	UGC	Cys	6	0.57	C
UUA	Leu	10	0.11	UCA	Ser	7	0.12	UAA	STOP	2	0.62	UGA	STOP	1	0.30	A
UUG	Leu	11	0.11	UCG	Ser	8	0.13	UAG	STOP	0.3	0.09	UGG	Trp	14	1.00	G
CUU	Leu	10	0.10	CCU	Pro	7	0.16	CAU	His	12	0.52	CGU	Arg	24	0.42	U
CUC	Leu	9	0.10	CCC	Pro	4	0.10	CAC	His	11	0.48	CGC	Arg	22	0.37	C
CUA	Leu	3	0.03	CCA	Pro	8	0.20	CAA	Gln	13	0.31	CGA	Arg	3	0.05	A
CUG	Leu	52	0.55	CCG	Pro	24	0.55	CAG	Gln	29	0.69	CGG	Arg	5	0.08	G
AUU	Ile	27	0.47	ACU	Thr	12	0.21	AAU	Asn	16	0.39	AGU	Ser	7	0.13	U
AUC	Ile	27	0.46	ACC	Thr	24	0.43	AAC	Asn	26	0.61	AGC	Ser	15	0.27	C
AUA	Ile	4	0.07	ACA	Thr	1	0.30	AAA	Lys	38	0.76	AGA	Arg	2	0.04	A
AUG	Met	26	1.00	ACG	Thr	13	0.23	AAG	Lys	12	0.24	AGG	Arg	2	0.03	G
GUU	Val	20	0.29	GCU	Ala	18	0.19	GAU	Asp	33	0.59	GGU	Gly	28	0.38	U
GUC	Val	14	0.20	GCC	Ala	23	0.25	GAC	Asp	23	0.41	GGC	Gly	30	0.40	C
GUA	Val	12	0.17	GCA	Ala	21	0.22	GAA	Glu	44	0.70	GGA	Gly	7	0.09	A
GUG	Val	24	0.34	GCG	Ala	32	0.34	GAG	Glu	19	0.30	GGG	Gly	9	0.13	G
U				C				A				G				

Data from the Arabidopsis Research Companion on the World Wide Web (//weeds/mgh.harvard.edu). For codon frequencies for other bacteria, see Miller (1992).
[a] Frequency represents the average frequency that this codon is used per 1000 codons.
[b] Ratio represents the abundance of that codon relative to all codons for a particular amino acid.

A

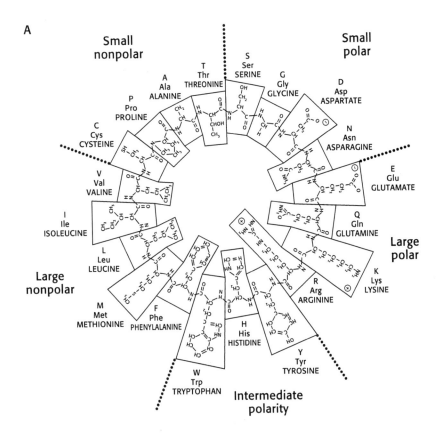

B

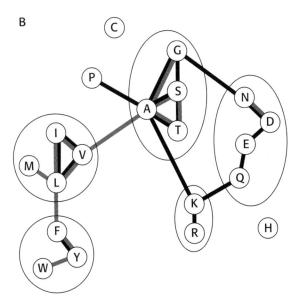

Figure 4.3 (*See following page for legend.*)

selection or screen, the important residues can be readily identified using random mutagenesis in vivo. Third, site-specific mutagenesis experiments often involve genes cloned on multicopy plasmids which may result in phenotypes that are quite different from the single chromosomal copy of a gene. Thus, random mutagenesis is extremely valuable during the initial stages of genetic analysis. However, once a gene has been characterized by this approach, site-specific mutagenesis in vitro may allow testing of critical predictions made from the in vivo experiments. Fourth, there may fortunately be no known homologs for the gene, and thus the investigator will not know what to change.

Figure 4.3 (*A*) Amino acid wheel. (Redrawn by L. Danley, with permission, from Doolittle 1985; copyright by Scientific American, Inc. All rights reserved.) Chemically similar residues have been organized into the following groups: K, R, H; D, E; Q, N; I, L, V, M; S, T; F, W, Y. C and P are not in any group (Grantham 1974; Margolin and Howe 1986). (*B*) Statistically preferred substitutions observed in buried residues (gray segments) and exposed residues (black segments) are shown. Residues roughly equivalent are grouped together in five subsets, which generally correlate with side chain physicochemical properties. (Redrawn, with permission, from Bordo and Argos 1991.)

Broad Host Range Allelic Exchange Systems

Many genetic manipulations require the exchange of mutations either between strains, from a plasmid into the appropriate chromosomal position, or from the chromosome onto plasmids. For example, such a strategy is needed for backcrossing mutations, which can be accomplished by generalized transduction or by plasmid exchange systems. A further use includes the isolation of mutations in a cloned wild-type gene with subsequent reintroduction of the altered gene into the chromosome. Allelic exchange is also generally used for in vivo cloning of chromosomally generated mutations. Several widely used allelic exchange systems have been developed to facilitate these tasks.

USE OF PLASMID INCOMPATIBILITY IN SELECTABLE MARKER EXCHANGE

The properties of plasmid incompatibility, particularly that of the IncP1 group, have been widely used as a means for marker exchange in a variety of gram-negative species. This method is largely unchanged from that first utilized by Ruvkun and Ausubel (1981), who replaced Tn5 insertions generated within fragments of *Rhizobium meliloti* DNA cloned in *Escherichia coli* into the corresponding wild-type sequences of the *R. meliloti* genome. The method can be used with any selectable marker. Generally, the fragment to be recombined is inserted into a broad host range IncP1 plasmid such as a Tc^r pLAFR derivative (Friedman et al. 1982) and maintained in an *E. coli* host. This plasmid is introduced into the strain in which the mutation is to be recombined. Although electroporation or mating can be used for this step, mating is usually preferred due to its high efficiency and the rather large size of the plasmid. The pLAFR plasmids require helper functions for their mobilization as provided by

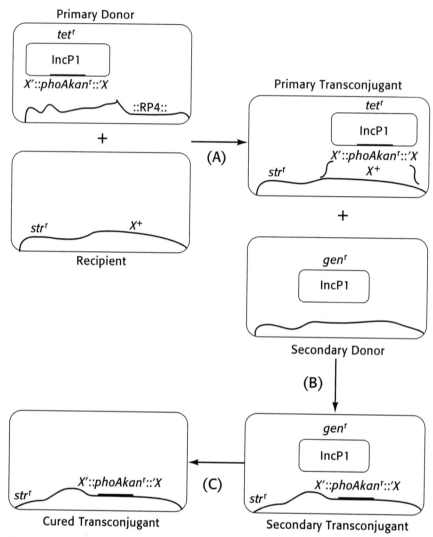

Figure 4.4 Allelic exchange of a Tn*phoA* fusion using the IncP1 system. The donor strain carries a pLAFR plasmid containing gene *x* that is disrupted with a *phoA* gene fusion marked by *kan*^r as would be the situation for a mutation generated by Tn*phoA* mutagenesis. The plasmid is present in a mobilizer strain such as SM10. This donor is mated with a recipient which carries a marker that will be used for counterselection against the donor, in this case *str*^r. (*A*) Transconjugants from *A* are mated with a second IncP1 plasmid (e.g., pPH1JI) that confers Gm^r. (*B*) Transconjugants from the second mating are selected on agar containing gentamicin and kanamycin for acquisition of the plasmid and retention of the *kan*^r marker. (*C*) Tc^s transconjugants are grown under conditions that promote curing of the Gm^r plasmid.

strains such as SM10 or S17-1 (Simon et al. 1983). Recipients carrying this plasmid are then mated with a strain carrying a second transmissible IncP1 plasmid with a different selectable marker. Plasmid pPH1JI confers resistance to gentamicin and is often used for this step, sometimes referred to as the "kickout" step. Plasmid pPH1JI is a large self-transmissible plasmid, which is again most efficiently introduced into the recipient cell by conjugation. By selection for Gm^r and Km^r (e.g., Tn5 insertion) and appropriate counterselection against the pPH1JI donor, exconjugants arise that have lost the original IncP1 pLAFR plasmid; however, they retain the *kan^r* marker that is now integrated into genomic DNA via the homology surrounding the insertion. Loss of the pLAFR vector is confirmed by scoring for Tc^s. Plasmid pPH1JI is somewhat unstable at high temperature, such that it can be cured from the final strain at a frequency of between 5% and 20% by overnight growth at 42°C in the absence of antibiotic selection.

In Figure 4.4, the plasmid incompatibility method is used to recombine a Tn*phoA* fusion into the chromosome. In this case, where the insertion can be visualized on colorimetric indicator plates, the marked insertion provides a means to make further unmarked constructions with the plasmid incompatibility method by screening for loss of the *phoA* phenotype. For example, the Tn*phoA* insertion could subsequently be replaced with a deletion that spanned or was nearby the insertion. The deletion would be inserted into a pLAFR plasmid and the procedure repeated, except that the recipient in this instance would carry the Tn*phoA* in the chromosome and one would score for its replacement by including XP in the selective agar and screening for white exconjugants. In addition, the method can be used to swap gene fusions or markers via homology at the ends of, for example, Tn5-derived elements (Wilmes-Riesenberg and Wanner 1992) (see Section 5, Gene Fusions with *phoA*).

Often, constructions that incorporate the antibiotic resistance marker within the genomic fragment of interest are first performed on a small, defined plasmid and then cloned into pLAFR for the allelic exchange into the appropriate recipient species. To avoid subcloning each construction, multiple mutations within a given region can be analyzed using a newly developed method that takes advantage of the *cos* site on pLAFR plasmids (McIver et al. 1995). Growing bacteriophage λ on a *rec$^+$* strain of *E. coli* that contains both the marked construct plasmid and pLAFR carrying the cloned corresponding wild-type region results in transducing particles that have recombined the insert from the smaller plasmid into the pLAFR plasmid. These recombinants are isolated by transduction with simultaneous selection for the Tc^r marker of pLAFR and the resistance marker carried on the insert. Cointegrates between the two plasmids can be avoided by subsequently scoring for lack of coinheritance of the plasmid used to make the initial construction. This technique is especially useful when studying a series of point mutations closely linked to a selectable marker such as a transposon insertion.

USE OF CONDITIONAL REPLICONS ("SUICIDE PLASMIDS") AS A MEANS OF CONSTRUCTING GENETIC DUPLICATIONS AND NULL ALLELES

Plasmids that are conditional for their replication can be used to create defined duplications within a target genome. In such instances, a specific DNA fragment is inserted into a plasmid which is then introduced into a recipient strain and placed under conditions where the plasmid cannot replicate. Since the plasmid cannot replicate, selection for its presence, for example, by an antibiotic marker, results in isolates that have integrated the plasmid into the host chromosome via homology between the cloned fragment and the corresponding region of the recipient chromosome. This system has also been used to obtain gene disruptions by cloning an internal portion of a gene sequence into such a suicide plasmid, thus generating two incomplete gene copies upon integration of the plasmid into the chromosome (Miller and Mekalanos 1988) (see Experiment 14).

Plasmid vectors used as the backbone for such insertional mutagenesis must have several properties desirable for this type of gene inactivation. First, the plasmid must be conditional for replication to allow for selection of insertion into the chromosome. This can be achieved by using a plasmid able to replicate autonomously only in permissive hosts or by using temperature-sensitive conditional replicons. Second, the plasmid must carry a selectable marker and, for broad host range work, should be transferable to, and selectable in, a variety of bacterial species. Plasmids that can be transferred by conjugation are preferable for situations in which other means of transfer such as transformation or electroporation are not efficient. Finally, it is useful to have an array of unique cloning sites. This system is typified by plasmid pGP704 (see Appendix G), which is a derivative of pBR322 that has a deletion of the pBR322 origin of replication (*oriE1*) but carries a cloned fragment containing the origin of replication of plasmid R6K. The R6K origin of replication (*oriR6K*) requires for its function a protein called π that is encoded by the *pir* gene. π is often supplied in trans in *E. coli* by a prophage (λpir) carrying a cloned copy of the *pir* gene. The plasmid also contains a 1.9-kb *Bam*HI fragment encoding the *mob* region of RP4. Plasmid pGP704 can be mobilized into recipient strains by transfer functions provided by a derivative of RP4 integrated in the chromosome of *E. coli* strain SM10 or S17-1 (Simon et al. 1983) carrying λpir, but it is unable to replicate in recipients because they do not provide the essential π protein function.

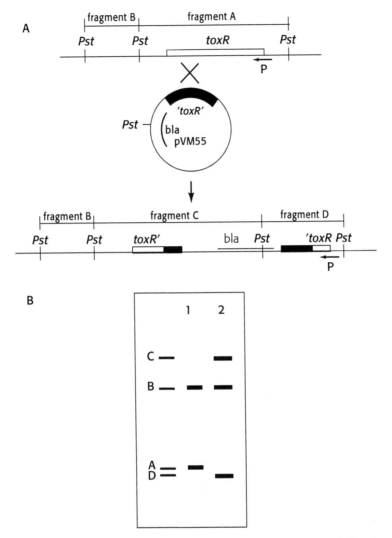

Figure 4.5 Chromosomal knockout of *toxR*. (*A*) Structure of the chromosome before and after recombination of plasmid pVM55; (*B*) representation of Southern analysis of parent (lane *1*) and recombinant (lane *2*) strains. (Adapted from Miller and Mekalanos 1988.)

Insertion mutations in a chromosomal gene are isolated by first sub-cloning DNA fragments carrying sequences internal to the coding sequence into pGP704. Because these derivatives cannot replicate in the recipient (e.g., *Vibrio cholerae*), Ap^r transconjugants should contain the mobilized plasmid integrated into the genome by homologous recombination between the internal gene sequence present on the plasmid and the corresponding intact gene in the chromosome. Generally, after putative insertion mutants have been purified, the structure of the chromosomal insertion is verified by Southern analysis (Fig. 4.5) or colony polymerase chain reaction (PCR).

USE OF SUICIDE VECTORS FOR ALLELIC EXCHANGE OF NONSELECTABLE MUTATIONS

Genetic interpretation is usually optimal when the mutation or genetic construction, such as a gene fusion, employed for the analysis is present in unit copy in its natural chromosomal context. If the mutation or construction is initially engineered on a plasmid, allelic exchange can be utilized to replace the wild-type chromosomal copy of the gene with the altered gene. Such gene replacements are usually easily achieved only if the mutation or fusion is either the result of, or closely linked to, a selectable marker such as an antibiotic resistance gene. This problem has been alleviated by modifying the suicide plasmid vectors such as those discussed above to provide a means to take advantage of the merodiploid nature of the plasmid integration to introduce nonselectable mutations into the chromosome. The modification provides a selection for excision of the plasmid by recombination between the flanking direct repeats of chromosomal homology that results in removal of one copy of the duplicated region. If one of the duplicated sequences carries a mutation, a certain percentage of the resolved products will remove the wild-type sequence and leave the mutation in the chromosome as seen in Figure 4.6. The frequency with which the mutation becomes incorporated into the chromosome is influenced by the position of the mutation within the region of homology. Thus, if the mutation is perfectly centered in the DNA fragment, approximately 50% of the selected colonies often carry the mutation in the chromosome. In the absence of a method to select for loss of the integrated plasmid, a large number of colonies would first have to be screened just to find those that had excised the plasmid sequences.

One marker that has been utilized to select for loss of the integrated plasmid is the *sacB* gene from *Bacillus subtilis* (Ried and Collmer 1987). Such vectors have been developed using either *pir*-dependent (Donnenberg and Kaper 1991) or temperature-dependent (Blomfield et al. 1991) plasmid replicons. The *sacB* gene encodes the enzyme levan sucrase, the expression of which is toxic for gram-negative bacteria grown in the presence of 5% sucrose (Gay et al. 1985). Thus, only colonies that have removed *sacB* by recombining out the integrated plasmid sequences can survive in the presence of sucrose. In practice, some sucrose-resistant colonies can arise from alterations in either the *sacB* gene itself or its expression, rather than by deletion of the vector from the chromosome.

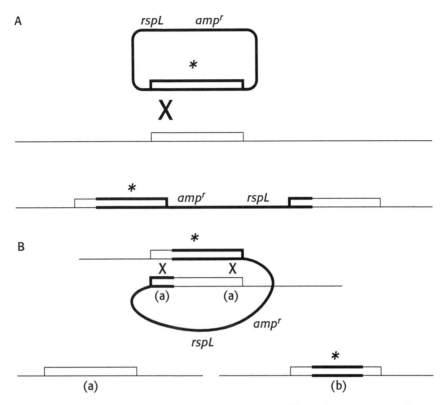

Figure 4.6 (*A*) Chromosomal integration of a plasmid carrying a mutation (*) on a cloned fragment by homologous recombination on either side of the mutation generates a heterodiploid strain. (*B*) Two possible excision events (a or b). Excision of the plasmid by homologous recombination on the opposite side of the mutation from which the integration occurred (excision b) results in allelic exchange of the mutation from the plasmid to the chromosome. The excised, non-replicating plasmid is lost upon cell division.

Thus, colonies arising from this selection are subsequently scored for the concomitant loss of Ap[r] to ensure that the sucrose-resistant phenotype is due to loss of the integrated plasmid.

Another strategy to select for loss of the integrated plasmid utilizes the *E. coli rpsL* gene. The *rpsL* gene encodes ribosomal protein S12. Integration of the plasmid carrying wild-type *rpsL* into a streptomycin-resistant strain carrying a chromosomal mutation in *rpsL* results in a merodiploid strain that is streptomycin-sensitive. This occurs because the mutation is recessive to the wild-type protein. Subsequent selection for streptomycin resistance results in colonies that have recombined out the vector, similar to the situation described for the *sacB* vectors. Plasmid pRTP1 is a *colE1* replicon carrying *rpsL* that has been used successfully for allelic exchange in *Bordetella pertussis* (Stibiz et al. 1986; Stibitz 1994). However, since *colE1* plasmids replicate efficiently in *V. cholerae* and a number of other bacterial species, the host range utility of pRTP1 is limited. To take advantage of the streptomycin selection to introduce mutations into *V. cholerae*, the *rpsL* gene has been inserted into the plasmid pGP704 discussed above (Skorupski and Taylor 1995). One such derivative, pKAS32, is utilized in Experiment 14. A map for this plasmid and a version carrying a kanamycin resistance gene (pKAS46) are shown in Appendix G (Plasmids pKAS32 and pKAS46). These vectors should prove useful for bacterial species in which the *E. coli rpsL* gene is expressed and provides a streptomycin-sensitive phenotype.

The inability of *colE1*-based plasmids to replicate in some bacterial species has also been exploited for allelic exchange in *Pseudomonas aeruginosa*. The pEMR series of vectors (Flynn and Ohman 1988) combine many of the properties for cloning and mobilization of the vectors described above and take advantage of the inherent instability of Tn5 in *Pseudomonas*, which makes it practical to screen transconjugants that have undergone gene replacement.

The high efficiences of allelic exchange, especially as achieved with the positive selection vectors, make it feasible to physically screen chromosomal integration of mutations for which there is no phenotypic screen. Such approaches are also useful for confirming mutations for which there is a phenotypic screen or selection. They generally involve analysis of an amplified product generated by rapid colony polymerase chain reaction (PCR) using primers that flank the region of interest (see Experiment 14). The presence of insertion or deletion mutations that alter the size of a particular DNA fragment can be determined directly by gel electrophoresis of the amplified product. For base substitutions, the colony PCR product can be sequenced directly or subjected to RFLP analysis if the mutation alters the restriction digest pattern of the amplified fragment.

IN VIVO CLONING OF MUTANT CHROMOSOMAL ALLELES ONTO PLASMID VECTORS

The principles discussed above for the recombination of mutant alleles from plasmids into the chromosome can also be used to recover chromosomal mutations onto plasmids (Muro-Pastor and Maloy 1995). This approach can be extremely useful when a number of alleles of a given region are to be analyzed since it circumvents the need for repeated time-consuming in vitro cloning manipulations or PCR amplification of a region that can be subject to the introduction of secondary mutations or inefficient recovery of large fragments. The diagram in Figure 4.7 shows this system as it is used with pSC101 in a *polA⁻/polA⁺* strain pair to pro-

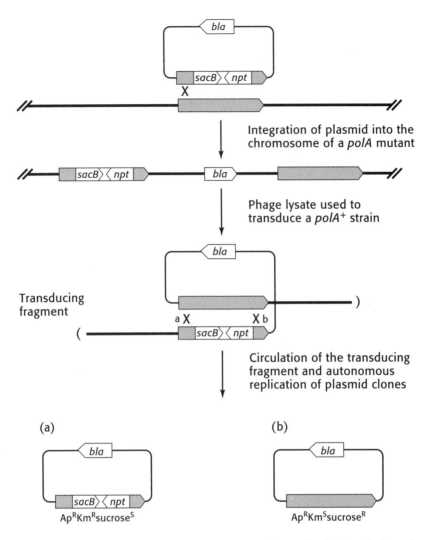

Figure 4.7 Scheme for in vivo cloning using positive selection for allelic exchange. A gene of interest with its internal portion substituted by a cassette containing the *sacB* gene and the *npt* gene (Km[r]) is inserted into the replication-conditional plasmid vector to be integrated into the chromosome. The construct is introduced into the nonpermissive host (*polA* mutant in the case of the pSC101-based vector in this example) with selection for Ap[r] and Km[r], which results in cells that harbor the construct integrated into the chromosome at a point of homology with the wild-type gene. A generalized transducing phage lysate is prepared on the integrated strain and used to transduce a host that is permissive for plasmid replication (*polA+*) with selection for Ap[r]. In the permissive host, the transducing fragment efficiently recircularizes and replicates. Resolution of the partial duplication results in one of the two possible plasmid segregants designated as *a* or *b*. Incorporating sucrose into the transduction plate provides a selection for plasmid *b* containing the gene of interest which is scored as Km[s]. In practice, the sucrose[s] phenotype of constructs carrying *sacB* must be monitored at each step to avoid accumulation of mutations leading to sucrose[r]. (Modified, with permission, from Muro-Pastor and Maloy 1995.)

vide nonpermissive and permissive hosts for plasmid replication. The scheme can also be applied using a *pir⁻/pir⁺* strain pair for plasmids requiring π protein for replication. A strain containing a mutant allele is transduced or transformed with the plasmid containing a copy of the wild-type gene that is disrupted by a DNA fragment carrying the *sacB* gene from *Bacillus subtilis* and a Km^r gene. A direct selection for integration of the plasmid into the chromosome produces a tandem duplication. Upon recircularization of the plasmid in a permissive host, the conditionally lethal phenotype associated with the *sacB* gene is used to select for the desired recombinant plasmids. Alternative conditional lethal phenotypes such as expression of the wild-type *rpsL* gene in an Sm^r *rpsL* mutant host should also be useful for this step.

Suppression

The characterization of genetic suppressors provides a powerful method for determining mechanisms of protein function and gene regulation, elucidating biological pathways, and dissecting complex processes. Suppression is particularly useful for determining molecular interactions and mechanisms of function within a single protein, between proteins, or between proteins and nucleic acids. Suppression studies can be thought of as a kind of reversion analysis. Most mutants, particularly conditional lethals, readily revert to the wild-type phenotype (i.e., growth, in the case of a conditional lethal). This can occur by reversion of the original mutation back to the wild-type DNA sequence, termed a true revertant, or by a secondary event that restores a wild-type phenotype to the cell, termed a pseudorevertant. Such a secondary mutation resulting in a pseudorevertant is termed a suppressor; i.e., the second event suppresses the original mutant phenotype. For the clever geneticist, suppressor analysis is a very powerful tool.

General mechanisms of suppression with some specific examples of how they have been applied are given below. In addition, two excellent reviews have been written by Hartman and Roth (1973) and Botstein and Maurer (1982).

INTRAGENIC SUPPRESSION

When pseudorevertants are due to a second alteration within the same gene as the original mutation, they are termed intragenic suppressors (Yanofsky et al. 1964). Such suppressors can provide the following useful information. If the change is a different nucleotide in the original codon that results in a substituted amino acid, information is gained regarding the stringency or constraints within a given position of a protein. Changes at a second site within the same gene can restore function by

135

several means. For example, altering an amino acid at a different site can allow a protein to regain function by restoring its conformation or stability. Such mutations provide information regarding the structure of a protein and the interactions that stabilize it. In a more confined regional analysis, these types of suppressors can shed light on the structure and mechanism of action of enzymatic sites. Another example of intragenic suppression is restoration of the reading frame in a frameshift mutant. Such a suppressor analysis was in fact instrumental in elucidating the nature of the genetic code (Crick et al. 1961). In this case, suppression is due to a compensating frameshift or, rarely, to the formation of a new initiation codon (i.e., create an appropriately spaced ribosome-binding site and ATG that still provide enough coding sequence for a functional protein). A mutation such as a frameshift may occasionally produce a polypeptide that is toxic to the cell. In such cases, intragenic suppression may occur by a second mutation that eliminates the toxic polypeptide. An example of such a suppressor might be an early termination mutation.

INTERGENIC SUPPRESSION

Each of the cases described above deals with a pseudorevertant resulting from an alteration within the same gene as the original mutation. Suppression can also occur when the compensating mutation is located in a different gene. There are many ways by which such intergenic suppression can occur.

Informational Suppression

It is possible to isolate suppressors that change the cell's translational machinery in such a way that the mutation is misread, making a functional protein. Such suppressor mutations are called informational suppressors because they change the way the cell reads the "information" in the mRNA (Bossi 1985). Most informational suppressors are the result of altered tRNA genes.

The most common class of informational suppressors are *nonsense suppressors*. Nonsense mutations are base substitutions that introduce a premature translational stop codon—UAG (amber), UAA (ochre), or UGA (opal)—into a gene. Since there are no tRNA molecules in a wild-type cell with anticodons to match these sequences, a truncated protein that usually has little or no activity is produced. Informational suppression can occur via a mutant tRNA that places an amino acid in place of a stop, or nonsense, codon. Such mutant tRNAs result from a change in the anticodon. For example, a tyrosine codon UAC could mutate to a UAG amber codon. If a tRNA^{Tyr} gene were mutated so that its normal anti-codon (GUA) were changed to CUA, it would be able to hydrogen-bond with UAG and place a tyrosine residue back at the original position, thus restoring the wild-type amino acid sequence to the mutant protein. Such a suppressor tRNA^{Tyr} could then place tyrosine residues at any UAG nonsense mutation occurring at any position in any gene, whether that position had originally encoded tyrosine or not. Some restored proteins could tolerate tyrosine at this position and regain activity, whereas other restored proteins might not. Thus, an amber nonsense suppressor can suppress many, but not all, amber mutations to various degrees. Cells that are wild type with respect to nonsense suppressors are denoted as *sup*^{0} and those carrying suppressors are denoted as *sup*. The following are important features of nonsense suppression: Not every nonsense codon is suppressed equally well by a particular nonsense suppressor and suppression is not always complete. Because of wobble, ochre suppressors may also be capable of suppressing amber mutations. Nonsense suppressors can only occur in cases where the cell contains two or more copies of the particular tRNA. For example, if there was not more than one copy of the tRNA^{Tyr} gene noted above, there would not be any tRNA to place tyrosine at UAC codons in the suppressor mutant. Finally, the majority of termination is still correct in *sup* strains and thus they are viable. Two features of the normal termination process contribute to this. First, protein termination factors respond to chain termination codons even when a tRNA that recognizes the codon is present. Thus, suppression of termination is not 100%. Second, many normal open reading frames end with more than one stop codon, with a second stop codon located several bases downstream and in the same reading frame as the first. Third, the most common termination is by a single UAA codon, and UAA suppressors are generally very inefficient.

Several other classes of informational suppressors are very rare compared to nonsense suppressors. One is *missense suppressors*, which can suppress missense mutations by amino acid substitution (Hill 1975). Such a substitution occurs by one of three ways: (1) a mutant tRNA molecule may recognize two codons, (2) a mutant tRNA molecule may be incorrectly recognized by an aminoacyl synthetase so that it carries the wrong amino acid, or (3) a mutant aminoacyl synthetase may charge an incorrect tRNA molecule. As one might imagine, this mechanism of suppression is very detrimental to the cell and all known suppressors of this type suppress with an efficiency of less than 1%. It is also possible to isolate tRNA suppressors that suppress certain frameshift mutations. *Frameshift suppressors* often change the tRNA so that the anticodon recognizes four bases instead of three bases as normal. These tRNA suppressors shift the translational reading frame at a low frequency, thus restoring translation of the correct amino acid sequence downstream from the frameshift mutation. A last type of informational suppression is due to a ribosomal alteration instead of one affecting tRNA. An example of one of these mutations, called *ram* for ribosomal ambiguity (Rosset and Gorini 1969; Vincent and Liebman 1992), results in ribosomes that incorrectly translate mRNA at high frequency. Because such ribosomal mutations cause the accumulation of errors in all proteins, such mutants are very unhealthy.

Interaction Suppressors

In addition to informational suppression, a number of other mechanisms exist by which intergenic suppression can occur. One of the most important and widely used in genetic analysis is a class called interaction suppressors. For example, when two proteins interact, a mutation in one of the genes may confer a mutant phenotype simply because it disrupts proper protein-protein interactions. In such cases, it may be possible to isolate interaction suppressors in the second gene, enabling its product to restore interaction with the first altered polypeptide. Because only specific alterations in the second polypeptide will interact with a given alteration in the first polypeptide, such interaction suppressors are usually allele-specific; i.e., a specific interaction suppressor will only suppress a small subset of mutations in a gene. This type of suppression can identify interacting polypeptides in a holoenzyme or directly interacting components of a biochemical pathway (Jarvik and Botstein 1975; Parkinson and Parker 1979). Synthetic lethality is another type of analysis that can infer direct interaction between proteins (Flower et al. 1995).

Dosage Compensation Suppression

Sometimes a negative phenotype is due not to a complete loss of activity, but rather to a defective protein with greatly decreased activity, or a decrease in expression of an otherwise wild-type gene. In such cases, the wild-type phenotype might be restored by increasing the amount of gene product, or dosage compensation suppression. This could occur by fusion of a gene to another promoter, insertion of a genetic element containing an outward promoter nearby, or more informatively by a promoter-up or altered regulation mutation. Altered regulation mutations might occur in regulatory genes or in cis regulatory sites that increase synthesis of the defective protein. Another way to suppress a phenotype due to a partially defective molecule is by increasing the effective gene dosage as might occur through a gene duplication or by high-copy overexpression suppression (Dammel and Noller 1995).

Bypass Suppressors

Although none of the suppressor mechanisms discussed so far will suppress every mutation in a given gene, bypass suppressors have this property. A bypass suppressor provides a new activity in the cell that can substitute for a normal activity that has been prevented by mutation. One way this suppression can occur is when a second mutation alters a second protein's activity so that it can substitute for the first. Alternatively, some bypass suppressors activate or derepress a previously inactive gene that has a similar function or whose function leads to accumulation of metabolic intermediates that are lacking in the mutant pathway (Kuo and Stocker 1969). Thus, in the case of bypass suppression, all mutations in a given gene, and sometimes in a complete metabolic pathway, are suppressed.

Physiological Suppressors

The final class of suppressors are termed physiological suppressors. For example, a missense mutation might produce a protein that is still functional but is unstable. Second mutations that alter the physiology of the cell, for example, by changing an intracellular ion concentration, could result in restoration of function (Suskind and Kurek 1959). Such a suppression mechanism might also work indirectly through chaperone proteins, which are a class of proteins that help other proteins to fold correctly.

Genetic Complementation

Geneticists often define genes operationally as *complementation groups* (or as cistrons; Benzer 1959). What do we mean by this? A strain that carries a nonpolar mutation in the *trpE* gene, which encodes component I of anthranilate synthase, is designated Trp⁻ (requires tryptophan for growth). The Trp⁺ phenotype can be restored by introducing the *trpE⁺* gene in trans, for example, on a λ specialized transducing phage. Although this strain retains the original *trpE* mutant allele, TrpE function is provided by the complementing gene on λ. This is distinct from recombination, where *trpE⁺* information physically replaces the region including the *trpE* mutant allele. Thus, *trp* mutants that are restored to the Trp⁺ phenotype by the λ *trpE⁺* phage are said to define the *trpE* complementation group (i.e., they are *trpE* mutants; Fig. 4.8).

Complementation tests should be carried out in *recA* hosts to eliminate possible recombination between the two copies of the gene being tested, as such recombination could restore a wild-type gene to give a spurious result. In *Escherichia coli*, complementation analysis is conveniently performed with F′ episomes or with λ specialized transducing phage. In *Salmonella typhimurium*, merodiploid complementation analysis is best performed with tandem duplication strains (see below).

INTRAGENIC COMPLEMENTATION

In general, the terms complementation group and gene define the same entity. However, there are special-case exceptions. For example, the *trpC* gene encodes a bifunctional enzyme, phosphoribosyl-anthranilate iso-

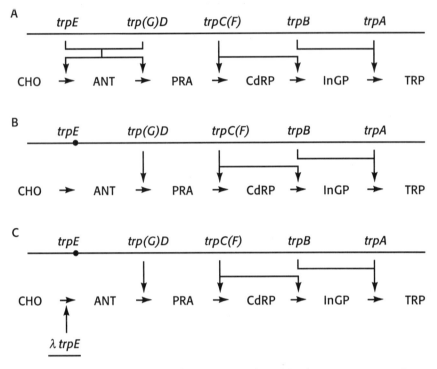

Figure 4.8 Complementation of a *trpE* mutation. (*A*) The *trp* operon and Trp synthetic pathway. Compounds are as follows: CHO, chorismate; ANT, anthranilate; PRA, phosphoribosylanthranilate; CdRP, carboxy-phenylamino-1-deoxyribulose-5-phosphate; InGP, indole 3-glycerol phosphate; TRP, tryptophan. (*B*) A *trpE* missense mutation blocks synthesis of ANT, but the remaining enzymes of the Trp pathway are synthesized. (C) A λ *trpE* specialized transducing phage provides TrpE function, thereby complementing the *trpE* mutation.

merase/indole glycerol phosphate synthetase (TrpF/TrpC functions). These two enzymatic activities are located in independent domains. Certain missense mutations affecting the TrpF domain will complement certain other missense mutations affecting the TrpC domain (Fig. 4.9) (Yanofsky et al. 1971). Thus, be vigilant when interpreting the results of complementation analyses. Indeed, the *hisI* and *hisE* genes were defined as such on the basis of complementation analysis, but subsequent analysis showed that both functions were encoded by a single gene *hisI* (Carlomagno et al. 1988).

A second form of intragenic complementation is sometimes observed for genes that encode multimeric proteins. Two distinct mutant alleles, each affecting assembly of the oligomer, may complement each other if the resulting mutant polypeptides are able to assemble into the correct multimeric structure (Crick and Orgel 1964; Gordon and King 1994).

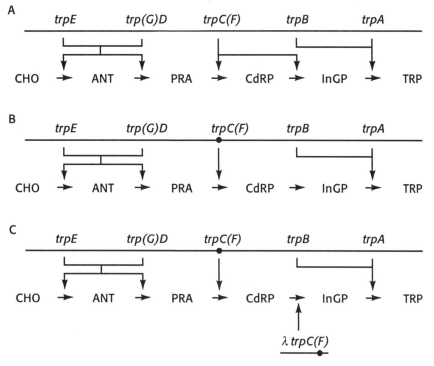

Figure 4.9 Intragenic complementation. (*A*) The *trp* operon and Trp synthetic pathway. (*B*) A missense mutation in the *trpC* domain blocks conversion of CdRP to InGP. (*C*) A λ *trpC* specialized transducing phage carrying a missense mutation in the *trpF* domain provides TrpC function.

METHODS FOR COMPLEMENTATION

Although multicopy plasmids are often used to perform complementation analysis, complications may arise in interpreting such results. In particular, an imbalance between the products of the single chromosomal allele and the multiple plasmid-borne alleles may lead to spurious results. However, the context of the complementation analysis is also a factor. For example, a null allele of a biosynthetic enzyme can usually be complemented by a multicopy plasmid carrying the wild-type allele if the purpose is simply to demonstrate that the cloned fragment indeed corrects the mutant defect. In contrast, the relative levels of regulatory gene products are often critical in establishing proper regulation, and imbalances in their levels of synthesis, due to expression from a multicopy plasmid, can yield different complementation patterns when compared to the same tests performed with single-copy systems (see, e.g., Collins et al. 1992).

The ideal system for conducting genetic complementation tests should consist of the following features: (1) both alleles of the gene in question should be present in single copy, (2) strains that are merodiploid for any predetermined region of the chromosome should be easily constructed, and (3) haploid segregants carrying either of the alleles should be easily isolated to confirm that the merodiploid strain has the correct genotype.

Tandem Duplications

All of the above features are provided by tandem duplications, which are easily constructed in *S. typhimurium* by using Mud elements located at known locations in the chromosome (Hughes and Roth 1985). One begins with P22 HT *int* lysates grown on each of two different MudA strains, with the positions of the MudA insertions defining the endpoints of the duplication. The MudA insertions serve as portable regions of homology for recombination. Thus, both MudA insertions must be in the same orientation with respect to each other. Since P22 only rarely packages an entire MudA (because the MudA genome is almost as big as the P22 genome), most transducing particles carrying chromosomal DNA from the region around the MudA insertion contain only a portion of MudA. Thus, the parental MudA insertions will be inherited with low frequency in transductional crosses. To generate duplications, one infects the recipient with high multiplicities of both MudA transducing lysates and selects for Ap^r. One way for Ap^r to be inherited in such a cross involves homologous recombination between the transduced MudA fragments, along with recombination in the flanking chromosomal regions (Fig. 4.10). Thus, to form a duplication between *nadB* and *cysH*, for example, one selects for Ap^r colonies on rich medium and replicates the colonies to defined medium lacking nicotinamide and cysteine. Prototrophic Ap^r colonies are candidate tandem duplication strains (see Experiment 3).

The newly generated intact MudA defines the duplication join-point. As long as the duplication strain is cultured in the presence of Ap, the duplication will be stably inherited. However, if Ap selection is removed, then Ap^s haploid segregants will arise at relatively high frequency. Thus, a true duplication strain will have an unstable Ap^r phenotype.

For complementation, appropriate alleles are transduced into the two halves of the duplication. For example, a *srl*::MudJ insertion transduced into one half of the duplication will generate a diploid *srl$^+$/srl*::MudJ strain. (The MudA at the duplication join-point is chosen to be in the opposite orientation from the MudJ in *srl*, to minimize potential problems from MudA-MudJ recombination.) Ap^s haploid segregants of this strain will be of two types: Srl$^+$ Km^s and Srl$^-$ Km^r. This type of segregation analysis provides final verification that the parental strain carries the desired duplication and the desired alleles in each of the two duplicated regions (Fig. 4.11).

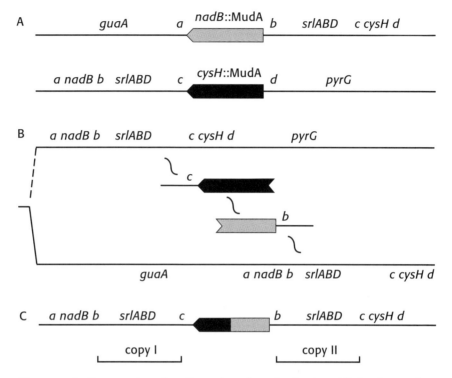

Figure 4.10 Construction of a directed tandem duplication. (*A*) Starting strains. (*B*) Recombination events following transduction with P22 HT *int* lysates grown on both starting strains. (*C*) Structure of the resulting Ap^r duplication strain, which is duplicated for the region *b* - *srl* - *c*.

Integrative Plasmids

The advent of nonreplicating suicide plasmids provides an analogous avenue for conducting complementation analysis, particularly in organisms other than *S. typhimurium* and *E. coli*. For example, replication factor π-dependent plasmids (e.g., pGP704) only replicate in strains carrying the π structural gene, *pir* (see Experiment 14). However, a π-dependent plasmid that carries a segment of bacterial DNA can integrate into the chromosome of a π$^-$ recipient by homologous recombination. Such integrants can be isolated by selecting for the drug resistance phenotype encoded by the plasmid. The resultant strain carries a tandem duplication, the boundaries of which correspond to the boundaries of the cloned DNA. Thus, complementation analysis simply requires that the donor plasmid and recipient bacterium contain different alleles of the gene in question. This method was used for a simple complementation analysis of a *Klebsiella pneumoniae* regulatory gene (Fig. 4.12) (Goldman et al. 1994).

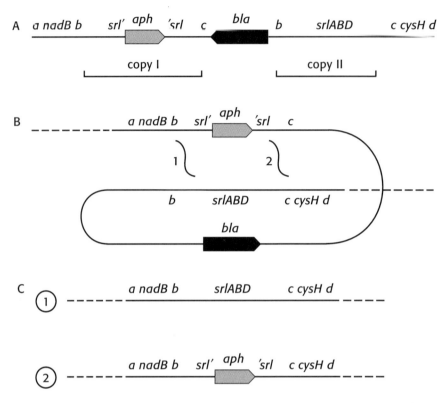

Figure 4.11 Segregation of a tandem duplication. (*A*) starting strain, merodiploid for *srl*. (*B*) Segregation. (*C*) Ap⁵ haploid segregants. Strain (1) arises from recombination at site 1; strain (2) arises from recombination at site 2.

One drawback of this method in comparison to the tandem duplication method described above is that it is sometimes difficult to isolate haploid segregants because of the small region of homology available for recombination. However, introduction of a counterselectable marker, such as *sacB* (which confers the sucrose-sensitive phenotype), allows for enrichment of segregants.

F' Episomes

A common method for conducting complementation analysis is to use F' episomes, which are derivatives of the F factor that carry a portion of the bacterial chromosome. A variety of F' episomes have been isolated for both *E. coli* and *S. typhimurium* (Holloway and Low 1987; Sanderson and Roth 1988). The advantages of F' episomes are that they are maintained at approximately one copy per chromosome, mutant alleles can be introduced by transduction or by homogenization, and F' episomes are easily

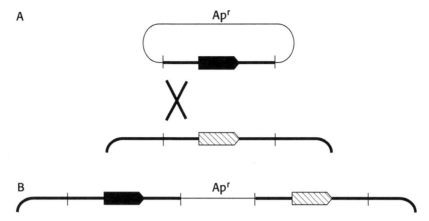

Figure 4.12 Construction of a defined tandem duplication by using an integrative plasmid. (*A*) Recombination events following introduction of a replication-defective plasmid into the recipient strain. (*B*) Structure of the resulting Ap^r duplication strain, which is duplicated for the gene of interest.

transferred from one strain to another by conjugation. However, some F′ episomes are relatively unstable and tend to lose the integrated bacterial DNA at a high frequency. A good example of using F′ episomes for complementation analysis is provided by Roof and Roth (1989).

Specialized Transducing Phage

Integration-proficient specialized transducing phage also provide a good means for conducting complementation analysis. For *E. coli*, a variety of integration-proficient (*att*$^+$ *int*$^+$) derivatives of bacteriophage λ have been developed as molecular cloning vehicles (Chauthaiwale et al. 1992). A transducing phage carrying the gene(s) of interest can be isolated either by direct cloning or by genetic screening of a library. The development of a λ specialized transducing phage system for complementation analysis is described by Egan and Stewart (1990, 1991) and by Collins et al. (1992).

COMPLEMENTATION ANALYSIS OF GENE REGULATION

One important use of complementation analysis is to elucidate mechanisms of gene regulation. A null allele of a gene encoding a trans-acting regulatory factor can be complemented by a wild-type copy of that gene (trans-recessive). However, a mutation in a cis-acting control sequence cannot be complemented (cis-dominant). Figure 4.13A–D describes two general mechanisms for gene regulation, repression and activation, and shows how complementation tests can define the genes encoding the trans-acting regulatory factors.

A   In the absence of inducer, repressor binds to the operator. This
blocks RNA polymerase from binding to the promoter and
initiating transcription.

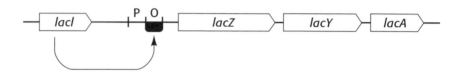

Inducer binds to and inactivates repressor, preventing it from binding
to the operator. RNA polymerase is now free to bind to the promoter
and initiate transcription.

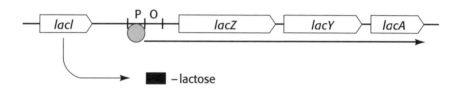

B   A null (e.g., insertion) mutation in the repressor gene causes
constitutive expression of the operon, irrespective of the
presence of inducer.

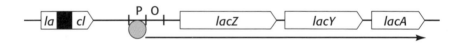

This null allele of the repressor gene is *recessive.* A
wild-type copy of the repressor gene, provided in trans, will
restore inducibility.

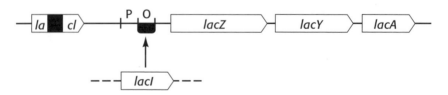

Figure 4.13 (*See facing page for parts C and D and legend.*)

C   In the absence of inducer, activator fails to bind to the activator
binding site. Thus, RNA polymerase cannot bind to the promoter
and initiate transcription.

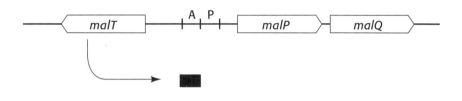

In the presence of inducer, activator binds to the activator binding site.
This facilitates RNA polymerase binding to the promoter, allowing it to
initiate transcription.

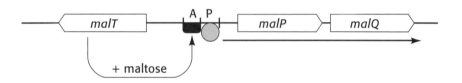

D   A null (e.g., insertion) mutation in the activator gene causes
expression of the operon to be uninducible, irrespective of
the presence of inducer.

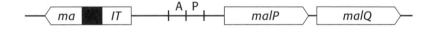

This null allele of the activator gene is *recessive.* A wild-type
copy of the activator gene, provided in trans, will restore
inducibility.

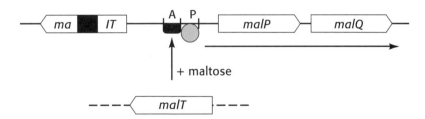

Figure 4.13 (*A*) The *lac* operon: negative regulation. (*B*) Complementation of a
lacI null allele. (*C*) The *malPQ* operon: positive regulation. (*D*) Complementation
of a *malT* null allele.

Polarity

Transcriptional polarity occurs when a mutation in an upstream gene of an operon causes reduced expression of the downstream genes in the same operon. Polarity is a consequence of premature transcription termination and thus fewer RNA polymerase molecules reach the downstream portion of the operon.

The transcriptional polarity caused by upstream nonsense or insertion mutations is well-documented in a variety of systems (see, e.g., Imamoto et al. 1966; Zipser et al. 1970; Yanofsky et al. 1971; Kleckner et al. 1977), but the *Salmonella typhimurium his* operon provides a particularly rich source of material for discussion (Martin et al. 1966; Fink and Martin 1967). The *hisGDCBHAFE* operon encodes the enzymes required for histidine synthesis. The *hisG* gene encodes the first enzyme in this pathway, and the *hisD* gene encodes the final enzyme, histidinol dehydrogenase. *S. typhimurium hisD*$^+$ strains can use histidinol as the sole source of histidine (Hol$^+$ phenotype), even if other *his* genes (such as *hisG*) are inactivated by mutation. Thus, selections or screens involving Hol phenotypes provide a good means for examining the effects of polar mutations in the *hisG* gene.

RHO-DEPENDENT TRANSCRIPTION TERMINATION

Prokaryotic transcription and translation are coupled. This simple statement is usually taken to mean that both processes occur in the same cellular compartment (as opposed to eukaryotes), but here the meaning of the term *coupled* is somewhat more profound. In fact, transcription and translation are intimately connected, and a failure to complete translation quickly leads to termination of transcription. This mechanism has likely evolved to prevent wasteful mRNA synthesis in cells that are sufficiently starved for amino acids that they cannot effectively translate mRNA chains into polypeptide chains.

151

Transcribing RNA polymerase pauses at certain locations within each structural gene. An oncoming ribosome, translating the mRNA into protein, causes release of RNA polymerase from the pause site so that it can continue transcription until it encounters the next pause site. Thus, RNA polymerase waits at specific sites within a transcription unit while translating ribosomes catch up. The coupling of transcription and translation therefore represents an active form of communication between these two forms of macromolecular synthesis.

If mRNA translation is slowed or blocked, then transcription will be terminated at a pause site. The transcription termination factor is the Rho protein (Greek letter ρ, for release factor; Roberts 1969). This hexameric ATPase loads onto untranslated mRNA and is thought to move cartwheel-like along the mRNA until it encounters paused RNA polymerase, whence it catalyzes dissociation of the RNA polymerase/DNA/mRNA ternary complex (transcription termination; Richardson 1991). This process requires untranslated RNA because the presence of ribosomes on the mRNA blocks Rho action. Rho-dependent transcriptional terminators (pause sites) are unlike factor-independent terminators, which consist of easily recognizable GC-rich stem-loop structures immediately followed by a run of T residues. Rather, the DNA sequence features that characterize pause sites are less well-defined and are not easily detected by sequence inspection alone (Alifano et al. 1991).

NONSENSE POLARITY

Consider coupled transcription-translation of the *S. typhimurium his* operon (Fig. 4.14). RNA polymerase binds to the promoter and transcribes the DNA until it reaches a transcriptional pause site. Meanwhile, a ribosome loads onto the mRNA at the ribosome-binding site (Shine-Dalgarno sequence plus initiator codon) and translates the mRNA into protein. The encounter between the translating ribosome and the paused RNA polymerase releases the pause, and both processes continue. At the end of the first gene (*hisG*), the ribosome dissociates from the mRNA at the termination codon, and RNA polymerase continues transcription into the second gene (*hisD*). Once the *hisD* ribosome-binding site is transcribed into mRNA, a ribosome will load and commence translation. These events continue through the remainder of the operon.

Introduction of a nonsense mutation into the *hisG* gene causes premature translation termination (Fig. 4.14). Transcription continues, however, until RNA polymerase reaches the pause site. (As it happens, the *hisG* gene is unusual in that it seems to contain only one major pause site; Ciampi and Roth 1988; Ciampi et al. 1989.) However, there are no translating ribosomes to release RNA polymerase from its pause, and thus Rho

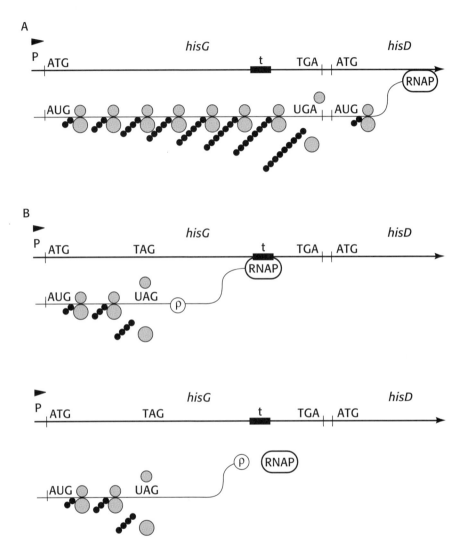

Figure 4.14 Nonsense polarity in the *S. typhimurium his* operon. The promoter, *hisG* gene, and the proximal portion of the *hisD* gene are shown. DNA and RNA are shown as thick and thin lines, respectively. RNA polymerase and Rho factor (ρ) are labeled. Ribosomes and nascent polypeptide chains are indicated by gray and black circles, respectively. The polarity site (t) within the *hisG* gene is indicated by a black rectangle. (*A*) Coupled transcription and translation of the *his* operon under normal circumstances. (*B*) Nonsense polarity. The UAG (amber) mutation within the *hisG* sequence causes premature translation termination. Rho factor recognizes the untranslated RNA, and catalyzes transcriptional termination of RNA polymerase, which has paused at the polarity site.

protein has sufficient time to bind to the mRNA and catalyze transcription termination. This RNA polymerase complex is unable to reach the *hisD* gene, and thus the nonsense mutation in *hisG* has caused decreased expression of downstream genes in the *his* operon.

The severity of the polarity effect depends on a variety of factors, including the time RNA polymerase spends at the pause site before spontaneously continuing transcription, the efficiency of Rho protein action, the distance between the nonsense codon and the pause site, and the distance between the pause site and the beginning of the next gene. These factors indicate that more upstream nonsense mutations usually exhibit strong polarity, because few RNA polymerase complexes reach the next gene, whereas more downstream nonsense mutations are usually less polar, because more RNA polymerase complexes can reach the next gene before Rho has time to act. Once translation has initiated at the downstream gene mRNA, RNA polymerase is again immune to Rho protein action. Examination of several different nonsense mutations within a given gene reveals this gradient of polarity, where the magnitude of the polar effect correlates with the location of the mutation (Fink and Martin 1967; Zipser et al. 1970; Yanofsky et al. 1971).

INSERTIONAL POLARITY

Polarity can also be caused by an insertion in an upstream gene (Fig. 4.15). Most of the commonly used transposons contain strong terminators, and transposon insertions are generally considered to be strongly polar (although special-case exceptions have been documented; see below). Indeed, transposon insertions can be used to probe the structure of operons. If a series of both transposon insertions and *lacZ* operon fusions are isolated within a newly studied gene cluster, then both types of insertions can be combined to examine polarity. An upstream transposon insertion (e.g., Tn*10*) in an operon will block transcription from reaching a downstream *lacZ* fusion (e.g., MudJ). If the two elements are in separate operons, the upstream insertion will have no effect on *lacZ* expression, and if the two elements are in a complex operon with internal promoters, then the upstream insertion will cause decreased *lacZ* expression. Examples of this type of analysis are provided by Roof and Roth (1988) for the simple *eut* operon and by Jeter and Roth (1987) and Escalante-Semerena et al. (1992) for the complex *cob* operon.

Not all insertions by a given transposon are completely polar, however (Berg et al. 1980; Ciampi et al. 1982). An explanation for this phenomenon comes from studies of the *S. typhimurium hisG* gene (Ciampi and Roth 1988). Genetic analysis suggests that the *hisG* gene is unusual in that it contains only one strong polarity site, located in the downstream por-

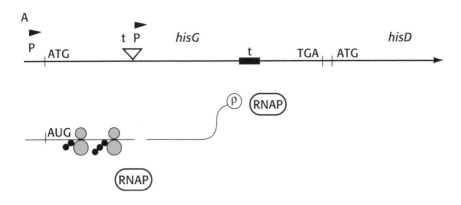

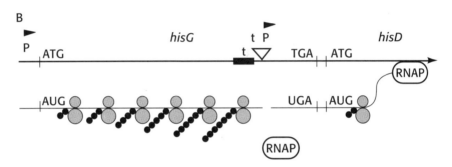

Figure 4.15 Insertional polarity in the *S. typhimurium his* operon. The promoter, *hisG* gene, and the proximal portion of the *hisD* gene are shown. DNA and RNA are shown as thick and thin lines, respectively. RNA polymerase and Rho factor (ρ) are labeled. Ribosomes and nascent polypeptide chains are indicated by gray and black circles, respectively. The polarity site (t) within the *hisG* gene is indicated by a black rectangle. Insertions are shown as open triangles, and contain both a terminator (t) and an outward-directed promoter (P). (*A*) An insertion upstream of the polarity site. Transcription from the *his* promoter is blocked by the insertion, and outward-directed transcription from the insertion is blocked at the polarity site. (*B*) An insertion downstream of the polarity site. Transcription from the *his* promoter is blocked by the insertion, but outward-directed transcription from the insertion is not blocked, leading to low-level constitutive expression of the downstream genes.

tion of the coding sequence. Insertions of Tn*5* and Tn*10* that lie upstream of this site are completely polar on downstream gene expression (Hol⁻ phenotype). However, Tn*5* and Tn*10* insertions downstream from this polarity site retain the Hol⁺ phenotype due to relatively low-level constitutive expression of the *hisD* gene. This low-level constitutive expression results from the outward-directed promoters at the ends of the transposons (Ciampi et al. 1982; Wang and Roth 1988). Analogous phenomena have been observed with other operons, including *lacZYA* (Berg et al. 1980). Thus, when using transposons for studies of genetic polarity, it is always a good idea to test several independent insertions in case some of the insertions exhibit incomplete polarity.

A different approach to the same problem involves construction of defined insertions by molecular cloning. In this case, the Ω (Greek letter Omega) elements are particularly well-suited. Ω elements consist of an antibiotic resistance gene flanked by strong transcription terminators derived from bacteriophage T4 (Fellay et al. 1987). The presence of these strong terminators prevents transcription from escaping the element, and these insertions are strongly polar. Introduction of an Ω element upstream of a MudJ insertion provides a means for determining operon structure (see, e.g., Berg and Stewart 1990; Lin et al. 1994).

COMPLICATIONS OF POLARITY

The phenomenon of polarity potentially complicates complementation analysis. A polar mutation in an upstream gene will not exhibit complementation when tested against mutations in downstream genes, despite the fact that the mutations being tested are indeed located in different genes. The alert geneticist will use several point mutations, each examined for potential polarity effects, to conduct rigorous complementation tests (see, e.g., Roof and Roth 1989; Hughes et al. 1991).

Many genes share overlapping start-stop signals (e.g., UGAUG, AUGA, and AUGNNUAA), which often specify translational coupling between the upstream and downstream genes. In such cases, efficient translation of the downstream coding region depends on translation of the upstream region, and a component(s) of the upstream ribosome may remain attached to the mRNA to initiate downstream translation. Such a mechanism ensures equimolar production of subunits in an enzyme complex (Oppenheim and Yanofsky 1980; Das and Yanofsky 1989; Adhin and van Duin 1990). Thus, a nonsense mutation in an upstream gene can potentially affect downstream gene expression not only by transcriptional polarity, but also potentially by disrupting translation of a translationally coupled downstream gene.

It is becoming increasingly popular to use so-called "nonpolar" insertions to probe gene function. These insertions generally consist of antibiotic resistance cassettes that contain a relatively strong outward-directed promoter. Placement of such a cassette within a gene of interest, using molecular cloning techniques, disrupts the gene of interest and is often viewed as eliminating any complications of polarity because of the strong outward-directed promoter. The function of the gene of interest can then be explored, without concern for expression of the downstream genes in the operon. However, this technique presents at least two potential complications. First, the existence of internal polarity sites means that the outward-directed transcription is not guaranteed to reach the downstream genes. Second, even if this transcription does lead to downstream expression, such expression is likely to be quite different from that of the wild type both in amount of transcript (reflecting the relative strengths of the normal and introduced promoters) and in regulation (the normal promoter is likely to be regulated, whereas the introduced promoter is constitutive). A variety of scenarios can be envisioned where this inappropriate expression of downstream genes could obscure or even alter the phenotype caused by the disruption of the upstream gene.

A more rigorous method to explore gene function, particularly in cases where it is difficult or impracticable to isolate null point alleles, is to construct a defined in-frame deletion within the gene of interest (this requires that the DNA sequence of the region of interest is known). By making a large in-frame deletion, the complications of nonsense (frameshift) polarity are eliminated, and the downstream genes will be normally expressed from the native promoter. Such deletions may be placed into the chromosome by using a linked selectable marker outside of the operon of interest (Egan and Stewart 1990), and a variety of allele-replacement methods are available for introducing silent (unmarked) deletions into the chromosome (Ried and Collmer 1987; Miller and Mekalanos 1988; Hamilton et al. 1989; Penfold and Pemberton 1990; Metcalf and Wanner 1993). In such cases, it is critical to ensure that the in-frame deletion does not disrupt translation initiation or termination signals in case the downstream gene is translationally coupled (see, e.g., Egan and Stewart 1990). Normal translation of a short deletion polypeptide encoded by the upstream gene deletion will help ensure normal synthesis of the downstream gene product.

Transposons and Fusions

"Just as Russell [Chan] was defending his thesis, Nancy Kleckner joined us as a postdoc. She took up an experiment that Russell had just begun: to show directly that the transduction of tetracycline resistance by P22Tc10 lysates at low multiplicity was indeed caused by transposition of the element. The idea was to screen for auxotrophic mutations caused when the element inserts itself into bacterial genes and then to show genetically that the two phenotypes, auxotrophy and drug resistance, were the consequence of the same insertion mutation. This line of research proved very fruitful indeed, and by the fall of 1974, we had submitted a very complete paper (Kleckner et al. 1975) to the *Journal of Molecular Biology* that clearly described the essential properties of the element now called Tn*10*.

"Just at this time, I took my sabbatical year at Cold Spring Harbor Laboratory. My intention was to work on yeast with Gerry Fink and John Roth. As it turned out, we did a lot with yeast that year, but transposons were a persistent subtext. John and I spent a great deal of time thinking about how insertion mutations might be used in bacterial genetics, ideas that occupied both our laboratories for many years thereafter."

David Botstein
Reprinted from Discovery of the Bacterial Transposon Tn*10*.
In *The Dynamic Genome: Barbara McClintock's Ideas in the Century of Genetics*
(ed. N. Federoff and D. Botstein), pp. 225–232.
Cold Spring Harbor Laboratory Press, Cold Spring Harbor, NY (1992)

Use of Transposons in Bacterial Genetics

Transposons are virtually essential for conducting contemporary investigations in bacterial genetics. The number and variety of applications that transposons find in bacterial genetics seem to be limited only by the imagination of the geneticists themselves. Considered here are some of the basic uses for transposons.

The first consideration in using transposons is the delivery system: how to get the transposon into the species of interest. The most convenient methods involve bacteriophage delivery systems. For example, transposon mutagenesis of *Escherichia coli* uses bacteriophage λ delivery vehicles constructed in the laboratory of N. Kleckner (Kleckner et al. 1991). These λ derivatives carry the transposon, the gene for transposase, conditional (amber) mutations in genes essential for phage replication, and mutations that prevent lysogeny. First, a stock of the phage on a suppressing host is prepared and then a suppressor-free strain is infected to isolate derivatives in which the transposon has moved into the chromosome. The delivery phage can neither replicate nor lysogenize, so each drug-resistant colony is the result of an independent transposition event. Specialized methods for transposition mutagenesis of *Salmonella typhimurium*, which is resistant to λ, are described below. Finally, conjugational delivery systems developed for nonenteric bacteria generally involve the use of plasmids that replicate in *E. coli* but not in the host of interest (Simon et al. 1983) or plasmids whose replication is conditional on functions provided by a specially engineered donor host (Taylor et al. 1989; see below).

The second consideration is the source of transposase. Transposons encode their own transposase, but many elements have been genetically engineered and thus they contain only the extreme ends of the transposon (the sites of transposase action) and a drug resistance marker.

These so-called defective transposons are incapable of transposing unless transposase is provided in trans (specific methods are described below). One advantage of defective transposons is that they can be induced to transpose more or less in synchrony by inducing transposase synthesis. Once isolated, insertions of defective transposons are stable and thus concern with secondary transposition events or transposase-promoted rearrangements is eliminated.

Kleckner et al. (1977), in a highly influential and important paper, described many of the ways in which transposons are useful in bacterial genetics. Sanderson and Roth (1988) provide a short update, with discussion of more recently developed methods and ideas. The following key considerations are annotated from Kleckner et al. (1977):

1. Transposons can be found inserted at a large number of sites on the bacterial chromosome. (In principle, it is possible to find transposon insertions in or near any gene of interest.)

2. Interrupted genes suffer complete loss of function. (Transposon-generated mutations are considered to be null alleles in all but rare exceptional cases.)

3. The phenotype of the insertion mutation is completely linked to drug resistance in genetic crosses. (It is possible to transfer mutations into new strain backgrounds simply by selecting for drug resistance.)

4. Insertion mutants can be recovered at high frequency after low-level "mutagenesis." (Strains with multiple mutations are rare, and backcrossing allows one to establish that the mutant phenotype is due exclusively to the transposon insertion.)

5. Insertion mutations revert by precise excision with concomitant loss of drug resistance. (This generally occurs at very low frequency; with some transposons, such as bacteriophage Mu, it is never observed.)

6. Insertions in operons are strongly polar. (Transposon insertions can be used to determine whether genes are in an operon. There are special-case exceptions to this rule of polarity [see Ciampi and Roth 1988].)

7. Transposons can generate deletions nearby. (This provides a convenient method for isolating deletions, and even determining genetic map position.)

8. Transposons can provide a portable region of homology. (Insertions can be used to construct deletions or duplications with defined endpoints or can serve as sites of recombinational integration of other genetic elements.)

9. Insertions behave as point mutations in fine-structure genetic mapping. (This is true only when the insertion serves as the recipient for transductional crosses.)

10. Insertions can be specifically obtained which are *near* but not *within* a gene of interest. (Such insertions are useful for generating deletions, and for genetic mapping.)

TRANSPOSON Tn*10* AND DERIVATIVES

Tn*10*, a Tc^r transposable element, is one of the most widely used transposons in bacterial genetics. It is a composite transposon: Its central region, encoding Tc^r, is flanked on both sides by nearly identical inverted copies of an insertion sequence, IS*10* (Fig. 5.1). Each IS*10* element contains short inverted repeats at each end. These repeats are the cis-acting sites of transposase action. The rightward copy of IS*10* (termed IS*10*R) encodes transposase, and it transposes independently of Tn*10* when transposase recognizes the two repeats of IS10R—the inside repeat and the outside repeat. Movement of the entire Tn*10* element occurs when transposase recognizes the two copies of the outside repeat, the one in IS*10*R and the one in IS*10*L, the leftward element. The IS*10* elements are each about 1.4 kb in length, and Tn*10* is about 9.3 kb.

The central region of Tn*10* contains at least four genes (Fig. 5.1). *tetA* encodes the tetracycline efflux protein and *tetR* is a repressor of *tetA* transcription (Beck et al. 1982); the functions of *tetC* and *tetD* are unknown (Braus et al. 1984). The gene for transposase, *tnpA*, is located entirely within the IS*10*R element; the *tnpA* gene in IS*10*L is nonfunctional. Expression of *tnpA* is maintained at an extremely low level by a variety of

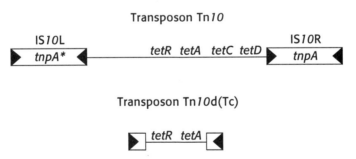

Transposon Tn*10*

Transposon Tn*10*d(Tc)

Figure 5.1 Schematic representation of transposons Tn*10* and Tn*10*d(Tc), not to scale. Tn*10* is flanked by inverted copies of insertion sequence IS*10*, in inverted orientation. The black triangles represent the inverted repeats at each end of IS*10*. The transposase gene, *tnpA*, is not functional in IS*10*L. The *tetA* gene encodes the Tc^r determinant. The *tetR* gene encodes the repressor of *tetA* gene expression. The functions of the *tetC* and *tetD* genes are unknown.

mechanisms that include infrequent transcription initiation and weak translation initiation (Kleckner 1990). The result is that Tn*10* transposition occurs very infrequently, about 10^{-8} per cell per generation or even lower, which is a drawback when isolating insertions in the first place but a great advantage in subsequent manipulations because Tn*10* insertions are relatively stable.

ATS Transposase

Tn*10* transposes by a nonreplicative mechanism: Tn*10* is excised from its flanking sequences in the donor DNA and is spliced into the DNA at the target site (Bender and Kleckner 1986). Selection of target sites shows some sequence preference, and hot spots for insertion are apparent. Thus, it is possible to find a Tn*10* insertion in virtually any gene if one searches hard enough. However, examination of a number of insertions in a given gene reveals that many independent insertions occur repeatedly at the same site (Kleckner et al. 1979a,b). Recently, Kleckner and co-workers have isolated mutants of *tnpA* termed *ats* (altered target specificity; Bender and Kleckner 1992). The most useful ATS transposase is a double mutant (*ats-1 ats-2*), with cysteine to tyrosine changes at positions 134 and 249. This ATS transposase exhibits only about a threefold decrease in transposition frequency, so it is still quite useful for insertion mutagenesis.

The relaxed target specificity of ATS versus wild-type transposase was revealed by an experiment comparing insertions into the *lacZ* gene. Fifty independent insertions each of wild-type and *ats-1 ats-2* versions of Tn*10* were analyzed by DNA sequence analysis. The wild-type insertions were distributed among 12 sites, with 24 of the insertions occurring at the same hot-spot site. In contrast, the ATS insertions were distributed among 23 sites, with no more than 5 insertions falling in a given site (Bender and Kleckner 1992).

Defective Mini-Tn*10* Elements

Tn*10* and IS*10* insertions promote adjacent DNA rearrangements including deletions and inversions (Shen et al. 1987). In wild-type Tn*10*, these rearrangements occur spontaneously with relatively high frequency (10^{-4} per cell per generation). Furthermore, IS*10*R can transpose at a high frequency relative to Tn*10*; because IS*10*R transpositions are not marked by Tc^r, they are silent and difficult to detect (Sanderson and Roth 1988). These observations have catalyzed the construction of a variety of defective mini-Tn*10* (Tn*10*d) elements (Way et al. 1984; Elliott and Roth 1988;

Kleckner et al. 1991). These elements consist of the extreme ends of IS*10*R (the sites for transposase action) flanking a drug resistance marker such as Tc^r and Cm^r. Thus, in the absence of transposase, Tn*10*d elements are stable: They do not promote rearrangements of adjacent DNA, and they do not transpose.

A variety of delivery systems for Tn*10*d transposition have been developed. All rely on high-level regulated expression of transposase. λ vectors are the method of choice for mutagenizing *E. coli*. A variety of λ vectors carrying Tn*10*d elements and the gene for ATS transposase have been constructed by Kleckner et al. (1991). Upon infection, the *tnpA(ats)* gene product is synthesized at high level, and it catalyzes excision and transposition of the Tn*10*d element. The donor vehicle is destroyed by nuclease, and thus the Tn*10*d element, having transposed once, is now locked in place.

Other methods have been developed for other organisms, two of which are described here. The first, trans-acting complementation, is illustrated in Experiment 2. The second, regulated transposase expression, is illustrated in Experiment 3. Both of these methods are available for use with Tn*10*d(Tc), Tn*10*d(Cm), and Tn*10*d(Km).

Trans-acting Complementation

This method uses a *tnpA(ats)* expression plasmid in the recipient cell and a P22 HT *int* lysate grown on a strain carrying F' *lacZ*::Tn*10*d as the delivery vehicle (Elliott and Roth 1988). Because *S. typhimurium* does not have the *lac* genes, a recipient cell that receives a transducing particle carrying *lacZ*::Tn*10*d cannot inherit the Tn*10*d insertion by homologous recombination. However, the recipient cell expresses ATS transposase at high levels and thus the Tn*10*d element can be excised from the donor *lacZ*::Tn*10*d transducing fragment and transposed to the recipient chromosome. Since the recipient expresses transposase, the Tn*10*d insertions may not be stable. This method is therefore most useful for constructing pools of Tn*10*d insertions; P22 HT *int* grown on the pools can be used to transduce other strains to drug resistance in conjunction with screens or selections for the desired phenotype.

Regulated Transposase Expression

This method uses a plasmid that carries three elements: a Tn*10*d insertion, the *tnpA(ats)* gene under the expression of the *tac* promoter, and *lacI*^{Q}, a highly expressed version of the *lac* (*tac*) repressor (Kleckner et al. 1991). In the absence of the *lac* inducer IPTG, transposase expression is virtually nil, and there is little transposition of the Tn*10*d element. However, the addition of IPTG results in high-level transposase expression and efficient Tn*10*d transposition to the chromosome. Again, this method suffers from the fact that the recipient cell retains an active transposase gene, and thus desired Tn*10*d insertions must be backcrossed by transduction before further analysis.

Tn*10*-promoted Deletions and Rearrangements

As described above, wild-type Tn*10* catalyzes the formation of adjacent deletions and inversions. The development of a positive selection for Tc^s strains (Bochner et al. 1980; Maloy and Nunn 1981) allowed bacterial geneticists to exploit this property. Tn*10*-promoted rearrangements occur as a result of abortive transposition involving an outside or inside end of IS*10*. Often, this abortive transposition results in the deletion of the central *tet* region of Tn*10*. Thus, selection for Tc^s strains greatly enriches for those that have undergone a Tn*10*-promoted rearrangement. However, there are significant drawbacks to the use of this method for generating deletions. The first is that it is often difficult, in a given system, to distinguish a deletion from an inversion by purely genetic criteria. The second is that rearrangement formation, a type of transposition, is subject to the effects of hot spots. Thus, even a large collection of independent Tn*10*-promoted deletions contain many examples of the identical deletion (Kleckner et al. 1979a,b). Finally, most of the Tn*10*-promoted deletions are unidirectional and contain an IS*10* outside or inside end as one endpoint.

The Tn*10*d(Tc) element cannot transpose and cannot promote adjacent deletions and inversions. Thus, selection for Tc^s derivatives of Tn*10*d(Tc) elements enriches for those strains in which spontaneous deletions have removed the Tn*10*d(Tc) element and flanking DNA. Such deletions occur at a greatly reduced frequency compared with those promoted by wild-type Tn*10*, but they are more likely to occur randomly, with varying endpoints, compared with the hot-spot-prone deletions promoted by the wild type.

TRANSPOSON Tn*5*

Tn*5*, like Tn*10*, is a composite transposon (Reznikoff 1993). The central region of Tn*5* encodes three proteins that confer resistance to kanamycin and neomycin, streptomycin, and bleomycin (Mazodier et al. 1985). The latter two resistance determinants are not expressed in enteric bacteria. The central region is flanked by nearly identical inverted copies of the insertion sequence IS*50*. Each IS*50* element contains the cis-acting sites of transposase action as short inverted repeats at each end. The rightward copy of IS*50* (IS*50*R) encodes transposase and an inhibitor of transposase. IS*50*L differs by a single nucleotide, but this single difference has two consequences: It inactivates the IS*50*L-encoded transposase by creating a nonsense codon, and it allows expression of the drug resistance elements by creating a promoter for the transcription of their genes. The IS*50* elements are each about 1.5 kb in length, and Tn*5* is about 5.7 kb.

Tn*5* has been widely used in bacterial genetics because its high transposition frequency (~10^{-5} per cell per generation) makes it relatively easy to isolate a large number of insertions. In addition, Tn*5* is active in a wide variety of gram-negative bacteria, adding to its popularity. However, the high frequency of Tn*5* transposition can cause complications in genetic analysis, particularly when Tn*5* insertions are used as donors in genetic crosses (Biek and Roth 1980). One solution to this problem has been the development of diverse Tn*5*d derivatives carrying a variety of antibiotic resistance determinants (de Lorenzo et al. 1990; Herrero et al. 1990).

TRANSPOSON FOR SIGNATURE-TAGGED MUTAGENESIS

The application of transposon mutagenesis to identify genes involved in bacterial pathogenesis has been limited from the standpoint of in vivo screens since insertion into most genes of interest would result in loss of viability of the mutant in the in vivo situation such as survival in an animal model. Thus, a large number of animals would be required to screen for mutants that are unable to survive. To circumvent this problem, a transposon derivative has been developed that allows one to retrieve clones of specific bacteria represented in a pooled inoculum that are avirulent and hence are not recoverable from the animal. This method, termed signature-tagged mutagenesis, relies on a mini-Tn*5*d harboring a cassette that contains a variable region DNA sequence tag. The variable region is designed such that each bacterium present in a transposon insertion pool carries a unique DNA sequence (Hensel et al. 1995) (see Section 1, General Concepts in Bacterial Pathogenesis). After infection of an appropriate animal model, DNA hybridization is used to identify clones of those bacteria in the inoculum that are not represented in the population of bacteria recovered from the animal. Transposon mutagenesis incorporating this technique should be useful for identifying genes that encode products essential in any given growth situation.

Bacteriophage Mu

Bacteriophage Mu (Mu for Mutator) is a temperate phage with a double-stranded DNA (dsDNA) genome of approximately 37 kb (Symonds et al. 1987). Mu lysogenizes susceptible hosts at high frequency, but the mechanism of prophage integration involves transposition rather than site-specific recombination. Thus, independent lysogens contain Mu prophages at distinct chromosomal locations. Upon lysogenic induction, prophage DNA synthesis is accomplished by replicative transposition (Fig. 5.2). Thus, during induction, a single prophage becomes two, then four, then eight, and so on until lysis. Packaging occurs as the prohead/terminase complex recognizes the left end of the prophage DNA. Terminase cuts roughly 150 bp to the left of the prophage end, and headful packaging proceeds for approximately 38.5 kb. Thus, phage genomes consist of phage DNA embedded within host DNA, about 150 bp at the left end and about 1.5 kb at the right end.

Mu transposase is encoded by genes A and B, located at the left end of the genetic map. Expression of the AB operon is controlled by the product of the repressor gene, c, which is at the leftmost of Mu. The remaining phage genes encode head, tail, and assembly proteins and host killing and lysis functions; their expression is regulated by a positive regulator encoded by gene C. Martha Howe (1972) isolated tight temperature-sensitive alleles of c, termed $c$ts, so thermoinduction of a Mu $c$ts lysogen results in phage production and lysis.

Mu particles adsorb to the cell surface through a specific interaction between the lipopolysaccharide side chain and the Mu tail fiber. Phage DNA is injected into the cell cytoplasm, and transcription initiates at the AB operon promoter. Transposase is synthesized, and the phage DNA is integrated into the host chromosome by nonreplicative transposition (Harshey 1983). By this time, sufficient c repressor has been synthesized to repress the AB operon, preventing the synthesis of more transposase.

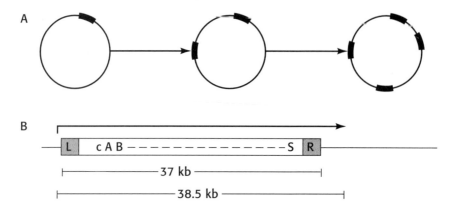

Figure 5.2 Bacteriophage Mu lytic growth. (*A*) Mu replication. Upon induction, a single prophage duplicates itself by replicative transposition, ad infinitum. (*B*) Mu encapsulation. Terminase cuts a short distance proximal to the left (L) end of Mu, and packaging proceeds from left to right. Thus, the vegetative phage DNA is bounded by host chromosomal DNA on both sides. Gene *c* encodes Mu repressor; genes *A* and *B* encode transposase; the remaining genes encode lytic functions and structural components.

One aspect of this integration deserves attention: Mu integration is often (10%) accompanied by a deletion of adjacent host DNA sequences. This can lead to complications when using Mu for genetic analysis and should be remembered when confronted with unexpected results.

Mud DERIVATIVES

Defective Mu genomes (Mud) have been genetically engineered as deletion-substitution derivatives of Mu that are suitable for genetic analysis. Mud elements are propagated as phage particles with the aid of a wild-type helper prophage. A strain for making Mud lysates carries a single copy of a Mud prophage and a single copy of a Mu *cts* prophage. Upon induction (temperature shift of a *cts* lysogen), the helper phage provides transposase, phage structural proteins, and assembly proteins that may be required by the Mud element. The Mud genomes are replicated and packaged (via headful packaging) with the same efficiency as the wild-type helper genome. Thus, a lysate prepared from such a strain contains approximately equal proportions of particles carrying Mu *cts* genomes and Mud genomes.

A variety of Mud elements have been constructed, but only the most widely used are described here. Virtually all of these Mud elements were constructed by Malcolm Casadaban (see Table 5.1). The original Mud, Mud1(Ap^r, *lac*), was developed for constructing in vivo operon fusions. This defective Mu phage retains the genes for transposase, and many phage structural genes as well, but it still requires a helper phage for packaging. The promoterless *lacZYA* operon (the *trp-lac* W209 fusion) is located close to the right end of the Mud genome, so that transcription coming into the prophage from the right end will read the *lacZYA* sequence, forming an operon fusion. A derivative of this element, Mud301(Ap^r, *lac*), is identical, except that it forms gene fusions. Conditionally transposition-defective derivatives of these elements were constructed by placing amber mutations in genes *A* and *B*, so that transposition occurs only in amber suppressor hosts.

Currently, the best Mud elements for most purposes are mini-Mud elements constructed by Casadaban and associates, MudI1734 (MudJ) and MudII1734 (MudK) (Castilho et al. 1984). These small (10 kb) elements lack transposase altogether, so subsequent instability is not a concern. Furthermore, strains carrying them can be propagated at 37°C; strains carrying $A^+ B^+$ Mud elements must be grown at 30°C to prevent induction of transposase synthesis (due to inactivation of the *c*ts repressor).

Mud nomenclature is cumbersome and potentially confusing, so shorthand designations have been developed to describe some of the most widely used elements (Table 5.1). MudI is a generic term for operon fusion Mud elements, and MudII denotes gene fusion elements.

TRANSITORY CIS-ACTING COMPLEMENTATION

One requirement for using transposase-deficient Mud derivatives is to provide transposase upon demand for transposition. An extremely useful method for doing this with *S. typhimurium* is termed transitory cis-acting complementation (Hughes and Roth 1988). This method uses a strain carrying *hisD*::MudJ and *hisA*::Mud1 alleles; *hisD* and *hisA* are only a few kilobases apart (Fig. 5.3). Some particles in P22 HT *int* lysates grown on this strain carry chromosomal DNA, including *hisD*::MudJ and the *cAB* end of the *hisA*::Mud1 insertion. Since P22 can only package about 44 kb, a fragment carrying an intact MudJ element (11 kb) will also contain only a portion of the Mud1 element, from the *cAB* end. Thus, the genes for transposase are genetically linked to the MudJ element but remain outside of the transposed DNA. Once the *AB*-encoded transposase has catalyzed excision and transposition of the MudJ element, the remaining transduced DNA is degraded, and the MudJ element, which has transposed only once, is now locked in place with no source of transposase.

Table 5.1 Properties of Commonly Used Mud Derivatives

Element[a]	AKA[b]	Fusion[c]	kb[d]	Marker[e]	Genes[f]	Reference
Mud1(Ap^r, *lac*)	Mud1	O	37.2	Ap	cts $A^+ B^+$	1
Mud301(Ap^r, *lac*)	Mud2	G	35.6	Ap	cts $A^+ B^+$	2
Mud1-8	MudA	O	37.2	Ap	cts A(Am) B(Am)	3
Mud2-8	MudB	G	35.6	Ap	cts A(Am) B(Am)	3
MudI1734	MudJ	O	11.3	Km	cts $\Delta(AB)$	4
MudII1734	MudK	G	9.7	Km	cts $\Delta(AB)$	4

[a]Original designation.
[b]"Also known as"; alternate designation.
[c]O, operon (transcriptional) fusion; G, gene (translational; protein) fusion.
[d]Length of element DNA in kilobases.
[e]Selectable antibiotic resistance marker
[f]Genotype of repressor (*c*) and transposase (AB) genes.
References: (1) Casadaban and Cohen (1979); (2) Casadaban and Chou (1984); (3) Hughes and Roth (1984); (4) Castilho et al. (1984).

Mud AS A CLONING VECTOR

The random transposition of Mu genomes during lytic induction has been harnessed to develop a method for in vivo cloning. Groisman and Casadaban (1986) have constructed defective Mu elements carrying a selectable marker and a plasmid origin of replication (Fig. 5.4). During lytic growth, the Mud element transposes to many positions on the chromosome. Frequently, two Mud elements will have transposed to either side of a particular gene of interest. Headful packaging results in encapsulation of one Mud, the gene of interest, and at least part of the second Mud. The resulting lysate can be used to transduce an appropriate recipient, selecting for the antibiotic marker on the Mud element. Homologous recombination between the Mud elements forms a circular plasmid, in which the selectable marker and origin are provided by the Mud element, and the insert DNA consists of the chromosomal region that was flanked by the two Mud elements in the original phage particle.

Two aspects of Mud cloning vectors make them attractive for cloning. First, the virtually random transposition of Mu ensures that the gene bank will contain representatives of the entire genome at nearly equal frequency. This is tricky to achieve when constructing gene banks with restriction enzymes, which have fixed cutting positions. Second, the

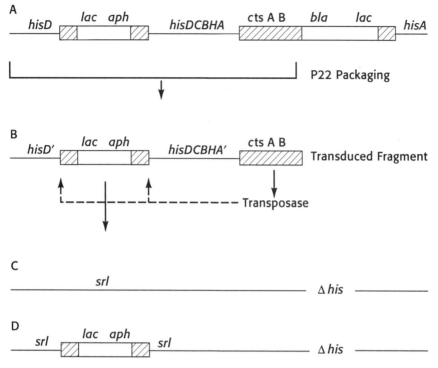

Figure 5.3 Transitory cis-acting complementation. (*A*) donor strain (TSM102) containing *hisD*::MudJ and *hisA*::Mud1. P22 HT *int* packages all of MudJ and the *cAB* end of Mud1. (*B*) Upon transduction, transposase is synthesized from the Mud1 fragment, and acts on MudJ. (*C*) Transposition in the recipient (TSM101) is the only way to inherit Km^r; the lack of *his* sequences in TSM101 prevents homologous recombination. (*D*) A MudJ insertion in the *srl* locus. Adapted from Hughes and Roth (1988).

headful packaging of Mu encapsulates about 38.5 kb of DNA, so inserts are large; with Mud5005, which is only 7.9 kb, inserts of 20–25 kb are routinely recovered. This facilitates molecular analysis of large regions of the genome.

Mud5005, the most widely used Mud cloning vector, contains genes *cAB* and thus encodes transposase. Recipient strains for cloning must therefore be Mu lysogens. Because the recipient Mu prophage produces *c* repressor, the incoming *AB* operon of the Mud5005 is immediately repressed, which prevents zygotic induction of the Mud5005 element and permits the isolation and propagation of stable clones.

Mud cloning can be used with virtually all enteric bacteria, by using appropriate hosts and vectors (see below). A related method, using the *Pseudomonas aeruginosa* transposable phage D3112, is equally successful with this organism (Darzins and Casadaban 1989).

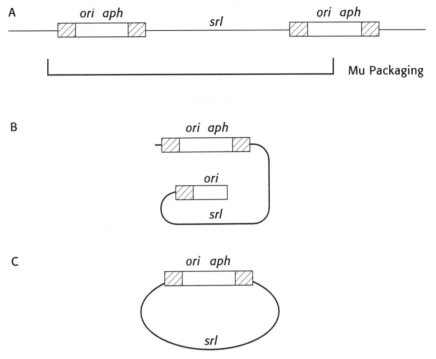

Figure 5.4 Cloning with Mud5005. (*A*) Mu transposition in one of the donor cells has placed direct repeats of Mud5005 on either side of the *srl* locus. All of the left Mud5005, *srl*, and part of the right Mud5005 are packaged into a Mu particle. (*B*) Homologous recombination between Mud5005 sequences in the recipient yields: (*C*) a circular plasmid containing one complete copy of Mud5005 and the chromosomal DNA that was between the two Mud5005 elements in the donor. (Adapted from Groisman and Casadaban 1986.)

PRACTICAL ASPECTS OF USING Mu IN BACTERIAL GENETICS

Mu lysates are quite unstable and lose significant titer even in 1 day. Thus, it is best to make the Mu lysate on the same day in which it is to be used. This of course does not apply to the use of P22 as a vector for transferring Mud elements (as in transitory cis-acting complementation), as P22 lysates are quite stable.

Hosts for Mu infection must have the appropriate lipopolysaccharide (LPS) composition or Mu will not adsorb. Only *Escherichia* spp. and the closely related *Citrobacter* spp. and *Shigella* spp. have the appropriate LPS. However, the genes for tail fiber, the component that recognizes LPS, are homologous in Mu and in coliphage P1. Martha Howe has isolated recombinants between Mu and P1 in which the P1 tail fiber gene has replaced the homologous gene of Mu. The resulting Mu phage, termed Mu *cts* hP1, thus has the host range of P1 (Csonka et al. 1981). It is possible to isolate P1-sensitive mutants of a wide variety of enteric bacteria (Goldberg et al. 1974), so Mu *c*ts hP1 can be used in these species. Likewise, Mud cloning vectors can also be used with P1-sensitive enterics (Groisman and Casadaban 1987).

Salmonella represents a special case. Mutant strains lacking galactose epimerase (*galE*) produce a truncated LPS that is effectively recognized by P1 tail fiber. These strains are resistant to P22. However, growth of a *galE* mutant in the presence of 1% glucose and 1% galactose restores the normal LPS. Thus, *S. typhimurium galE* mutants can be infected by Mu *c*ts hP1 or by P22, depending on the composition of the medium in which they are grown (Enomoto and Stocker 1974; Ornellas and Stocker 1974).

Operon and Gene Fusions

Two related developments in the mid 1970s provided a new approach to the practice of bacterial genetics. The first was the harnessing of transposable elements for genetic analysis. The second development, which depends on transposons, was the establishment of generalized methods to construct operon and gene fusions.

Any study of gene regulation quickly meets the serious challenge of how to measure expression from the gene whose regulation is being studied. In many cases, enzyme or other assays are suitable for studying the regulation of a particular gene, but often such assays are tedious, inconvenient, or simply not available. Malcolm Casadaban, a graduate student in the laboratory of Jon Beckwith, faced exactly this problem when he chose to study the regulation of the *ara* operon regulatory gene, *araC*: No assay was available for measuring AraC synthesis. Thus, Casadaban invented a general method for constructing *lacZ* operon fusions to any gene (Casadaban 1976a) and applied this method to demonstrate autoregulation of *araC* gene expression (Casadaban 1976b).

With a fusion in hand, genetic methods available for the reporter gene may be used to isolate regulatory mutants. This is particularly accessible with *lacZ* fusions because of the wealth of methods available for selecting and visualizing *lacZ* expression (Silhavy and Beckwith 1985). Other reporters (e.g., *phoA*) allow study of subcellular localization of a particular gene product.

An operon fusion (termed by some a transcriptional fusion) can be used to study transcriptional regulation of virtually any gene (Fig. 5.5). In an ideal operon fusion, translation initiation of the reporter gene is completely independent of translation initiation signals for the gene in question. Thus, the rate of reporter enzyme synthesis is directly proportional to the rate of transcription initiation over a wide range. A gene fusion (termed by some a translational fusion or protein fusion) results in the

A. The *srl* operon

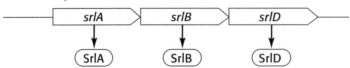

B. Operon fusion within *srlB* constructed with MudJ

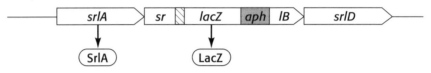

C. Gene fusion of *srlB* to *lacZ* constructed with MudK

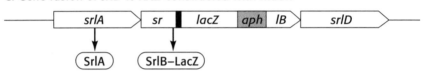

Figure 5.5 Schematic representation of operon and gene fusions. (*A*) The *srl* operon. (*B*) An insertion of MudJ in *srlB*, forming a Φ(*srlB-lacZ*) operon fusion. The hatched portion of MudJ represents *trpCBA* sequences. (*C*) An insertion of MudK in *srlB*, forming a Φ(*srlB-lacZ*) gene fusion. Note the production of a hybrid SrlB-LacZ fusion polypeptide.

production of a hybrid protein, with the translation initiation signals and the amino-terminal coding region derived from the gene of interest and the carboxy-terminal coding region derived from the reporter. Thus, for gene fusions, synthesis of the reporter (fusion) enzyme depends not only on transcription initiation, but also on translation initiation signals from the gene of interest. It is of course essential that gene fusions be constructed such that the coding region of the gene of interest be fused in-frame with the reporter gene coding region.

OPERON AND GENE FUSIONS WITH *lacZ*

By far the most successful and widely used reporter gene for fusions is the structural gene for *Escherichia coli* β-galactosidase, *lacZ*. Three features have contributed to this success (Silhavy and Beckwith 1985). First, because of the extensive array of β-galactosidase indicator and selective media, screens or selections can be tailored for (or to some extent against) β-galactosidase synthesis. MacConkey, tetrazolium, and XGal media all provide excellent differentiation of Lac$^+$ and Lac$^-$ colonies in a wide variety of physiological contexts. Direct selection for Lac$^+$ colonies is easily achieved by demanding growth with lactose as the sole carbon source, and selections for Lac$^-$ strains (resistance to tONPG, *o*-nitro-phenyl-β-D-thiogalactoside; resistance to galactose killing in *galE* strains) also find use under certain circumstances. Second, the enzyme assay for β-galactosidase is unusually sensitive and extremely convenient to per-form; many dozens or even hundreds of samples may be assayed in a single day with high precision and accuracy. Finally, β-galactosidase can tolerate virtually any replacement of its extreme amino terminus and retain full enzymatic activity, a feature that is invaluable for constructing gene fusions.

Many currently available operon fusion elements and vectors use a hybrid β-galactosidase gene (the *trp-lac* W209 allele), in which the first 59 codons of *trpA* (*E. coli* tryptophan synthase α subunit) are fused to codon 3 of *lacZ*. The resulting hybrid TrpA-LacZ protein has normal β-galactosidase activity and is quite stable. These fusion constructions also contain all of *trpB* and the extreme carboxy-terminal coding region of *trpC*. Thus, transcription reading into the fusion must traverse about 1.5 kb of *trp* DNA before reaching *lacZ*. Translation of the *trpA-lacZ* gene fu-sion initiates at the *trpA* translation initiation codon. (The reason for using a *trpA-lacZ* fusion is historical; this fusion was a well-documented source of functional *lacZ* that is expressed completely independently of the *lac* transcriptional regulatory elements.)

Operon and gene fusions with *lacZ* may be isolated in vivo using transposable elements that contain appropriate *lacZ* sequences near one end of the element. The most successful of these fusion elements are based on bacteriophage Mu and are described more fully in this section (see Bacteriophage Mu). The following are two general strategies for in vivo fusions: (1) isolate insertions in a particular gene(s) of interest and then use the resulting fusions to study gene regulation (see, e.g., Stewart 1982; Maloy and Roth 1983) and (2) screen a large number of fusions for response to a particular regulatory signal to define new genes that are members of a specific regulon (see e.g., Kenyon and Walker 1980; Wan-ner and McSharry 1982).

Operon and gene fusions with *lacZ* may also be constructed in vitro using molecular cloning techniques. A variety of vectors have been devel-oped for fusions constructed in vitro, the most versatile of which allow for recombination of the fusion onto a λ specialized transducing phage so that expression of the fusion can be studied in single copy. A particularly useful system has been developed by Simons et al. (1987). A somewhat analogous system for *Salmonella typhimurium* integrates the fusion as a single copy within the *putAP* locus (Elliott 1992).

OPERON AND GENE FUSIONS WITH *uidA* (*gus*)

The structural gene for β-glucuronidase, *uidA* (more generally known as *gus*), also provides a convenient marker for construction of operon fusions (Jefferson et al. 1986; Jefferson 1989). The enzyme β-glucuronidase cleaves a variety of substituted β-glucuronides, and commercially available substrates analogous to ONPG and XGal are used for measuring its activity and detecting its expression. Versatile *uidA* cassettes have also been constructed to facilitate construction of fusions (Metcalf and Wanner 1993). *uidA* fusions are particularly useful in contexts where high endogenous β-galactosidase activity precludes the use of *lacZ* fusions, for example, in studying bacterium-plant interactions.

GENE FUSIONS WITH *phoA*

The *phoA* gene fusion approach is based on the fact that for the normally periplasmic protein bacterial alkaline phosphatase to be active, it must be localized extracytoplasmically. Signal sequence mutations that block its export render the enzyme inactive. It has been shown that export and activity can be concomitantly restored by fusing a restriction fragment containing a truncated *phoA* gene (lacking secretion signals) to portions of genes encoding signal sequences of heterologous proteins such as OmpF or LamB (Hoffman and Wright 1985). Manoil and Beckwith (1985; Manoil et al. 1990) have extended the utility of this approach by inserting a similar *phoA* restriction fragment near one end of Tn5 to create a transposon, designated Tn*phoA*, that can randomly generate gene fusions to *phoA* upon insertion into a cloned target gene or the chromosome (Fig. 5.6). The hybrid proteins expressed by such gene fusions display alkaline phosphatase activity only if the target gene encodes a membrane, periplasmic, outer membrane, or extracellular protein. A mini-Tn5*phoA* that may be more stable because it separates the transposase gene from Tn*phoA* has also been constructed (de Lorenzo et al. 1990).

A major use of Tn*phoA* has been to study the topology of inner membrane proteins (Manoil and Beckwith 1985; Manoil et al. 1990). This utility has been broadened by its use in conjunction with β-galactosidase fusions that have the complementary property of being active only when cytoplasmic (Manoil 1990). Elements have been further developed to facilitate switching *phoA* fusions with other gene fusion or antibiotic resistance cassettes (Wilmes-Riesenberg and Wanner 1992) as well as for additional in vitro *phoA* fusion construction (Gutierrez and Devedjian 1989). Another major use of Tn*phoA* has been to identify genes that encode exported proteins. This utility has found widespread application in bacterial pathogenesis studies since exported proteins represent the major

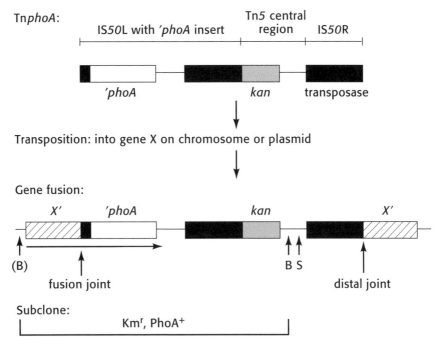

Figure 5.6 Events leading to the formation and cloning of an active *phoA* gene fusion by using Tn*phoA*. The transposon is a derivative of Tn*5* with the region encoding *E. coli* alkaline phosphatase, minus the signal sequence and expression signals, inserted into the left IS*50* element. Active insertions into gene *X* interrupt the gene and result in production of a hybrid protein from the *X-phoA* fusion. One scheme for isolating the gene fusion is to utilize the *Bam*HI (B) or *Sal*I (S) sites that lie distal to the Km^r gene and a hypothetical site (B) upstream of the gene to which *phoA* is fused. Restriction enzymes such as *Xba*I, *Stu*I, *Sac*I, and *Eco*RV, which do not cut within Tn*phoA*, are useful for isolating cloned fragments that carry the insertion plus DNA flanking both ends of the fusion joint. Also see transposon Tn*phoA* map in Appendix G (Plasmid and Transposon Restriction Maps). (Redrawn, with permission, from Taylor et al. 1989.)

class of virulence factors. Tn*phoA* provides a strong enrichment for insertion mutations in virulence genes (for a recent review, see Kaufman and Taylor 1994).

A number of Tn*phoA* delivery systems have been developed. A system widely used in *E. coli* is λTn*phoA*1 (Gutierrez et al. 1987). General systems to introduce Tn*phoA* into a variety of gram-negative species have been developed by placing the transposon onto mobilizable plasmids that allow for selection of transposition events after their introduction to the recipient (Taylor et al. 1989; de Lorenzo et al. 1990). Because of the broad host range of both the delivery plasmids and Tn*5* transposition, these systems are applicable to a wide range of gram-negative bacterial species. An overview of one of these systems is shown in Figure 5.7. The pRT291 sys-

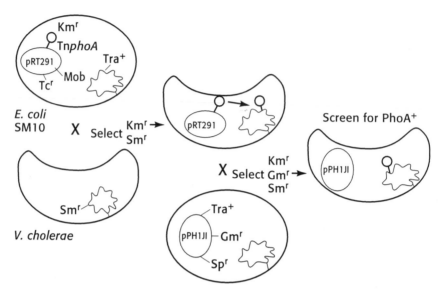

Figure 5.7 Method for delivery and selection of transposition of Tn*phoA* from *E. coli* to other gram-negative bacteria. On the left, pRT291, which carries Tn*phoA*, is mobilized into an Sm^r *V. cholerae* recipient strain. Exconjugants are then utilized as the recipients for a second plasmid, pPH1J1, of the same incompatibility group as pRT291 and encoding Gm^r. By maintaining selection for the presence of Tn*phoA* (Km^r), colonies are obtained in which Tn*phoA* has transposed and the vector plasmid has been lost. Colonies resulting from a fusion that expresses active alkaline phosphatase appear blue when XP is incorporated into the agar. (Redrawn, with permission, from Taylor et al. 1989.)

tem is utilized in Experiment 12 for isolating *phoA* fusions in *Vibrio cholerae*. This system is based on the use of incompatible plasmids of the IncP1 group as discussed in Section 4 (Broad Host Range Allelic Exchange Systems). It is noteworthy that when using this technique, different bacterial species may give various levels of background blue color on the XP indicator agar used to detect active fusions. Defined *phoA* deletion strains are available for experiments that utilize *E. coli* for the identification of fusions. In the case of the *V. cholerae* strains used here, inclusion of a fermentable carbon source such as glucose into the agar greatly diminishes background cleavage of XP due to endogenous phosphatase activity.

IN VIVO EXPRESSION TECHNOLOGY

A current area of investigation in bacterial pathogenesis is the identification of genes that are specifically expressed during the process of infection and might thus escape detection under typical laboratory growth condi-

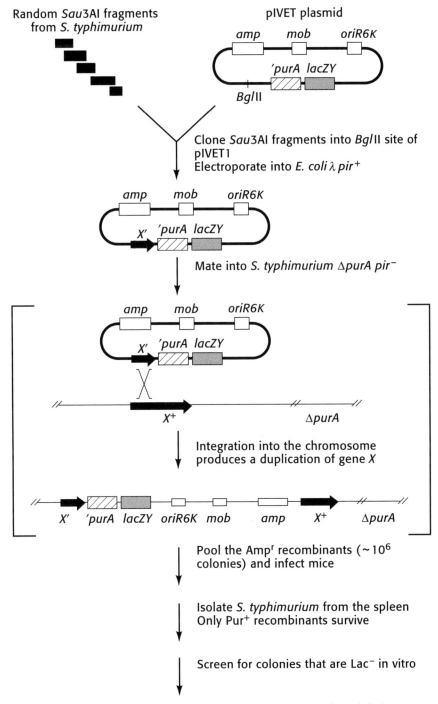

Figure 5.8 The gene fusion system termed IVET (in vivo expression technology). (Redrawn, with permission, from Mahan et al. 1993. Copyright 1983 by the AAAS.)

tions. A gene fusion system termed IVET (in vivo expression technology) has been developed to select for fusions to bacterial genes that are specifically induced when bacteria infect a host organism (Mahan et al. 1993; Slaugh et al. 1994). This system was first used in *S. typhimurium* and takes advantage of the requirement for purine synthesis by this organism to sustain in vivo growth. Thus, a *purA* deletion strain is greatly attenuated for in vivo persistence. The system, as diagrammed in Figure 5.8, is designed to form a series of gene fusions to a promoterless *purA* gene by cloning random DNA fragments into the pIVET1 plasmid vector. This plasmid functions essentially as the pGP704 vector described in Section 4 (Broad Host Range Allelic Exchange Systems) and thus can only replicate in a λ*pir* host. Upon introduction into a *purA S. tyhimurium* recipient, the plasmid integrates into the chromosome, resulting in a duplication of the cloned region. The duplication results in one copy of the target gene being fused to *purA* while retaining a complete wild-type copy of the gene to allow for recovery of organisms from the animal model. This technique was chosen to study the resultant fusions in single copy and to allow for isolation of strains carrying fusions to genes encoding products essential for in vivo growth. Fusions of *purA* to promoters that are expressed in vivo thus complement the *purA* deletion of the recipient strain and allow it to persist in vivo. To provide a convenient way to monitor the expression of the identified promoters in vitro as well as in vivo, the vector also includes a *lacZY* reporter downstream from *purA*. This allows rapid screening for promoters specifically expressed in vivo by plating the bacteria recovered from the host onto MacConkey lactose agar. It is expected that such systems will be useful for the identification of new virulence factors that may provide antigens for vaccine components or targets for new antimicrobial drug development.

SECTION 6
Gene Regulation

"As our knowledge of mechanisms of regulation of gene expression increases, virtually every event involved in macromolecule synthesis and function is seen to be modulated by one or more regulatory strategies. Common targets for regulatory intervention are the three stages of transcription: initiation, elongation, and termination. In addition, RNA processing and modification, messenger RNA translation and degradation, and protein modification, turnover, and activity are subject to regulatory control. Not only are many regulatory mechanisms in use, but often two or more, each responding to a different environmental signal or intracellular stress, control expression of a single gene or operon. Apparently regulation of each gene's expression is so vital that every event influencing that expression is optimized by exercise of one or more appropriate controls. However, granting the importance of regulation, one wonders why there is such diversity when a few regulatory mechanisms should suffice for each class of molecular events. One answer to this question is that cellular activities generate many types of signals and that these signals create opportunities for the evolution of new regulatory schemes."

Charles Yanofsky
Reprinted, with permission, from Transcription Attenuation.
J. Biol. Chem. 263: 609–612 (1988)

Regulation of Virulence Genes

WHAT IS A VIRULENCE GENE?

Virulence is the ability of an organism to infect a host and cause disease. Likewise, virulence factors are those properties of an organism that allow it to cause a disease. Virulence depends on numerous cell properties: A successful pathogen must be able to enter its host, find an appropriate niche, and multiply while continually subverting or avoiding a number of host defenses in the process (Bliska et al. 1993). Any bacterial property required for entry or growth of a bacterium in the host may be considered a virulence factor. For example, virulence factors may include a property of the bacterial cell surface that prevents killing by complement, properties that allow the bacterial cell to adhere to host cells, the ability to adapt to the temperature of the host, or the ability to synthesize nutrients that are limiting in the host. Some virulence factors are normal "housekeeping" functions of the bacterium, whereas others are only needed when the bacterium infects an appropriate host. For example, the *purA* and *aroA* gene products, which synthesize intermediates in purine and aromatic amino acid biosynthesis, are essential for virulence of *Salmonella* presumably because the product of these genes is limiting in the intracellular environment. (This clue led to the development of a live vaccine using *aroA* mutants of *Salmonella typhi*; Dougan et al. 1987.) However, the virulence factors required by one type of bacteria may not be required by another type of bacteria, even when the bacteria are closely related and the pathogenesis seems very similar. For example, although *Salmonella typhimurium* and *S. typhi* are very similar and share many virulence factors, each requires a unique set of virulence factors as well (Stocker and Makela 1986).

Genes that encode virulence factors are often called "virulence genes." When virulence genes are lost by mutation or segregation, there is a considerable decrease in virulence. It has been estimated that at least 4% of the genes in *S. typhimurium* encode virulence factors (Groisman and Saier 1990).

WHY ARE VIRULENCE GENES REGULATED?

A pathogenic bacterium may be exposed to a variety of harsh conditions: extremes of temperature and pH, strong detergents and digestive enzymes, and competition with an established community of microorganisms. To survive, bacteria have evolved mechanisms to tolerate these conditions. Often, these conditions act as signals to turn on or off specific genes needed for adaptation to the unique microenvironments.

The invading bacterium must be able to distinguish and rapidly adapt to differences in growth conditions between the host and the outside environment and within the host itself. Different microenvironments may have different temperatures, oxidation-reduction potentials, pH, and concentrations of inorganic and organic nutrients. For example, before ingestion, *Salmonella* is often present in water at a low temperature, low osmotic strength, near neutral pH, and a low concentration of organic nutrients. In contrast, once ingested by a mammalian host, *Salmonella* must adapt to a higher temperature and higher osmotic strength. It is transiently exposed to low pH in the stomach and then passes to the intestine where it encounters a higher pH, high concentrations of bile salts, anaerobiosis, and abundant organic nutrients but very low free iron concentrations. Within several hours after ingestion, *Salmonella* passes through the upper gastrointestinal tract and arrives in the lumen of the small bowel, where it must make intimate contact with appropriate host cells. To survive this sequential onslaught of offenses, the bacteria must rapidly express a variety of gene products: adhesions that keep it from washing out of the intestine, iron chelators that allow it to scavenge the limiting iron, and invasions that allow it to penetrate mucosal cells. Simultaneously, certain other gene products must be turned off under each different condition. This finely tuned ability to turn genes on and off is essential for pathogenesis.

HOW ARE VIRULENCE GENES REGULATED?

Not surprisingly, expression of virulence genes may be regulated by a variety of environmental signals (see Table 6.1). Furthermore, virulence genes are commonly regulated by multiple conditions, allowing different genes to respond to several different environmental signals individually or only when multiple signals occur at the same time (Mekalanos 1992). For example, all virulence genes may be temperature-regulated, but only a subset may be induced by low pH.

Expression of virulence genes may be regulated at many levels, including the initiation or termination of transcription, mRNA turnover, initiation of translation, posttranslational modification, protein turnover, and DNA structure. Many virulence genes are regulated by global regulatory

Table 6.1 Some Potential Regulatory Mechanisms for Virulence Genes

Signal	Potential mechanism
Carbon limitation	activates transcription by directly interacting with RNA polymerase, responds to cAMP levels (CRP-cAMP)
Iron starvation	represses transcription by interfering with RNA polymerase binding (Fur)
Supercoiling	responds to temperature or osmotic stress (HNS)
Temperature, ionic conditions, nicotinic acid, pH	two component systems, transmembrane sensor-mediated phosphorylation of regulator activates virulence gene expression by binding upstream of promoters (BvgAS in *Bordetella*; PhoPQ in *Salmonella*)
Temperature, pH, osmolarity, amino acids	ToxR transmembrane transcriptional activator stabilized in active conformation by ToxS (ToxRS in *Vibrio*)
Autoinducer	homoserine-lactone molecule detected by membrane sensors activate gene expression in response to bacterial population density
Gene rearrangement	leads to variations in the expression locus, some of which result in no product being produced (antigenic variation in *Neisseria*)
DNA inversion	leads to phase variation in gene expression (H-antigenic variation in *Salmonella*)
Replication slippage	leads to altered reading frames, some of which are frameshift mutations, and loss of protein expression (*Neisseria* P proteins)

systems, i.e., regulatory systems that control expression of many different genes simultaneously (Miller et al. 1989). For example, global regulatory systems essential for the virulence of *Salmonella* include activation of transcription by the CRP-cAMP complex in response to the availability of carbon sources and derepression of transcription by the Fur-Fe complex in response to iron availability. DNA supercoiling also seems to affect the expression of many virulence genes (Dorman 1991). The extent of supercoiling is maintained by DNA gyrase (a type II topoisomerase encoded by the *gyrA* and *gyrB* genes which adds negative supercoils into DNA in an ATP-dependent manner) and Topo I (a type I topoisomerase encoded by the *topA* gene which removes negative supercoils). Many important processes such as DNA replication, recombination, and transcription are affected by changes in DNA supercoiling, so the amount of supercoiling is normally maintained in a homeostatic balance by the competing action of these two topoisomerases. However, the extent of supercoiling may be modulated by a variety of environmental factors such as temperature, anaerobiosis, and osmolarity, which are likely to be experienced by pathogenic bacteria during infection. Thus, the level of supercoiling may be an important mechanism for coordinating expression of a variety of pleiotropic virulence genes: TopA is required for invasion and for normal expression of invasion genes in *S. typhimurium*, and many genes that have an important role in virulence have promoters that are sensitive to supercoiling. However, despite considerable circumstantial evidence that supercoiling is important for regulation of gene expression, there are many unanswered questions about the extent and mechanism of this regulation. One problem is that certain promoters are only sensitive to supercoiling when maintained in the correct DNA context, and transcription of adjacent genes can also affect the extent of local supercoiling. Thus, potential differences between local and global supercoiling may make it difficult to interpret results obtained when genes are cloned onto plasmids to study the effect of supercoiling. A second problem is that since supercoiling increases or decreases the expression of many genes (estimates suggest that about 50% of all genes are affected by supercoiling), it is necessary to determine whether the effect of supercoiling on virulence genes in vivo is direct or indirect. Bacterial histone-like proteins also have an important regulatory role in virulence. Integration host factor (IHF) regulates many genes by binding to DNA and causing DNA bending. HNS is a histone-like protein that binds to curved DNA and affects the expression of many bacterial genes. Mutations that eliminate IHF or HNS result in decreased virulence of *S. typhimurium* in mice. Although supercoiling and histone-like proteins may have an important role in the global regulation of virulence genes, most virulence genes are also regulated by other systems as well.

WHY STUDY REGULATION OF VIRULENCE GENES?

Understanding the regulation of virulence genes gives insight into conditions encountered upon infection and may provide an approach for identifying new virulence genes. On the basis of the prediction that many virulence factors may affect gene products on the cell surface that interact with other cells or the environment, other virulence genes have been identified by looking for gene fusions to membrane and secreted proteins (Kaufman and Taylor 1994). An alternative approach for identifying virulence genes is to search for new mutants that are regulated by environmental signals likely to be encountered upon infection of a host. A simple approach to isolate mutants regulated by specific conditions is to screen for gene or operon fusions that seem to be regulated appropriately in vitro (e.g., Lee et al. 1992). Recently, Mekalanos and colleagues (Mahan et al. 1993) developed a modification of this approach that allows the detection of genes that are specifically turned on during infection of a host. This approach, called the IVET selection for in vivo expression technology, is described in Section 5 (In Vivo Expression Technology).

Once the regulatory signals are identified, dissecting the details of the regulatory mechanism may provide further clues to the mechanism of virulence and may identify new targets for antimicrobial agents.

Challenge Phage: Genetic Analysis of DNA-Protein Interactions

Regulation of gene expression has an essential role in pathogenesis. Specific genes are turned on when bacteria enter a potential host, allowing the bacteria to survive host defenses and efficiently grow under conditions that differ drastically from the outside environment. Such regulatory responses commonly involve DNA-protein interactions controlled by specific physical or nutritional factors in the host. Understanding the DNA-protein interactions may give insight into the modification of these interactions and provide important clues to the molecular mechanisms of pathogenesis.

DNA-BINDING PROTEINS

The physiological role of DNA-binding proteins is determined by the affinity and specificity of the DNA-protein interaction. These properties depend on the precise interactions between amino acids in the DNA-binding protein and nucleotides in the DNA-binding site. Higher-order protein-protein interactions are often required for DNA-protein interactions as well.

DNA-BINDING SITES

Gene expression is often regulated by proteins that activate or repress transcription by binding to short, specific DNA sequences. Such cis-acting sites are usually located close to the promoter (RNA-polymerase-binding site) for the regulated gene, but sometimes they act from a considerable distance (e.g., by DNA looping). When positioned close enough to the promoter, almost any specific DNA-binding protein may act as a repressor. Repression may be caused by (1) competition with RNA polymerase binding to the promoter, (2) blocking isomerization of RNA polymerase required for initiation of transcription, (3) changing the local DNA structure (e.g., by bending the DNA), or (4) inhibiting the elongation of RNA polymerase after transcription has initiated. Repression due to binding to a single site is usually strongest when the binding site is close to the promoter (within ~20 bp) (Elledge and Davis 1989; Collado-Vides et al. 1991).

DNA-PROTEIN INTERACTIONS

How can features of a DNA-binding site or DNA-binding protein that determine DNA-protein interactions be identified? Many DNA-binding proteins have common structural motifs involved in DNA-protein interactions, but it is not possible to identify the amino acids involved in DNA binding by simple sequence gazing. For example, even though a protein may have an amino acid sequence with all the features of a consensus DNA-binding motif, that amino acid sequence may not be involved in DNA binding. Furthermore, different DNA-binding proteins that share common motifs may recognize DNA differently. Therefore, the critical nucleotides in a DNA-binding site and the interacting amino acids in DNA-binding protein must be determined empirically. DNA-protein interaction can be dissected using in vitro biochemical approaches or in vivo genetic approaches. Even when DNA-protein interactions have been characterized in vitro, genetic analysis of the interaction is needed to confirm that the specific DNA-protein contacts identified in vitro are necessary and sufficient for DNA binding under physiological conditions.

Any genetic approach for characterizing DNA-protein interactions which regulate gene expression requires the isolation of mutants that affect regulation. A general approach for isolating regulatory mutations is to place the expression of a selectable reporter gene under the control of the specific DNA-protein interaction. Transcription of the reporter gene is then regulated by binding of the protein to a specific DNA sequence upstream of the reporter gene. This approach can be applied to many different types of DNA-binding proteins.

Challenge phage provide a powerful method for genetically dissecting DNA-protein interactions by exploiting the *immI* regulatory region of bacteriophage P22 (Benson et al. 1986). Challenge phage use the phage P22 *ant* gene as the reporter gene. Under appropriate conditions, expression of the *ant* gene determines the lysis-lysogeny decision of P22. This provides a positive selection for and against DNA binding: Repression of *ant* can be selected by requiring growth of lysogens, and mutants that cannot repress *ant* can be selected by requiring lytic growth of the phage. Thus, placing *ant* gene expression under the control of a specific DNA-protein interaction provides very strong genetic selections for regulatory mutations in the DNA-binding protein and DNA-binding site that either increase or decrease the apparent strength of a DNA-protein interaction in vivo. The challenge phage selection provides a general method for identifying critical residues involved in DNA-protein interactions. Challenge phage selections have been used to genetically dissect many different prokaryotic and eukaryotic DNA-binding interactions (P. Youderian et al., in prep.).

P22 REGULATION

P22 has two regulatory regions that control the decision between lysis and lysogeny. The *immC* region encodes the *c2* repressor that regulates transcription from P_L and P_R. The regulation of P22 by the *c2* repressor is virtually identical to the regulation of bacteriophage λ by the *cI* repressor (Ptashne 1992). However, P22 has a second regulatory region, *immI*, that is not present on bacteriophage λ. The *immI* region indirectly regulates the lysis/lysogeny decision by modulating the activity of the *c2* repressor. The *immI* region contains three genes: *ant*, *arc*, and *mnt*. The *ant* gene encodes *ant*i-repressor, a protein that binds noncovalently to the *c2* repressor and inactivates it. Expression of the *ant* gene is negatively regulated by the other two *immI* gene products: the Mnt protein (*maintenance of lysogeny*) and the Arc protein (*antirepressor control*) (Fig. 6.1). Both *arc* and *ant* are expressed from the P_{ant} promoter.

How is this complex regulatory network coordinated? Early after infection, expression from P_{ant} results in a burst of Arc and Ant synthesis. Arc protein then binds the O_{arc} operator that overlaps P_{ant} and represses further transcription. Repression of P_{ant} by Arc activates P_{mnt}, turning on expression of Mnt. Mnt then binds to O_{mnt} and further represses *ant* expression during lysogeny. Because of this dual regulation by Arc and Mnt repressors, *ant* does not normally affect the decision between lysis and lysogeny during infection of a sensitive, nonlysogenic host (Susskind and Youderian 1983). However, in *arc* mutants, Ant is dramatically overproduced, preventing lysogeny of the infected cells. This overexpression of Ant can be prevented if the *arc⁻* phage infects a lysogen that is producing the Mnt repressor: Mnt produced by the lysogen can bind to O_{mnt} on the incoming phage, repressing *ant* expression and allowing the superinfecting phage to lysogenize the cell. Thus, in *arc⁻* phage, the binding of Mnt repressor to O_{mnt} regulates *ant* expression and thereby controls the decision between lysis and lysogeny (Weinstock et al. 1979).

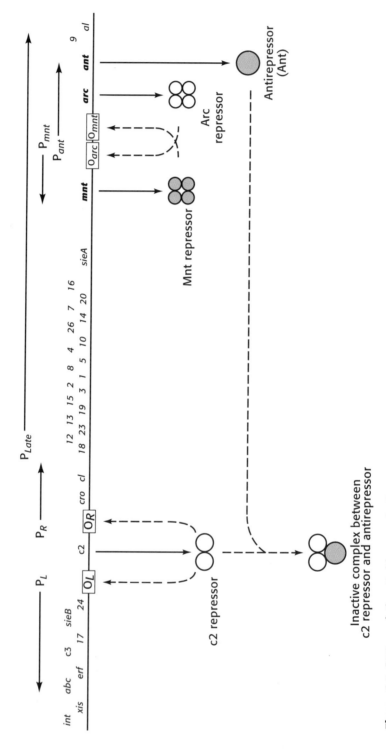

Figure 6.1 P22 regulation. Bold arrows indicate transcription. The map positions of major P22 genes are shown, but map distances are not drawn to scale. Important regulatory genes are shown in bold and DNA-binding sites are boxed.

In addition to Ant, other physiological factors can also affect the lysis/lysogeny decision mediated by the *c2* repressor. One important variable is the multiplicity of infection (moi). A high moi favors the lysogenic pathway: When the moi is greater than 10, more than 95% of the infecting phage form lysogens (Levine 1957). However, if the moi is too high, the incoming phage may titrate repressor binding to O_{mnt}, resulting in lytic growth of the phage.

CHALLENGE PHAGE

Challenge phage are derivatives of P22 with a kanamycin-resistant (Km^r) disruption of the *mnt* gene, an *arc*(Am) mutation, and a substitution of any desired DNA-binding site for O_{mnt}. Thus, binding of a protein to the substituted O_{mnt} site controls the decision between lysis and lysogeny for the challenge phage: If no protein is bound to the site, *ant* will be expressed and the host cells will lyse, but if a protein binds to the site, *ant* expression will be repressed, yielding Km^r lysogens. This provides a very tight, direct selection for studying specific DNA-protein interactions. The efficiency of lysogeny is typically between 10^{-1} and 10^{-2} if the host produces a protein that can bind to the substituted O_{mnt} site, but the efficiency of lysogeny is less than 10^{-7} if the protein cannot bind to the substituted O_{mnt} site.

CONSTRUCTION OF CHALLENGE PHAGE

Challenge phage contain a specific DNA-binding site substituted for O_{mnt}. The substitutions are obtained by cloning a DNA-binding site onto a plasmid that carries the P22 *immI* region, then crossing the plasmid clones with P22 *mnt*::Km9 *arc*(Am), and screening for recombinants that inherited the O_{mnt} substitution.

The plasmid vector we use for constructing challenge phage is pPY190 (see Appendix G, Plasmid and Transposon Restriction Maps). Plasmid pPY190 is a pBR322 derivative that carries a 500-bp *Eco*RI-*Hind*III fragment from the *immI* region of P22. The P22 fragment cloned on pPY190 includes the *mnt* gene, P_{ant}, and the 5′ end of the *arc* gene (including the *arc*⁺ allele of the *arc*(Am) mutation present on the challenge phage). The O_{mnt} site on pPY190 was replaced with a small fragment containing an *Sma*I site and an *Eco*RI site (Fig. 6.2). Blunt-ended DNA fragments can be directly cloned into the unique *Sma*I site on pPY190. Cloning into the *Sma*I site of pPY190 places the insert 3 bp upstream of the start of *ant* transcription.

The O_{mnt} substitution cloned into pPY190 can be moved onto P22 *mnt*::Km9 *arc*(Am) by recombination in vivo. To cross the phage with the

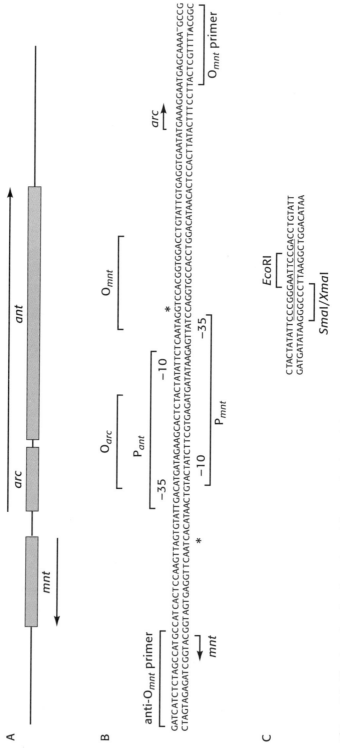

Figure 6.2 The *imml* region from P22. (*A*) Organization of the *mnt*, *arc*, and *ant* genes. (*B*) DNA sequence of the wild-type *imml* control region. The positions of the anti-O$_{mnt}$ and O$_{mnt}$ primers used for PCR amplification and sequencing are also shown. (*C*) Position of the O$_{mnt}$ replacement on pPY190 is shown directly below the corresponding sequence from the wild-type *imml* control region.

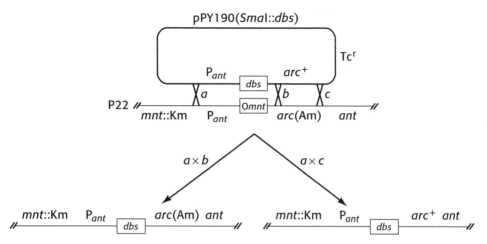

Figure 6.3 Construction of challenge phage by recombination with pPY190 in vivo. Recombinants between P22 *mnt*::Km9 *arc*(Am) and pPY190 can occur by two crossovers, one to the left of O$_{mnt}$ and one to the right of O$_{mnt}$ (indicated by a x b and a x c). Crossovers between O$_{mnt}$ and the *arc*(Am) mutation produce recombinant challenge phage that retain the *arc*(Am) mutation. Crossovers to the right of the *arc*(Am) mutation produce recombinant challenge phage that lack the *arc*(Am) mutation.

plasmid, the phage are grown on a permissive host carrying both an amber suppressor (*supE*) mutation and the pPY190 plasmid with the O$_{mnt}$ substitution. Since pPY190 has homology with P22 on both sides of the *Sma*I site, the plasmid can recombine with P22 and replace the P22 O$_{mnt}$ site with the substituted DNA-binding site.

Although the region of homology between the plasmid and phage is small (~500 bp), the frequency of recombination between the multicopy plasmid and the lytic phage is quite high because the plasmid and phage are present in multiple copies, and recombination is enhanced by phage-encoded proteins. The recombinant challenge phage can be selected by infecting a strain carrying a P22 *c2*+ *mnt*+ *sieA*− prophage. When the parental P22 *mnt*::Km9 *arc*(Am) phage infect a *c2*+ *mnt*+ lysogen, they are repressed by the *c2* and Mnt proteins and no plaques are formed. In contrast, when challenge phage with an O$_{mnt}$ substitution infect a *c2*+ *mnt*+ lysogen, they form plaques. *ant* expression is not repressed by the Mnt produced by the prophage because the challenge phage lack O$_{mnt}$; thus, the synthesis of antirepressor inactivates the *c2* repressor and lysis results.

Recombinants with pPY190 may carry either the *arc*+ allele from the plasmid or the *arc*(Am) allele from the phage (Fig. 6.3). Both types of recombinants form plaques on a P22 *c2*+ *mnt*+ lysogen: The *arc*(Am) mutants form large, clear plaques, whereas the *arc*+ recombinants form smaller, turbid plaques because Arc can partially repress P$_{ant}$. The *arc*+ recombinants are about twice as common as *arc*(Am) recombinants. The large, clear plaques containing the desired *arc*(Am) recombinant phage must be purified several times to eliminate contamination with phage from nearby plaques.

Large, clear plaques may also arise from spontaneous O_{mnt} constitutive mutants. The desired recombinants can be distinguished from O^c mutants by (1) directly assaying for repression in vivo, (2) testing for a restriction fragment length polymorphism (RFLP), or (3) determining the DNA sequence of the O_{mnt} region on the phage. Testing the phage that form clear plaques for RFLP analysis is a simple way of screening for recombinant phage. The O_{mnt} substitution on pPY190 contains an *Eco*RI site that is not present on P22. Thus, when digested with *Eco*RI, recombinant phage that inherited the O_{mnt} substitution from pPY190 will have lost one large fragment and gained two smaller fragments. In addition, an *Fnu*4HI site is destroyed by the *arc*(Am) mutation, so recombinants that retain the *arc*(Am) mutation can be screened by digesting the phage with *Fnu*4HI.

EXPRESSION OF DNA-BINDING PROTEINS

Most specific DNA-binding proteins are expressed at low levels in vivo. Overexpression of DNA-binding proteins at very high levels is often lethal. Therefore, to express sufficient levels of a DNA-binding protein for use with challenge phage, it is necessary to express the DNA-binding protein under control of a regulated promoter.

A useful approach is to clone the structural gene for the DNA-binding protein downstream from the tac promoter/operator (P_{tac}). P_{tac} is a strong hybrid promoter composed of the −35 region of the *trp* promoter and the −10 region of the *lacUV5* promoter/operator (Amman et al. 1983). Expression of P_{tac} is repressed by the LacI protein. The *lacI*^{Q} allele is a promoter mutation that expresses the LacI repressor at high levels, resulting in strong repression of P_{tac} unless the inducer IPTG (isopropyl-β-D-thio-galactopyranoside) is added. IPTG inactivates the LacI repressor. Thus, the amount of expression from P_{tac} is proportional to the concentration of IPTG added: Low concentrations of IPTG result in relatively low expression from P_{tac} and high concentrations of IPTG result in high

Figure 6.4 An example of regulation of challenge phage by induction of a DNA-binding protein (Nac) under the control of P_{tac}. (*A*) In the absence of IPTG, *nac* is repressed by LacI. In the absence of Nac protein, Ant is expressed and the phage grows lytically forming plaques. (*B*) When IPTG is added, LacI is inactive, so *nac* is expressed. Nac protein binds to the substituted DNA-binding site (O_{nac}) and represses *ant* expression. When Ant is repressed, the phage forms Km^r lysogens. (*C*) An IPTG challenge curve with O_{nac}+ challenge phage.

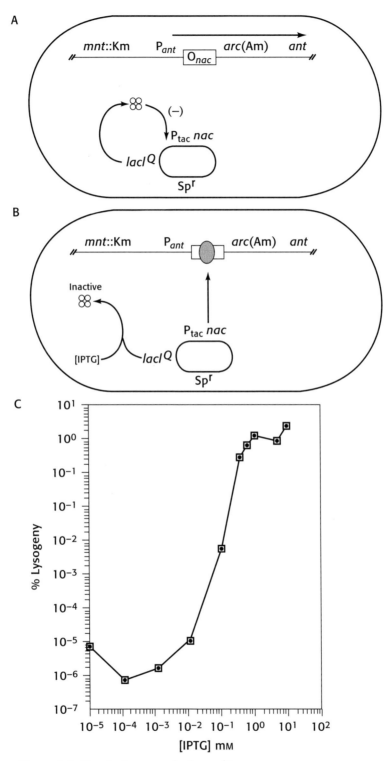

Figure 6.4 (*See facing page for legend.*)

expression from P_{tac}. By varying the IPTG concentration, the amount of a DNA-binding protein expressed from P_{tac} can be varied over several orders of magnitude.

When expression of a DNA-binding protein is regulated in this way, the relative affinity of a DNA-binding protein for the binding site on the challenge phage can be quantitated by measuring the frequency of lysogeny in media with different concentrations of IPTG (and hence different concentrations of the DNA-binding protein) (see Fig. 6.4). The relative DNA-binding affinity can be expressed by plotting the log(% lysogeny) versus the log[IPTG].

Several potential problems must be considered when expressing a DNA-binding protein from P_{tac}: (1) Ideally, *lacI*^{Q} should be cloned on the same plasmid as the regulated gene, because if *lacI*^{Q} is on the chromosome or on another plasmid, there may be insufficient LacI protein to fully repress the P_{tac} promoter in trans. (2) Cell viability should be measured at different concentrations of IPTG, because excessive overexpression of a DNA-binding protein may cause the protein to accumulate in inclusion bodies (Nilsson and Anderson 1991) or inhibit cell growth. (3) Even when maximally repressed, there is some expression from P_{tac}. If this leaky expression causes problems, it may be necessary to clone the gene into an alternative expression vector that is more tightly repressed. The plasmid vector pBAD is a useful alternative (Guzman et al. 1995). When cloned behind the P_{ara} promoter on pBAD, expression of the gene is regulated by the AraC repressor. Expression from the *ara* promoter is very tightly repressed on media with glucose and derepressed by arabinose (Schleif 1987).

USES OF CHALLENGE PHAGE

Challenge phage provide three tools for studying a DNA-binding site and the cognate DNA-binding protein: (1) an in vivo assay for determining the affinity of a DNA-binding protein for a specific DNA sequence, (2) a strong selection for mutations in a DNA-binding site that identify nucleotides which directly contact a DNA-binding protein, and (3) a strong selection for mutations in a DNA-binding protein that identify amino acids involved in recognition of a DNA-binding site.

Figure 6.5 Challenge phage provide strong selections for mutations that (*a*) affect critical nucleotides in a DNA-binding site and (*b*) affect amino acids in a DNA-binding protein that recognize a specific site.

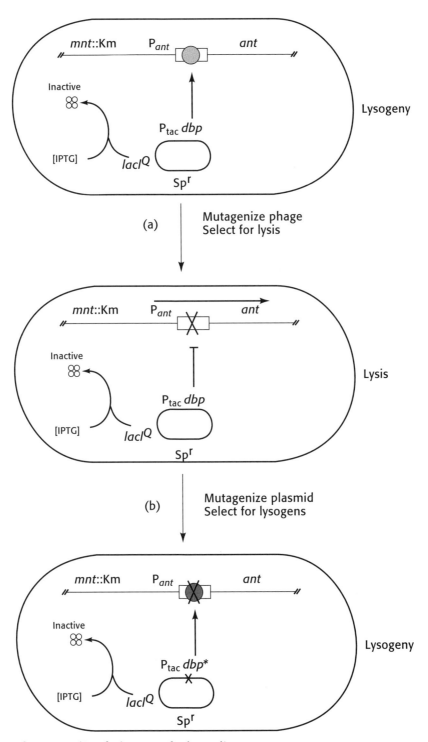

Figure 6.5 (*See facing page for legend.*)

In Vivo DNA-binding Assays

When a DNA-binding protein is expressed from P_{tac}, the relative affinity of a DNA-binding protein for the binding site on the challenge phage in vivo can be quantitated by measuring the frequency of lysogeny in media containing different concentrations of IPTG (and hence different concentrations of the DNA-binding protein). The relative DNA-binding affinity of different mutants can be determined by measuring the log(% lysogeny) versus the log[IPTG].

Selection for DNA-binding Site Mutations

Challenge phage form lysogens on a host that expresses a DNA-binding protein which binds to the substituted O_{mnt} site. It is possible to select for operator constitutive (O^c) mutations in the DNA-binding site by isolating challenge phage mutants that are not repressed by the DNA-binding protein and thus form plaques on a lawn of cells which express the DNA-binding protein (Fig. 6.5a). This provides a very strong selection for mutations that affect the DNA-protein interaction in vivo: Although mutations in a DNA-binding site are usually rare, the resulting rare plaques can be easily selected from a population of greater than 10^8 infecting phage.

Mutations in the DNA-binding site may result from spontaneous mutagenesis. However, spontaneous mutants often have deletion mutations that are less valuable for defining DNA-protein interactions than point mutations. The frequency of point mutations may be enhanced by mutagenizing the phage. Several mutagens with different specificities can be used to obtain a wide spectrum of mutations. Using this approach, it is possible to quickly isolate a large number of O^c mutations in the substituted DNA-binding site. The resulting mutations define critical residues of the DNA-binding site necessary for recognition by a DNA-binding protein.

Selection for DNA-binding Protein Mutants

Mutations in a DNA-binding protein that identify amino acids involved in recognition of a DNA-binding site can be obtained by (1) selecting for Km^r lysogens at a suboptimal concentration of IPTG, making it possible to select for mutants of DNA-binding proteins with increased affinity for DNA, and (2) selecting for Km^r lysogens from O^c challenge phage mutants, making it possible to isolate second-site suppressor mutant DNA-binding proteins that recognize the mutant site (Fig. 6.5b). Such second-site suppressor mutations may either increase the affinity of the DNA-binding protein for each of the DNA-binding sites (extended specificity mutants) or alter the site specificity of the DNA-binding protein (altered specificity mutants) (S. Maloy and P. Youderian, in prep.).

Many DNA-binding sites are palindromes that are bound by identical subunits of dimeric regulatory proteins: Each monomer recognizes one of the half-sites. If only one of the half-sites has a mutation, selection for DNA-binding protein suppressor mutants that recognize the palindrome will only result in enhanced specificity mutants—the mutant proteins must recognize the wild-type half-site in addition to the mutant half-site. Such enhanced specificity mutants are often due to a general increase in the DNA-binding affinity and do not define specific amino acid–nucleotide contacts. In contrast, altered specificity mutants recognize a mutant half-site much better than the wild-type half-site. Such altered specificity mutants often directly affect specific amino acid–nucleotide contacts. To isolate altered specificity mutants in a DNA-binding protein that recognizes a palindromic site, it may be necessary to construct symmetric double mutants by site-directed mutagenesis or direct synthesis.

AN EXAMPLE: THE NAC PROTEIN

The challenge phage experiments in this manual focus on the binding of the Nac protein to a specific DNA site located upstream of the *putA* gene in *Klebsiella aerogenes*. Nac is a LysR family regulatory protein that activates gene expression (Schwacha and Bender 1993; Goss and Bender 1995). When fixed nitrogen is limiting, expression of the *nac* gene is induced by the Ntr system (Fig. 6.6). The Nac protein then activates several other operons involved in nitrogen utilization, including the proline utilization (*put*) operon. Although the normal function of Nac is to activate *putA* gene expression, when the Nac-binding site is cloned into the *Sma*I site on the challenge phage, Nac binding prevents RNA polymerase from binding to P_{ant}. Thus, in the challenge phage, Nac functions as a repressor.

Although Nac had been shown to bind to DNA in vitro, the precise DNA sequences recognized by Nac were not known until mutations in the DNA-binding site were isolated using challenge phage (L.-M. Chen et al., in prep.). The wild-type Nac protein cannot bind to these mutant DNA-binding sites. To determine how Nac protein binds to DNA, suppressor mutations were isolated that allow the Nac protein to bind to the mutant DNA sites. The amino acid substitutions in allele-specific suppressor mutants identified a helix-turn-helix domain in the Nac protein that directly interacts with the DNA-binding site (L.M. Chen and S. Maloy, in prep.).

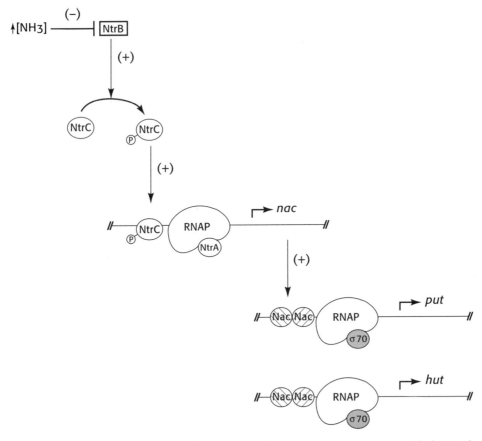

Figure 6.6 Nitrogen starvation induces a cascade of regulatory responses. When NH_3 is limiting, the kinase activity of NtrB protein is stimulated, causing NtrB protein to phosphorylate NtrC protein. The phosphorylated NtrC protein binds to specific enhancer sites on the DNA and activates transcription from NtrA (σ^{54})-dependent promoters, including the *nac* promoter. Nac protein subsequently activates transcription at certain σ^{70}-dependent promoters, including the *putA* gene in *Klebsiella*.

Challenge phage may also be used to study directly DNA-protein interactions that have important roles in bacterial virulence. For example, challenge phage have been used to determine how ToxR binds to sites that regulate expression of the *ctx* operon from *Vibrio cholerae* (Pfau and Taylor 1995).

RELEVANT P22 MUTATIONS

In addition to the modification of the *immI* region, the properties of several other P22 genes have critical roles in the challenge phage selections. The properties of a few relevant P22 mutations are described below.

Superinfection Exclusion

P22 has three mechanisms to avoid superinfection of a lysogen. The *a1* locus of P22 modifies the O-antigen receptor of the host cell, reducing (but not preventing) adsorption of superinfecting phage. The *a1* locus consists of three genes, *conABC* (M. Susskind, pers. comm.). The *a1* O-antigen modification also decreases adsorption of P22 to cell debris after lysis, allowing the preparation of very stable, high-titer P22 lysates. The two additional superinfection exclusion systems are encoded by the *sieA* and *sieB* genes. The function of these gene products is to prevent superinfection of P22 lysogens with other phage. The *sieA* system prevents superinfection by both P22 and related phage, but the *sieB* system only prevents superinfection by heteroimmune phage, not P22 derivatives (Susskind et al. 1971; Ranade and Poteete 1993). Hence, P22 *sieA* lysogens are sensitive to superinfection by P22.

arc(Am)

Repression of P_{ant} by Arc would complicate the analysis of P_{ant} repression caused by binding of a protein to the substituted O_{mnt} site. Therefore, challenge phage have an *arc*(Am) mutation. P22 *arc⁻ ant⁺* phage have a temperature-sensitive lethal phenotype because the overexpression of *ant* prevents proper expression of other phage late proteins (Susskind and Youderian 1983). This is less of a problem when *arc⁻* *arc*(Am) phage are grown on a *Salmonella typhimurium* host with a *supE* mutation because suppression of the amber mutation is only partial, so the *arc⁻* *arc*(Am) phage can be grown at 37°C.

Gene 9

P22 gene *9* encodes the tail protein. Therefore, gene *9* mutants are unable to adsorb to sensitive cells. However, P22 tail protein can be purified and will combine with intact phage heads and assemble into functional phage in vitro (Berget and Poteete 1980). Many of the P22 prophages have insertions of the Ap^r transposon Tn*1* in gene *9*. Since P22 circularizes by recombination between the terminal redundancy produced during the "headful" packaging mechanism, P22 mutants with large transposon insertions cannot form plaques because they cannot circularize after infection. However, it is possible to isolate revertants with compensating deletions that remove the nonessential genes between *mnt* and *a1* (Weinstock et al. 1979). Although these revertants still carry a mutation in gene *9*, they cannot form plaques unless tails are added. Hence, they are called tail-dependent plaque-forming revertants, designated Tpfr.

PHAGE PLAQUES

Plaques are clear zones formed in a lawn of cells due to lysis by phage. At a low moi, a cell is infected with a single phage and lysed, releasing progeny phage that can diffuse to neighboring cells and infect them, and so on. The morphology of the plaque depends on the phage, the host, and the growth conditions (Adams 1959). Phage infection is usually studied in a layer of soft agar (or "top agar"), which allows the phage to diffuse rapidly. The size of the plaque is proportional to the efficiency of adsorption, the length of the latent period, and the burst size of the phage. A diversity of plaque sizes can result if the phage infect cells at different times during the bacterial growth phase: Phage that adsorb early make larger plaques than those that adsorb later. To avoid this, phage are often preadsorbed to cells under conditions that do not allow phage growth (e.g., low temperature) and then shifted to permissive conditions to induce all of the phage to develop at the same time. A clear plaque will be formed if the host is completely susceptible to the phage. (Often, a clear plaque will be slightly turbid at the edge because the cells at the edge of the plaque are not yet fully lysed.) If the host is partially resistant to the phage, then the plaque may be uniformly turbid (e.g., if 10% of the cells survive phage infection).

Why do plaque assays accurately measure the titer of temperate phage? If a phage lysogenized a host cell immediately upon infection, it would never form a plaque. Instead, when temperate phage infects a population of exponentially growing cells, each phage produces a plaque with a "bull's-eye" plaque morphology, a turbid center surrounded by a ring of clearing. This characteristic plaque morphology is due to the role of the moi and cell physiology on the lysis-lysogeny decision. Lytic growth is favored when cells are growing rapidly and the moi is low, whereas lysogeny is favored when cells are growing slowly and the moi is high. This is the reason temperate phage typically have plaques with turbid centers. When the cells are initially infected with phage, the ratio of phage to cells is usually about 1 phage per 10^6 cells. Initially, the nutrients are plentiful and the bacteria grow rapidly and, since the moi is low, the phage grow lytically. After several lytic cycles, the local moi increases and most of the cells are lysed, producing a plaque in the lawn of cells. As the cell lawn becomes saturated, the rate of cell growth slows down, and since lysis requires rapid metabolism, the plaque stops increasing in size. However, any lysogens that formed in the center of the plaque are immune to lysis and can continue to grow because they do not have to compete with nearby cells for nutrients. Thus, lysogens begin to grow in the center of the plaque, giving the plaque a turbid, bull's-eye appearance.

SECTION 7
Recombinant DNA

". . .protein sequences derived from DNA sequences are subject to many possible errors. *(a)* The actual initiation codon is often conjectural, based on a putative ribosome binding site and analogy with the initiation codon in other, often distantly related organisms. *(b)* Undetected frameshift errors, usually in pairs, are sometimes present; a sequence segment may differ markedly from the rest when sequences from a variety of organisms are being aligned. If by adding and deleting single base pairs the odd sequence can be brought into conformity, a technical error in sequence determination (most often as a result of compression) is suspected. . . *(c)* DNA sequences are not error free; many of these errors are not obvious by comparison with other sequences. *(d)* Finally, the possibility exists that errors were introduced during manipulation for alignment or data reduction and missed in proofreading."

Irving P. Crawford
Reprinted, with permission, from (©1989 by Annual Reviews, Inc.) Evolution of a Biosynthetic Pathway: The Tryptophan Paradigm.
Annu. Rev. Microbiol. 43: 567–600 (1989)

Plasmids and Conjugation

Movement of plasmids among bacteria and even across species barriers occurs naturally and has a variety of applications in genetic analysis. A major mechanism by which this movement occurs is bacterial conjugation, which is the process of DNA transfer requiring direct contact between the donor and recipient cells. Nearly all naturally occurring plasmids either are self-transmissible or can be mobilized in trans. Most conjugation systems are capable of mediating transfer to related species, and this is particularly true of the IncP plasmids, which demonstrate an extremely broad host range and are capable of transfer to virtually all gram-negative species. This property of IncP-based plasmids is an extremely useful tool in the genetic analysis of diverse bacterial species (Guiney and Lanka 1989). Broad host range plasmids (see representative list in Table 7.1) provide vectors for which cloned DNA fragments can be manipulated in *Escherichia coli* and then transferred back to the original organism for functional studies (Christopher et al. 1989). They have also been utilized to mobilize chromosomal markers in various species (Haas and Reimann 1989) and are useful as vectors for the introduction of transposons into a variety of bacterial species (Simon 1989). For a number of extensive reviews on bacterial conjugation and the use of broad host range plasmids in genetic analysis, see Thomas (1989), Clewell (1993), and Hardy (1993).

MECHANISM OF CONJUGATION

The molecular mechanism of bacterial conjugation is perhaps best understood in the case of the F fertility plasmid of *E. coli*. The F plasmid is about 100 kb in size, of which 33 kb are relegated to encoding functions

Table 7.1 Some Common Broad Host Range Plasmids and Vectors

Incompatibility group/ Replicon	Plasmid	Phenotype characteristics	References
IncFI	F	Tra$^+$	Bukhari et al. (1977)
IncP1	RK2/RP4/RP1/R68	Ap^r, Km^r, Tc^r, Tra$^+$	Burkhardt et al. (1979)
	pRK290	Tc^r, Mob$^+$	Ditta et al. (1980)
	pLAFR series	Tc^r, Mob$^+$ cosmids, multiple cloning sites, lacZα for pLAFR3	Friedman et al. (1982)
	pPH1JI	Gm^r, Sp^r, low level Sm^r, Tra$^+$	Ruvkun and Ausubel (1981)
	pR68.45	Tc^r, Km^r, Ap^r, tandem duplication of IS21, broad host range un-stable Hfr formation	Haas and Halloway (1976, 1978)
	pME487	pR68.45ts	Haas et al. (1987)
IncP1/ ColE1	pRK2013	Km^r, Tra$^+$, mobilizer replicon	Figurski and Helinski (1979)
IncQ	RSF1010	Su^r, Sm^r, Mob$^+$	Guerry et al. (1974)
	pMMB66EH/HE	Ap^r, tac promoter expression vector	Furste et al. (1987)
IncX	R6K	Ap^r, Sm^r	Kontomichalou et al. (1970)
	pGP704	OriR6K cis region, MobRP4, Ap^r, multiple cloning site, suicide vector requires R6Kπ	Miller and Mekalanos (1988)
pMB1	pBR322	Ap^r, Tc^r, Mob$^+$	Bolivar et al. (1977)
	pBR327	Ap^r, Tc^r, Mob$^-$	Bolivar (1978)
	pUC series	Ap^r, lacZα, Mob$^-$	Yanisch-Perron et al. (1985)
P15A	pACYC177	Ap^r, Km^r	Chang and Cohen (1978)
	pACYC184	Tc^r, Cm^r	

Abbreviations: Tra$^+$, self-transmissible; Mob$^+$, can be mobilized with a helper plasmid or strain; Ap, ampicillin; Cm, chloramphenicol; Gm, gentamicin; Km, kanamycin, Sm, streptomycin; Su, sulfonamides; Tc, tetracycline.

involved in plasmid transfer. Most of the genes encoding these functions are organized in a large *tra* operon of 37 genes designated as *tra* and *trb* genes (for review, see Willetts and Skurray 1987; Ippen-Ihler and Skurray 1993). Expression of the *tra* operon is positively regulated at the transcriptional level by the product of *traJ*, which is the first gene of the operon. A second level of regulation is mediated at the level of fertility inhibition by *finP*. The *finP* gene encodes an RNA that is antisense to the untranslated mRNA leader region of *traJ* and acts in a negative manner by occluding the *traJ* ribosome-binding site. The *finP* message is stabilized by the protein product of the *finO* gene. In the case of F, there is an IS3 element in the *finO* gene and thus F is derepressed for conjugation. Most conjugal plasmids exhibit fertility inhibition, although the mechanisms by which it occurs may differ. When a plasmid under such negative control does happen to enter a recipient cell, a phenomenon similar to zygotic induction can occur. Since the recipient does not already contain the molecules necessary to prevent *tra* gene expression, expression of these genes from the newly incoming plasmid is not immediately repressed, and a burst of plasmid transfer can occur. This mechanism may ensure rapid dispersal of a plasmid under appropriate conditions. Signals that may promote plasmid transfer have not yet been elucidated. The IncP plasmid, RP4, is also naturally derepressed for conjugation.

Once the *tra* operon is expressed, the initial stages of cell contact required for conjugation are mediated by the F pilus. The F pilus is composed of a homopolymer of TraA pilin subunit. The 70-amino-acid mature pilin is initially synthesized as a 121-amino-acid precursor molecule. During export, prepilin is processed to its mature form, which requires TraQ. In addition, the amino-terminal residue of the mature pilin must be acetylated by the product of *traX* prior to its incorporation into a pilus structure. Interestingly, the *traA*, *traQ*, and *traX* genes are located at the beginning, middle, and distal end of the *tra* operon, respectively. This may ensure that the complete operon is expressed prior to any pilin processing. The transport and assembly of the mature pilin into a pilus structure require the coordinated activity of many gene products. These products are located in the inner membrane, periplasm, and outer membrane of the cell. In addition to genes required for pilus assembly, some F mutants make pili but lack certain outer membrane proteins and are defective in conjugation. The corresponding genes, *traN* and *traG*, are thought to encode products that stabilize the aggregate formation between donor and recipient cells. Other membrane proteins are involved in a property of F-containing cells termed surface exclusion. These proteins appear to inhibit interaction and uptake of DNA from other F-containing cells.

Once a stable complex or aggregate is established between the donor and recipient cells, DNA transfer can proceed. The DNA transfer origin on the plasmid is designated *oriT*. The DNA is nicked on a single strand and unwound by a plasmid-encoded helicase enzyme. A single strand is displaced and transferred in a 5' to 3' direction, with new synthesis by a rolling circle mechanism in the donor cell. A number of proteins of unknown function and predicted to be localized to various cellular locations are required for the actual process of DNA transfer. The identities and functions of some proteins involved in this process are known. For example, F encodes its own single-stranded DNA-binding protein that may protect the DNA should it enter a recipient with insufficient amounts of host-encoded Ssb protein. The *psi* gene products inhibit SOS response and RecA protease activity and may thus also serve a protective function for the incoming DNA. In addition, a plasmid-encoded primase enters the recipient cell, thus providing RNA primers for discontinuous synthesis of the complement to the transferred plasmid DNA strand by resident host polymerase. These general requirements are likely to prove true for the majority of conjugal plasmids, although the mechanisms by which they are achieved can differ.

MOBILIZATION OF CHROMOSOMES

The discovery of mobilization of the *E. coli* chromosome initiated genetic mapping studies of the bacterial genome (Lederberg and Tatum 1946). Transfer of the chromosome by F occurs at low frequency compared with F transfer itself (at $\sim 10^{-7}$ for a given marker). Integration occurs via two mechanisms, one utilizing the host recombination system and the other utilizing transposition of resident elements. In the case of F, the plasmid carries IS2, Tn*1000*, and two IS3 elements (Willets and Skurray 1987). These provide homology with the *E. coli* chromosome since IS2 and IS3 are present in many strains. A single crossover event will integrate the entire F plasmid into the chromosome in a RecA-dependent fashion. Origins of chromosomal transfer can be specifically engineered in this manner by providing specific homology between a self-transmissible plasmid such as an F' and a chromosomal insertion (see Fig. 7.1). A second method for integration is by IS-mediated transposition of F into the chromosome. Strains in which F has become stably integrated into the chromosome are termed Hfr strains (high frequency of transfer). Such strains greatly facilitate genetic mapping using crosses with F⁻ recipients. In this case, *oriT* and the other genes encoding necessary transfer functions have become chromosomally encoded, allowing transfer of portions or the entire chromosome to recipient cells (see Section 3, Genetic Mapping of Chromosomal Genes). Aberrant excision of F from the chromosome can result in an F' plasmid that carries a portion of chromosomal DNA adjacent to the original insertion (for review, see Low

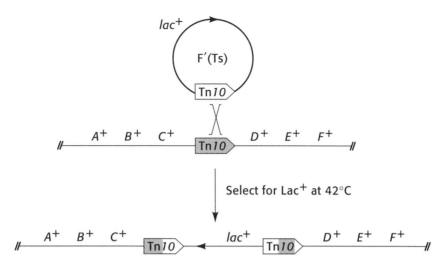

Figure 7.1 Method to create an Hfr strain with a defined origin of transfer. A gene of interest might first be identified by a chromosomal transposon insertion mutation. In this case, an F′ that contains homology with the transposon (Tn*10*), an additional selectable marker (Lac), and is temperature-sensitive for replication is introduced into the transposon mutant. Selection for Lac⁺ colonies at nonpermissive temperature results in isolation of colonies that contain the F′ integrated into the chromosome. To determine whether integration occurred at the point of transposon homology, as well as to determine orientation of integration, the time of transfer of adjacent markers can be examined in an interrupted mating experiment.

1972). F′ plasmids are useful for complementation analyses because they are self-transmissible and are maintained at one to two copies per cell. Thus, the interpretation of complementation results is not complicated by the increased gene dosage that would occur using a multicopy plasmid.

Plasmids other than F are useful for chromosome mobilization in a broader range of bacterial species. The broad host range of IncP-based plasmids makes them particularly suited for this purpose. As in the case of F, at least transient integration into the chromosome via transposition or homologous recombination must occur for mobilization to take place. One method of mobilization by the RK2/RP4/R68 family of plasmids is through replicative transposition of their resident Tn*1* element. However, since the Tn*1* resolvase functions very efficiently, chromosomal transfer by this mechanism is very infrequent. This group of plasmids also contains an IS*8* (designated IS*21* on R68) element. A derivative of R68 (R68.45) has been isolated that mediates increased frequency of chromosomal transfer. This plasmid has been shown to contain two tandem copies of IS*21*, which then allow the plasmid to act as a composite transposon resulting in integration of the entire R68 into the chromosome

(Leemans et al. 1980; Reimmann and Haas 1987). Although these structures are more stable than those mediated by Tn*1*, they are not stable enough to allow isolation of Hfr-like strains. One method used to form stable Hfr strains is to insert the RK2/RP4 *oriT* into the chromosome of potential donor strains. This has been accomplished through the construction of Tn*5-oriT* (or Tn5-Mob) (Simon 1984; Yakobson and Guiney 1984). Chromosomal mobilization from strains carrying the transposon is accomplished in the presence of an RK2/RP4-derived helper plasmid. This system can be used to mobilize plasmids as well.

One mechanism of Hfr formation in the case of F is by integration into the chromosome via homologous sequences present on both molecules. IncP and other plasmids do not naturally carry regions of homology with the chromosome as in the case of F in *E. coli*. There are several ways to provide homology artificially between a plasmid and the chromosome. For example, a similar transposon can be placed onto the plasmid and into the chromosome of a potential donor cell. Such instances can result in transient integration and low frequency of chromosomal transfer, or stable donors can be identified using means such as that outlined in Figure 7.1. Such a system has been developed for the resident sex factor, P, of *Vibrio cholerae*, which does not promote chromosomal transfer very efficiently on its own (Newland et al. 1987).

MOBILIZATION OF PLASMIDS

Plasmids that are not self-transmissible but can be transferred with the aid of other plasmids are said to be mobilizable (Mob$^+$). There are two general mechanisms by which mobilization can occur. In the first case, the mobilizable plasmid contains an *oriT* region but does not express a complete complement of the other functions necessary for transfer. The conjugation apparatus and transfer functions must therefore be provided by a second plasmid in the cell. A series of vectors and transfer strains have been developed by Simon et al. (1983) based on this system. In this case, the RP4 *oriT* sequence was inserted into plasmids to be mobilized. Mobilization is accomplished by using a helper strain such as SM10 as the donor. SM10 has a defective RP4 plasmid integrated into its chromosome that provides all the necessary transfer functions to promote DNA transfer initiated at *oriT* present on the plasmid to be mobilized. This system is used to transfer the plasmids used for allelic exchange in Experiment 14. A second mechanism of mobilizing plasmids requires transient fusion of the mobilizable plasmid with a self-transmissible plasmid. As in the case of chromosome transfer, this event can be mediated by homologous recombination or promoted by IS elements (Haas and Reimmann 1989).

INCOMPATIBILITY AND GENETIC ANALYSIS

Two plasmids closely related in their mechanisms of replication or maintenance cannot stably coexist in the same cell and are termed incompatible. Incompatibility groups can be determined by introducing a plasmid into a cell that already has a plasmid and selecting for acquisition of the new plasmid. If all of the cells lose the original plasmid, then the two are in the same incompatibility group. For example, two pMB9 replicons, say pBR322 and pUC18, cannot be maintained in the same cell because they share a similar repressor protein involved in replication control. Thus, even if they are marked with different antibiotic resistances, their copy numbers with respect to each other cannot be reliably maintained. When conducting experiments where two plasmids are required, the two plasmids should be of different incompatibility groups. For example, pACYC184 (*ori* P15A) and pBR322 (*ori* pMB1) can be maintained in the same cell. Incompatibility can also be used as a method to get rid of a plasmid vector from a cell by introducing and selecting for a second plasmid of the same incompatibility group. This concept is used for broad host range IncPα transposon delivery systems (Taylor et al. 1989) (see Section 5, Operon and Gene Fusions) and as a method of allelic exchange (Ruvkun and Ausubel 1981) (see Section 4, Broad Host Range Allelic Exchange Systems).

CONJUGATION IN PRACTICE

The considerations for using conjugation as a genetic tool are relatively simple and inexpensive. The first is to choose genetic characteristics for the donor and recipient strains such that a condition can be found to select for acquisition of the plasmid by the recipient strain without carryover of the donor. This condition is termed a counterselection. For example, an Sm^r recipient strain can be mated with an Sm^s donor carrying a transmissible plasmid marked with Tc^r. Plating the mixture on agar containing streptomycin and tetracycline will select for both the presence of the plasmid and the growth of the recipient. The donor cannot grow in the presence of streptomycin (see Fig. 4.4 in Section 4). Streptomycin resistance is a common counterselection since it is relatively easy to isolate Sm^r mutants that are essentially unaffected in growth properties or other characteristics. Another good counterselection is the use of an auxotroph as the donor. For example, the donor strain might be Pro$^-$ while the recipient is prototrophic. Plating on minimal agar lacking proline will select for the recipient. Selective growth medium may provide another useful counterselection when mating between species. For example, an *E. coli* donor cannot grow on TCBS agar which supports the growth of *V. cholerae*. It should be noted that selective media often contain compounds not usually found in more typical growth media that may interfere with the action or stability of certain antibiotics. A second consideration is the method of mating. The highest frequency of transfer for IncP plasmids occurs on solid medium (Bradley et al. 1980). In the experiments in this manual, IncP matings are done simply by cross-streaking fresh donor and recipient strains on a rich medium such as LB agar and allowing incubation for 8 hours to overnight. Filter matings are also commonly used for many broad host range plasmids. Donors containing some plasmids, such as F, seem to mate more efficiently in static liquid culture using cells from the exponential phase of growth.

Transformation and Electrotransformation

Transformation is the uptake of naked DNA from the environment. The ability to be transformed depends on the physiological state of the cell. When cells are in a physiological state suitable for transformation, they are called competent. Some bacteria become naturally competent for transformation at some stage of cell growth, but transformation of many bacteria requires treatment with chemicals or an electrical shock.

TRANSFORMATION OF *E. COLI* AND *S. TYPHIMURIUM*

Escherichia coli and *Salmonella typhimurium* do not take up exogenous DNA efficiently unless appropriately treated with chemicals (Hanahan 1987; Hanahan et al. 1991). Hypotonic Ca^{++} shock is a common method for preparing competent cells. This method involves centrifuging an early exponential phase culture and then resuspending it in a cold hypotonic $CaCl_2$ solution. DNA added to these cells forms a calcium-DNA complex that adsorbs to the cell surface. The cells are then briefly warmed (heat-shocked), which allows transport of the DNA into the cell. When selecting for antibiotic resistance, the cells are often grown briefly in a non-selective medium to allow expression of the antibiotic resistance gene prior to plating on selective medium. A quicker and easier method for preparing competent cells uses polyethylene glycol (PEG) and dimethyl-sulfoxide (DMSO) (Chung and Miller 1988). In this technique, a PEG-DNA complex adsorbs to the cell surface. DMSO apparently facilitates entry of the DNA into the cell. Both of these techniques yield transformation efficiencies of approximately 2×10^8 transformants per microgram of DNA.

219

Several factors affect the transformation frequency. (1) Rapidly growing early exponential phase cells must be used. Using a culture that is not well aerated or cells grown to high density decreases the transformation efficiency. (2) Competent cells are very fragile, and thus it is important to keep the cells cold and avoid vigorous mixing. Even briefly allowing the cells to warm up during the treatment will decrease the transformation efficiency. (3) For unknown reasons, different strains have different inherent transformation efficiencies. (4) Competent cells can be stored frozen at −70°C. Freezing decreases the transformation frequency of competent cells prepared by the $CaCl_2$ protocol at least tenfold, but it does not significantly affect the transformation efficiency of competent cells prepared by the PEG-DMSO protocol.

ELECTROTRANSFORMATION

Cells can also take up exogenous DNA by electroporation (for review, see Chassey et al. 1988; Shigekawa and Dower 1988). When cells are exposed to an electric field, the membrane becomes polarized and a voltage potential develops across the membrane. If the voltage potential exceeds a threshold level, small pores transiently form in the membrane that make the cells permeable to exogenous macromolecules, including DNA. Uptake of exogenous DNA by electroporation is called electrotransformation.

Although electroporation requires a special apparatus, it has several advantages over chemically induced transformation. (1) Preparation of the cells is simple and quick. A culture of cells is thoroughly washed (to remove salts from the medium) and resuspended in a low-ionic-strength solution (e.g., deionized H_2O) with 10% glycerol. (2) After preparing cells for electroporation, they can be stored at −70°C without loss of electrotransformation efficiency. (3) The efficiency of electrotransformation is often several orders of magnitude greater than that of other methods of transformation. Electroporation of E. coli or S. typhimurium typically yields 10^9 to 10^{10} electrotransformants per microgram of plasmid DNA. (4) The efficiency of electrotransformation of E. coli and S. typhimurium is so high that plasmids can be directly transferred between cells without purifying the DNA. (5) Electrotransformation is much less persnickety about the donor DNA than many other transformation methods. Electrotransformation of supercoiled and relaxed plasmids from 2 to 44 kb occurs at high efficiency. In addition, electroporation of linear DNA from phage P1 or P22 occurs with an efficiency of about 0.1% that of small plasmids. (6) A wide variety of bacteria can be transformed by electroporation, including many bacteria that are transformed poorly or not at all by other methods.

ELECTROPORATION

The electroporator made by Bio-Rad is described here. Other electroporators may differ somewhat but the basic principles are similar. The electroporation apparatus generates a rapid, high-voltage electrical pulse across a cuvette that holds the cells and DNA. The voltage is stored in a capacitor and released as an electrical pulse that decays exponentially. The initial voltage (V_0) decays over time according to the equation:

$$V_t = V_0 \ e^{-t/\tau}, \text{ where } \tau = R \times C$$

R is the resistance in ohms, C is the capacitance in farads, and τ is the time constant in seconds. τ is the time required for the voltage to decay to $1/e$ (~37%) of the initial voltage. Thus, τ is proportional to the length of the pulse. τ is optimally 4–5 msec for electroporation of enteric bacteria.

The pulse controller contains several variable sized resistors that can be placed in parallel with the sample, as well as a fixed 20-ohm resistor that is placed in series with the sample. Typically, a 200-ohm resistor is placed in parallel for enteric bacteria. This resistor effectively swamps out the contribution of the resistance of the sample to the total resistance of the circuit and thus determines the total resistance across the capacitor and controls the time constant, τ. The 20-ohm resistor protects the machine by limiting the current if arcing occurs.

The voltage potential across the cell membrane is proportional to the voltage gradient between the electrodes, also called the electric field (E):

$$E = V/d$$

where d is the distance between the electrodes (i.e., the path length of the cuvette). E and τ are the most important electrical variables affecting electroporation. If E is too low, the voltage potential across the membrane is too small to disrupt the membrane and electroporation does not occur. If E is too high, the cell membrane is irreversibly disrupted. E can be adjusted by varying the voltage or by changing the path length of the cuvette. For electroporation of *E. coli* and *S. typhimurium*, the optimal E is between 12.5 and 20 kV/cm, but the optimal conditions for other bacteria may be different. Under these electroporation conditions (with cells in a 0.1-mm cuvette shocked in an electroporation apparatus set at 1.5 kV, 200 ohms, and 25 μF, $E = 12.5$ kV/cm and $\tau = 4$ msec), approximately 30% of the cells remain viable.

The resistance across the cuvette is inversely proportional to the ionic strength of the sample. Discharge of a capacitor into a solution of higher ionic strength (lower resistance) produces a pulse with a shorter τ. Under some conditions, the voltage may be discharged across the cuvette very rapidly, causing arcing. Arcing drastically reduces the efficiency of electroporation and the number of surviving cells. Several possible causes of arcing arc (1) carry over of salts in the cell suspension due to inadequate washing, (2) carry over of salts in the DNA solution, (3) inadequate mixing of the cells and DNA, (4) failure of the sample to contact both sides of the cuvette (e.g., if the volume of cells is too small), (5) too many bubbles in the sample, (6) cracks in the electroporation cuvette, and (7) wet surfaces on the cuvette or slide chamber. However, the most common cause of arcing is too much salt in the DNA solution. As a rule of thumb, the salt concentration of the DNA should be reduced to less than 10 mM before dilution into the cell suspension. Drop dialysis is a simple, effective way to decrease the salt concentration of DNA samples before electroporation (see Appendix F, Drop Dialysis).

In addition to the electrical variables, several other factors may affect the efficiency of electrotransformation. (1) The frequency of electrotransformation is directly proportional to the DNA and cell concentration. For *E. coli* and *S. typhimurium*, the number of transformants obtained typically increases linearly with DNA concentrations from 10 pg/ml to greater than 5 μg/ml and is proportional to cell concentration between 10^9 and 10^{10} cells/ml. (2) The frequency of electrotransformation is dramatically affected by temperature. For *E. coli*, the number of transformants obtained is at least 100-fold higher when the cells and cuvette are cooled to 0–4ºC before electroporation than when used at room temperature. (The temperature effect may be due to joule heating during the pulse. Joule heating from a single high-voltage pulse can cause a 15–25ºC rise in temperature.) (3) The frequency of electrotransformation is higher for exponential phase cells than for stationary phase cells. (4) Not all strains are transformed with the same efficiency. (5) The DNA does not need to interact directly with the cell membrane for electrotransformation, and thus preincubation of the cells and DNA before pulsing is not required.

In addition to allowing transformation of cells with purified DNA, electroporation can also be used to transfer plasmids directly between cells. Both the donor and recipient cells must be prepared for electroporation. If live donor and recipient cells are used, a counterselection against the donor cells is needed to isolate recipient colonies that have taken up the plasmid (Pfau and Youderian 1990; Summers and Withers 1990). However, if the donor cells are heat-killed prior to electroporation, plasmids can be efficiently transferred to the recipient without a counterselection (A. Lorincz and P. Youderian, unpubl.).

Basic Molecular Biology Techniques

There are endless variations and modifications of essentially every molecular biology technique. For several excellent sources, see Perbal (1988), Sambrook et al. (1989), and Ausubel (1994). Adding to the number of choices are commercial "kits" that are advertised to be easier and faster (but not necessarily cheaper). Given the variety of choices available, the temptation is strong to quickly switch to a different method if one method does not work right away. However, if the biochemical basis for the methods is understood, this time-consuming and expensive "random-walk" through numerous molecular biology manuals and catalogs can be avoided by simple trouble-shooting. Therefore, the rationale for a few generally useful techniques are described below.

SMALL-SCALE PLASMID PURIFICATION ("MINIPREPS")

High-copy-number plasmids can be easily purified from bacteria simply grown in rich medium. However, under these conditions, a poor yield of plasmid DNA is often obtained from low- or medium-copy-number plasmids. Therefore, it may be necessary to amplify the plasmid before purification to increase the yield of plasmid DNA relative to chromosomal DNA. One common method for plasmid amplification is to grow the plasmid-containing strain to mid exponential phase in rich medium, then add chloramphenicol, and continue incubating the culture overnight. Chloramphenicol inhibits initiation of chromosomal replication, but since initiation of replication of many multicopy plasmids does not require protein synthesis, the plasmids continue to replicate (Hershfield et al. 1974). However, this procedure requires monitoring the growth phase of the cells, and it does not work for chloramphenicol-resistant plasmids. An alternative procedure is to grow the plasmid-containing strain to late stationary phase in plasmid broth. Plasmid broth is a very rich medium that allows cells to grow to a very high density. In addition, the high concentration of yeast extract promotes the continued replication of plasmids after chromosomal DNA replication stops in stationary phase, yielding a large number of plasmids per cell. The culture may then be stored at 4°C for 1–2 days before purifying the plasmid (for certain plasmids, leaving the culture overnight at 4°C increases the plasmid yield somewhat).

The alkaline lysis procedure is the most common method for purifying plasmid DNA from gram-negative bacteria (Birnboim and Doly 1979). The bacteria are treated with EDTA, which chelates divalent cations, disrupting the outer membrane. (Many procedures include lysozyme to degrade the cell wall peptidoglycan, but for most Enterobacteriaceae, the yield of DNA is just as high without lysozyme.) The cells are then lysed with a solution of sodium dodecyl sulfate (SDS) and NaOH. Once the cells are lysed, the solution must be mixed gently to avoid shearing the chromosomal DNA into small fragments that might copurify with the plasmid DNA. SDS disrupts the cytoplasmic membrane and denatures much of the cell protein. The NaOH raises the pH, which denatures double-stranded DNA. Acetate is then added to neutralize the pH, allowing the denatured single-stranded DNA to reanneal. The rate of reassociation is proportional to the length of the DNA. Since the plasmid DNA is small and intertwined, it reanneals much faster than the chromosomal DNA. When the mixture is centrifuged, the reannealed plasmid DNA remains in solution, but large aggregates of denatured chromosomal DNA, RNA, and protein are pelleted. The plasmid DNA is then precipitated with isopropanol or ethanol. A procedure for alkaline lysis minipreps is included in Appendix F (Plasmid DNA Minipreps).

Some RNA copurifies with the plasmid DNA. The contaminating RNA does not interfere with many procedures, such as restriction digests, but it can cause problems. For example, the RNA may obscure small DNA fragments on ethidium-bromide-stained gels, the free ends of RNA fragments may compete with DNA during end-labeling reactions, and short RNA fragments may act as nonspecific primers during DNA sequencing reactions. Therefore, it is sometimes necessary to eliminate contaminating RNA from the plasmid DNA. This can be done by degradation of the RNA with DNase-free RNase or differential precipitation of the RNA with LiCl (Pelham 1985). RNA is insoluble in 2.5 M LiCl, but DNA remains in the supernatant. A procedure for eliminating RNA from plasmid DNA preparations using LiCl is described in Appendix F (Plasmid DNA Minipreps).

Typically, 5–10 µg of plasmid DNA is obtained from a 1.5-ml culture of *Escherichia coli* containing a pBR322 derivative. The purified DNA is usually stored in Tris buffer containing EDTA (TE). The EDTA chelates heavy metals that could cause nicking of the DNA and divalent cations that are required by nucleases. The DNA may be stored at −20°C or at 4°C. It was previously thought that storage at −20°C induces nicks in DNA, but even after many freeze-thaw cycles, the number of nicks that accumulate in supercoiled DNA is negligible. One advantage to storing DNA at 4°C is that it is not necessary to wait for it to thaw before use.

LARGE-SCALE PLASMID PURIFICATION

The DNA from a small-scale plasmid purification is usually sufficient for restriction analysis or cloning. However, if a large amount of plasmid DNA is needed, a large-scale plasmid purification may be necessary. The initial steps of the large-scale plasmid preparation are similar to that of the small-scale preparation described above. The resulting DNA is much more concentrated than DNA from small-scale plasmid preparations; however, since the concentration of contaminants such as chromosomal DNA and RNA is much greater, the DNA must be further purified. For many years, purification of plasmid DNA from large-scale plasmid preparations was done by CsCl density gradient centrifugation in the presence of ethidium bromide. Covalently closed, circular plasmid DNA can be separated from any contaminating chromosomal DNA, nicked plasmid DNA, and RNA in a CsCl–ethidium bromide density gradient. When centrifuged to equilibrium, CsCl forms a density gradient, and any macromolecules in the CsCl gradient will band at the equivalent density of CsCl. To separate the plasmid DNA from contaminating chromosomal DNA, include ethidium bromide in the CsCl gradient. Ethidium bromide binds to DNA by intercalating between the bases, causing the DNA to unwind. Since covalently closed, circular plasmid DNA has no free ends, it can only unwind a limited amount and thus only binds a limited amount of ethidium bromide. In contrast, linear DNA and nicked circles do not have these topological constraints and thus they bind much more ethidium bromide. Ethidium bromide is less dense than DNA; therefore, the density of DNA decreases as more ethidium bromide is bound. Linear and nicked circular DNAs have a lower density than closed circular DNA. RNA is denser than DNA and pellets in a CsCl gradient. Thus, after centrifugation to equilibrium in the CsCl–ethidium bromide gradient, the lower DNA band contains purified, closed circular plasmid DNA. The plasmid DNA band is removed from the CsCl gradient, the ethidium bromide is extracted with butanol, the DNA is dialyzed to remove the CsCl, and the purified DNA is then concentrated by ethanol precipitation. Although CsCl purification works well, it is expensive and quite time-consuming. Several alternative methods have been developed recently for purification of plasmid DNA from large-scale plasmid preparations. Two general approaches that work well are (1) differential polyethylene glycol precipitation (Sambrook et al. 1989) and (2) selective binding to DNA affinity agents (available as a wide variety of commercial kits[1]).

PHENOL EXTRACTION

Phenol extraction is often used to remove proteins from aqueous DNA samples. Extraction with phenol followed by phenol:chloroform (1:1) efficiently removes most contaminating proteins. Phenol and chloroform denature proteins. The denatured proteins partition into the organic phase or remain at the interphase, but the DNA remains in the aqueous phase. Many procedures recommend adding isoamyl alcohol to the chloroform to improve the separation of the aqueous and organic phases, but the extractions work equally well without isoamyl alcohol.

Oxidation of phenol forms quinones, and the quinones can be further oxidized to yellow- and pink-colored diacids and phenoxide radicals. These oxidation products interact with primary amines and can cause breakdown or cross-linking of DNA. Therefore, when preparing pure phenol solutions, always check the color and do not use the solution if it has a yellow or pink color. The oxidation products can be removed by distilling the phenol, but this is a potentially hazardous and time-consuming job. It is safer, easier, and relatively inexpensive to purchase ultrapure phenol. Pure phenol melts at 43°C.

When phenol is saturated with water, the aqueous phase often has a low pH due to small amounts of contaminating diacids. Therefore, phenol is usually extracted with buffer before use. The antioxidant 8-hydroxyquinoline can be added to the buffer-saturated phenol to decrease the breakdown of phenol. Hydroxyquinoline has a yellow color, which also helps identify the phenol phase. Buffer-saturated phenol is only good for about 1 month if stored at 4°C, but small aliquots stored at –20°C seem to be stable for many months. If the phenol is not saturated with an aqueous solution, the aqueous phase of the DNA sample may completely mix with the organic phase. If this happens, addition of a small amount of TE or chloroform usually resolves the two phases. In addition, if the salt concentration of the sample is too high, the aqueous and organic phases may be inverted, but the phenol phase can be easily identified by the yellow color of the hydroxyquinoline.

---

[1]Several kits that work well and are easy to use include Gene-Clean® (Bio-101), US-Bioclean® Kit (United States Biochemical Corporation), Quiagen® Plasmid Kits (Quiagen, Inc.), and Wizard® DNA Purification Systems (Promega Corporation).

A procedure for phenol extraction of small aqueous nucleic acid samples is included in Appendix F (Phenol Extraction). Since phenol is highly corrosive and can cause severe burns, it should be used cautiously. Always wear gloves, protective clothing, and safety glasses when handling phenol. All manipulations with phenol should be carried out in a chemical fume hood. Any areas of skin that come in contact with phenol should be rinsed with a large volume of water and washed with soap and water—do not rinse with ethanol!

ETHANOL PRECIPITATION

DNA and RNA precipitate in aqueous solutions with monovalent cations and 70% ethanol. Ammonium acetate, potassium acetate, sodium acetate, and NaCl are commonly used counterions for ethanol precipitation. The salts are usually added to aqueous solutions at the concentrations shown in the following table.

Salt	Stock solution	Dilution
Ammonium acetate	7.5 M	1:2
Potassium acetate	5.0 M	1:2
Sodium acetate	3.0 M	1:10
Sodium chloride	5.0 M	1:50

In addition to nucleic acids, salts and small organic molecules may also be trapped in the precipitate. These contaminants may interfere with later reactions. Because DNA remains precipitated in ethanol but many salts are soluble in ethanol, some contaminating salts can be removed by "washing" the DNA pellet with 70% ethanol. Ammonium acetate is very soluble in ethanol and is effectively removed by a 70% ethanol wash, but other salts are less soluble in ethanol and are removed less effectively. Ammonium acetate also causes less coprecipitation of small organic contaminants (such as nucleotides and short oligonucleotides) than other salts. For these reasons, DNA precipitated in ammonium acetate seems to have fewer contaminants that inhibit restriction enzymes. For many purposes, ammonium acetate is thus the preferred salt for ethanol precipitation of DNA. However, because some DNA-modifying enzymes (e.g., T4 polynucleotide kinase) are strongly inhibited by ammonium, ammonium acetate precipitation should not be used if the DNA is to be treated with such enzymes (see Sambrook et al. 1989).

Most protocols for ethanol precipitation call for cooling the solution to –20ºC or –70ºC. However, even low concentrations of DNA (<5 ng/ml) are efficiently precipitated at 0–4°C (see Perbal 1988). After cooling, the precipitate is sedimented by centrifugation. Centrifugation at 4°C does not improve the recovery of DNA compared to centrifugation at room temperature.

Samples with DNA concentrations as low as 0.1 µg/ml can be effectively precipitated in cold 70% ethanol with an appropriate monovalent cation. For samples with lower DNA concentrations, adding a small amount of glycogen helps to precipitate the DNA quantitatively. Glycogen acts as a carrier that coprecipitates with nucleic acids but does not interfere with most subsequent manipulations. A general procedure for ethanol precipitation of DNA from small volumes of aqueous solution is included in Appendix F (Ethanol Precipitation).

Nucleic acids can also be precipitated by adding an equal volume of isopropanol (instead of ethanol) to aqueous solutions. However, more salt and other contaminants coprecipitate in isopropanol than in ethanol, especially if the solution is chilled. In addition, since isopropanol is less volatile than ethanol, it is much more difficult to remove residual isopropanol after precipitation. Any residual isopropanol may inhibit subsequent enzyme reactions. Therefore, isopropanol is typically used only if it is necessary to keep the volume of the solution to a minimum, and isopropanol precipitations are usually done at room temperature. After isopropanol precipitation, the DNA pellet can be rinsed with cold 70% ethanol to remove the residual isopropanol and precipitated salt.

REMOVAL OF CONTAMINATING SMALL MOLECULES FROM DNA

DNA preparations sometimes contain contaminants that inhibit DNA modification enzymes or salts that interfere with electroporation. Two simple approaches that remove many of these contaminants are drop dialysis and micro-gel filtration columns (or spin columns). Drop dialysis is a quick, simple procedure to remove inhibitors such as SDS or high salt from a small volume of a DNA preparation (Silhavy et al. 1984). A procedure for drop dialysis is included in Appendix F (Drop Dialysis). Spin columns are more effective for removing larger contaminants such as nucleotides or short RNA fragments. A procedure for the preparation and use of spin columns is included in Appendix F (Spin Columns).

ESTIMATING DNA CONCENTRATIONS

The concentration of a pure DNA solution can be determined by measuring the absorbance of UV light. A 50 µg/ml solution of double-stranded DNA has an $OD_{260} = 1$. (The value varies slightly with the ratio of G:C to A:T, but the error is usually small enough to ignore.) The ratio of absorbance to DNA concentration is linear to approximately OD = 2; it is thus possible to determine the concentration of a pure DNA solution from the OD_{260}. The purity of a DNA solution can also be estimated by measuring the absorbance of UV light. Due to the characteristic absorption spectra of DNA, the ratio of absorbance at 260 nm/280 nm for a pure solution of double-stranded DNA should be between 1.7 and 1.9. Higher ratios are often due to RNA contamination and lower ratios may be due to protein or phenol contamination.

If the DNA solution is not pure, the OD_{260} will not reflect the actual DNA concentration. Thus, spectrophotometric methods will not give a reliable estimate of DNA concentration in many DNA preparations, such as plasmid minipreps. Contaminating chromosomal DNA, protein, RNA and organic chemicals may all contribute to absorbance at 260 nm, so spectrophotometric methods frequently yield false high estimates of DNA concentration. Treatment of DNA preparations with RNase degrades contaminating RNA to ribonucleosides. Although these ribonucleosides are not observed on an ethidium-bromide-stained gel, they contribute to the absorbance at 260 nm. In such preparations, the DNA concentration can be estimated from the relative intensity of ethidium bromine staining as described in Appendix F (Purification and Quantitation of PCR Products).

DIDEOXY DNA SEQUENCING

Dideoxy sequence analysis is based on the random incorporation of analogs of deoxynucleoside triphosphates (dNTPs) into a growing DNA chain by DNA polymerase (Sanger et al. 1977). Dideoxynucleoside triphosphate (ddNTP) analogs lack the 3′-OH group on the deoxyribose moiety of the dNTP. Because the 3′-OH group is necessary for the formation of the next phosphodiester bond, incorporation of a ddNTP into a growing DNA chain causes termination of chain elongation. To determine the sequence of a DNA template, four separate reactions are run. Each reaction contains all four dNTPs but only one of the four ddNTPs. For every nucleotide on the template, DNA polymerase inserts the complementary nucleotide during synthesis of the new DNA strand. If a dNTP is inserted, chain elongation continues, but if a ddNTP is inserted, synthesis stops at that position. For example, consider the ddGTP reaction. When DNA polymerase encounters a cytosine on the template, it needs to incorporate an guanine nucleotide into the growing chain. Depending on the ratio of dGTP to ddGTP in the reaction, DNA polymerase has the choice between the substrates dGTP and ddGTP. If it incorporates the ddGTP, then the chain terminates. If it incorporates the dGTP, the reaction continues until another guanine nucleotide is needed, and then the enzyme again has a choice between the dGTP or ddGTP. Thus, at each position, there is a certain probability that the dGTP will be incorporated and DNA synthesis will continue or ddGTP will be incorporated and DNA synthesis will stop. This results in a nested set of DNA fragments of different lengths, each terminated at a different guanine residue. By determining the length of fragments produced with each of the four ddNTPs, it is possible to deduce the nucleotide sequence of the template DNA.

Initiation of DNA synthesis requires an oligonucleotide primer with a free 3'-OH group. The primer is typically a short oligonucleotide (usually an 18–24-mer) that hybridizes to the DNA just upstream of the region to be sequenced. DNA polymerase begins synthesis from the 3' end of the primer and adds dNTPs in the 5' to 3' direction through the adjacent DNA. Thus, every DNA fragment has the same 5' end but different 3' ends (terminated wherever a ddNTP was inserted).

The template for dideoxy sequencing can be either single-stranded DNA (e.g., from M13 or phagemid vectors) or double-stranded DNA (e.g., from phage or plasmid vectors). The primer can be directly annealed to single-stranded DNA templates, but for double-stranded DNA templates, the DNA must first be denatured to allow the primer to anneal. One approach is to denature the DNA with NaOH and then prevent the DNA from reannealing until the primer is added (see Sambrook et al. 1989) as described in Experiment 10. An alternative approach is to convert linear double-stranded DNA into single-stranded DNA using a double-strand-specific exonuclease (such as T7 gene 6 exonuclease before annealing the primer (Straus and Zagursky 1991).

The DNA polymerases used for dideoxy sequencing lack the 5' to 3' exonuclease activity that would degrade the common 5' end of the DNA fragments, making interpretation of the DNA sequence impossible. Many different DNA polymerases may be used for DNA sequencing. Presently, the most popular DNA polymerase for dideoxy sequencing is Sequenase.[2] Sequenase 2.0 is a modified form of phage T7 DNA polymerase (Tabor and Richardson 1989a,b) that lacks 3' to 5' exonuclease activity, is highly processive, and efficiently incorporates nucleotide analogs. Another popular approach that has several advantages is cycle-sequencing using a thermostable DNA polymerase (such as *Taq* polymerase) as described in Experiment 10.

After the reactions are stopped, the DNA fragments are denatured from the template and resolved according to size by polyacrylamide gel electrophoresis. The gels contain urea and are run at high voltages to keep

---

[2]Sequenase® is available from United States Biochemical Corporation.

the DNA denatured. The bands in the gel corresponding to the DNA fragments must then be visualized to determine the sequence. The most common approach is to radioactively label the DNA during the sequencing reactions and then expose the gel to film to visualize the bands by autoradiography. The DNA can be radioactively labeled by using a primer that has been end-labeled with ^{32}P or by incorporating [α-^{32}P]dATP or [α-^{35}S]dATP into the DNA during synthesis. Several approaches are available for DNA sequencing without using radioactivity. Incorporation of fluorescent derivatives of the four nucleotides can be used for automatic DNA sequence analysis. Alternatively, two nonradioactive approaches that are useful for manual DNA sequencing are (1) to use standard nucleotides and visualize DNA bands by silver staining[3] and (2) to incorporate chemiluminescent nucleotide derivatives into the DNA and visualize the bands by the resulting chemiluminescence. Commercial kits are available for both of these methods.

The resulting bands form a ladder corresponding to the size of the DNA fragments. The first band at the bottom of the gel represents the shortest fragment synthesized from the sequencing primer that terminated with the particular ddNTP used, the next band is one nucleotide longer, etc. (typically, the smallest readable band is 5–10 bp from the end of the primer). The sequence is determined by reading up the four lanes on the gel (corresponding to the four ddNTP reactions) in order of the occurrence of the bands on the ladder. It is important to note the band intensity and spacing as well as the presence or absence of bands. Compressions and sequencing artifacts often result in bands of unequal intensity or spacing.

DNA sequence analysis is described in more detail in Sambrook et al. (1989) and Ausubel et al. (1994). Some potential problems and suggested solutions are shown in Table 7.2 (Blakesley 1983; Ornstein and Kashdan 1985).

---

[3]A kit for silver-staining DNA sequencing gels is available from Promega Corporation.

Table 7.2 Troubleshooting Dideoxy DNA Sequencing Reactions

Problem	Possible causes	Possible solutions
Bands in all four lanes		
at high molecular weights (>50 bp)	secondary structure in the template DNA terminates elongation	run the reactions at a higher temperature; use thermostable DNA polymerase
at low molecular weights (25–45 bp)	specific activity of dATP used is too high; low [dATP] limits elongation	use lower specific activity (400–800 Ci/mmole) dATP
at all molecular weights	template DNA contains impurities[a]	repurify template DNA
	poor quality DNA polymerase	use new DNA polymerase
	DNA polymerase lost activity during storage	store at –20°C; do not dilute until immediately before use
	primer degraded	use fresh primer
	second-site priming	use a more specific primer
Few large fragments	ddNTP:dNTP ratio too high	reduce the [ddNTP][b]
Few small fragments	ddNTP:dNTP ratio too low	increase the [ddNTP][b]
Autoradiogram too light	poor annealing of primer to template	increase denaturing temperature and reanneal more slowly; use a longer primer
	film developer exhausted	prepare new developer
	exposure time too short	increase the exposure time or use intensifying screens
High background in all lanes	template DNA contains impurities[a]	repurify template DNA
	^{32}P-labeled DNA too old	prepare fresh ^{32}P-sequencing reactions or use [^{35}S]dNTP
Blurred bands in some parts of the autoradiogram	poor contact between gel and film during autoradiography	make sure film is pressed flat against the gel during autoradiography
Bands curve up near the outside edges (smiling)	nonuniform heating of gel during electrophoresis	heat gel before loading by pre-running for at least 30 minutes
	nonuniform thickness of the gel	use uniform spacers; place clamps over all spacers during polymerization
Gel sticks to film	contact of wet gel with film	dry gel completely before autoradiography
Nonspecific spots on the film	radioactive contamination on gel dryer, intensifying screens, gloves, or film cassettes	check for contamination with a Geiger counter
	stray radiation near film during autoradiography	protect film from other radiation sources during autoradiography

[a]Contamination of template DNA is probably the most common cause of sequencing problems.
[b]Avoid this problem by using commercially available sequencing kits with pretested reagents.

Electrophoresis of DNA

DNA fragments have a constant charge/length ratio due to the net negative charge of the phosphate backbone. Therefore, the rate of migration during electrophoresis mainly depends on the following three parameters.

Pore Size of the Gel and Length of the DNA. During electrophoresis, DNA molecules must pass through the pores in the gel. If the pores are closely spaced, then the DNA molecules must squeeze through the pores with one end first, followed by the rest of the molecule (like a snake might squirm through a pile of sticks). Sometimes as the end of the DNA molecule passes through one pore, it will be in the correct orientation to pass through the next pore, but sometimes the end of the DNA molecule will become temporarily trapped in a pore in the wrong orientation. The DNA molecule must then reorient before an end of the molecule can pass through the pore. Smaller DNA fragments can reorient much faster and hence migrate faster than larger DNA fragments. As the pore size decreases, it is more difficult for longer DNA molecules to orient properly to "snake" through the pores. By using gels with different pore sizes, a wide range of DNA fragment sizes can be separated. Within an appropriate gel matrix, the rate of migration of linear double-stranded DNA is inversely proportional to the $\log_{10}$ of the number of base pairs.

Conformation of the DNA. The shape of the DNA molecule also affects its ability to snake through the pores in the gel. In general, features that make the DNA less flexible or less compact slow the migration in a gel. The rate of migration of different forms of plasmid DNA is usually supercoiled > linear > nicked circles. In addition, secondary structure and bends in linear DNA may affect the rate of migration.

Applied Voltage. The strength of an electric field depends on the voltage and the length of the gel. At low voltages, the rate of migration of linear DNA is directly proportional to the applied voltage. However, as the voltage gradient increases, the rate of migration of large DNA fragments increases relative to smaller fragments, decreasing the effective separation of different size DNA fragments.

CURRENT, VOLTAGE, AND POWER

Two equations describe the relationship between current (I), voltage (V), resistance (R), and power (P) during electrophoresis:

$$I = V / R$$

$$P = I^2 \times R = I \times V$$

Current is measured in milliamps, voltage in volts, and power in watts. During electrophoresis, either the current, voltage, or power is held constant. Usually, the resistance of agarose and acrylamide gels increases during electrophoresis. When run at constant current, the voltage increases to compensate for the increase in resistance. This results in an increase in power produced by the system, causing the gel to heat up. If the gel heats up too much, the bands become distorted, and thus the initial current must be low enough so that the gel does not heat up excessively. In contrast, when run at constant voltage, the current decreases with time to compensate for the increased resistance. Thus, the gel will not heat up, but the rate of migration of the DNA decreases as the current decreases. DNA sequencing gels are often run at constant power to maintain a constant high temperature, which helps to prevent formation of secondary structure in the single-stranded DNA.

AGAROSE

Agarose is an uncharged polysaccharide purified from agar. Almost all agarose contains some anionic impurities, mainly sulfate and pyruvate, that can affect the electrophoresis. Agarose with the least amount of these anionic impurities ("low EEO" agarose) is best for DNA gels. Agarose melts when heated to 100°C and resolidifies when cooled below approximately 50°C. When solidified, agarose forms a gel matrix. The size of the pores in the gel matrix can be varied by using different concentrations of agarose: the higher the concentration of agarose, the smaller the pore size. Below are some useful agarose concentrations for separating DNA fragments.

Agarose concentration (%)	Range of linear dsDNA fragments resolved (bp)
0.5	2000–20000
0.8	600–1500
1.5	200–4000
3.0	80–1500

Standard agarose does not dissolve well at concentrations greater than approximately 1.5%. However, blends containing low-melting-temperature agarose can be used to prepare higher concentration agarose gels. A blend of three parts low-melting-temperature agarose with one part standard agarose (called agarose 3:1) is easy to prepare and gives good separation of small DNA fragments. Agarose 3:1 is available from FMC BioProducts and Sigma Chemical Co.

Larger DNA molecules can be separated by pulsed-field gel electrophoresis as discussed in Section 3 (Genetic Mapping).

ACRYLAMIDE

Polyacrylamide gels are produced by polymerization of acrylamide into linear chains with bis-acrylamide (N,N'-methylene-bis-acrylamide) cross-links between the acrylamide chains. Polymerization is initiated by adding ammonium persulfate, and the reaction is accelerated by TEMED (N,N,N',N'-tetramethylethylenediamine), which catalyzes the formation of free radicals from ammonium persulfate. Oxygen radicals can interact with acrylamide and terminate chain elongation, and thus oxygen inhibits the polymerization of acrylamide. When solutions are prepared from solid acrylamide, oxygenation of the solution may occur when the solution is mixed to dissolve the acrylamide. The resulting acrylamide solution may need to be "degassed" under a vacuum to remove oxygen before pouring the gel. However, degassing is not usually necessary for acrylamide solutions stored at 4ºC. To exclude oxygen, acrylamide gels are poured between two glass plates.

The concentration of acrylamide and the ratio of acrylamide to bis-acrylamide determine the pore size of the gel. Below are some useful acrylamide concentrations for separating double-stranded DNA fragments.

Acrylamide concentration (%)	Range of linear dsDNA fragments resolved (bp)
5	80–500
12	40–200
20	10–100

VISUALIZING DNA WITH ETHIDIUM BROMIDE

Ethidium bromide intercalates between the stacked bases of DNA and RNA. When excited by UV light between 254 nm (short wave) and 366 nm (long wave), it emits fluorescent light at 590 nm. The DNA–ethidium bromide complex produces about 50 times more fluorescence than free ethidium bromide. Ethidium bromide can be used to detect both double-stranded and single-stranded nucleic acids, but the sensitivity is much greater for double-stranded nucleic acids. Ethidium-bromide-stained gels photographed on a UV transilluminator require a UV filter to screen out the background UV from the transilluminator.

Ethidium bromide is a mutagen and must be used with caution. Always wear gloves when working with solutions that contain ethidium bromide. After use, ethidium bromide solutions should be decontaminated and disposed of in accordance with the safety practices established by your institution's Safety Office.

Polymerase Chain Reaction

Polymerase chain reaction (PCR)[1] is a powerful tool with a variety of applications. PCR-based methods can be used to clone specific genes, generate mutations, map mutations, sequence genes, and study gene expression.

SYMMETRIC PCR

PCR amplification of DNA occurs by cycles of three temperature-dependent steps: DNA denaturation, primer-template annealing, and DNA synthesis by a thermostable DNA polymerase (Fig. 7.2). First, the double-stranded template DNA is denatured to single-stranded DNA. Second, primers complementary to the 3′ ends of opposite strands of the DNA are hybridized with the single-stranded template DNA. Third, DNA polymerase extends each primer, duplicating the segment of DNA between the primers. DNA synthesized during the first cycle has the 5′ end of the primers and a variable 3′ end. When these strands are denatured, the parental strand will rehybridize to the primer, and the product with a variable 3′ end will continue to be synthesized during subsequent cycles of PCR. However, these products only accumulate arithmetically. The second cycle of denaturation, annealing, and primer extension produces discrete products with the 5′ end of one primer and the 3′ end of the other primer. Each strand of this discrete product is complementary to one of the two primers and thus acts as a template in subsequent cycles. Therefore, these discrete products accumulate exponentially with each round of DNA amplification, resulting in 2^n-fold amplification of the DNA in n cycles of PCR.

The optimal conditions for PCR differ for different templates and primers. The purity and yield of the PCR products depend on each of the

[1] The PCR process for amplifying nucleic acids is covered by patents owned by Hoffmann-LaRoche.

237

parameters of the reaction: the temperature of each step, the DNA polymerase, the primers, the DNA template, the Mg^{++} concentration, the buffer used, and the number of cycles. Since there is no simple way of knowing what conditions will work best for each PCR, the optimal conditions must be determined empirically. However, some general guidelines for optimizing each of these parameters are described below (Gelfand 1989; Innis and Gelfand 1990).

TEMPERATURE

The optimal temperature for each of the steps of PCR may vary for different templates and primers. The time at each step must be long enough to be effective, but if the time is too long, the purity and yield of the PCR product may decrease. Most of the time spent at each step is simply to ensure that the automatic thermal cycler reaches the correct temperature. (the time required may vary considerably for different thermal cyclers).

Step 1: Denaturation. Since partially denatured DNA will quickly reanneal when the temperature decreases, the DNA strands must be completely separated during the denaturation step. Denaturation of the DNA is very fast (within a few seconds) once an adequate temperature is reached. If the temperature is too high, the DNA polymerase may lose activity during each denaturation step of each PCR cycle. Denaturation at 94°C seems to be a good compromise.

Step 2: Annealing. For subsequent primer extension, the primers must stably anneal to the template. The optimal annealing temperature is a function of the melting temperature of the primer-template hybrid (Rychlik et al. 1990). If the temperature is too high, the primers do not anneal efficiently, decreasing the yield of desired product. However, if the annealing temperature is too low, the primer may anneal nonspecifically. Typical annealing is done at 5°C below the melting temperature (T_m) of the primers. As a rule of thumb, the T_m of the primers can be estimated by adding 2°C for each A or T and 4°C for each G or C. The primers are in vast excess to the template, so the annealing reaction occurs very quickly once the proper temperature has been reached.

Step 3: Primer extension. The extension temperature must be low enough to prevent the primer-template hybrid from denaturing, but high enough

Figure 7.2 Symmetric PCR amplification of DNA. Multiple cycles of denaturation of the template DNA, primer annealing, and primer extension result in approximately 10^6-fold amplification of the region of DNA located between the two primers.

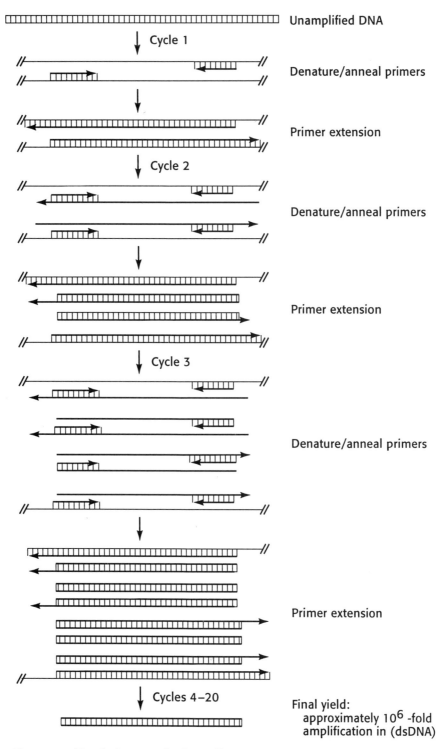

Unamplified DNA

Cycle 1

Denature/anneal primers

Primer extension

Cycle 2

Denature/anneal primers

Primer extension

Cycle 3

Denature/anneal primers

Primer extension

Cycles 4–20

Final yield:
approximately 10^6-fold
amplification in (dsDNA)

Figure 7.2 (*See facing page for legend.*)

for the thermophilic DNA polymerase to function efficiently. *Taq* polymerase is partially active even at typical annealing temperatures, so primer extension begins during the annealing step. Hence, the extended primers remain associated with the template at temperatures that are closer to the optimal temperature for *Taq* polymerase (75–80°C). Primer extension is usually done at 72°C. At this temperature, the rate of primer extension by *Taq* polymerase is approximately 50–100 nucleotides per second. The time required for primer extension depends mainly on the length of the sequence to be amplified.

DNA POLYMERASE

To withstand the cycles of high temperature, a thermostable DNA polymerase must be used (Gelfand and White 1990). *Taq* DNA polymerase (originally isolated from the thermophile, *Thermus aquaticus*) is most commonly used for PCR. *Taq* polymerase is a highly processive enzyme with an optimal temperature of 75–80°C, and it will extend primers up to 10 kb, but the efficiency of primer extension longer than 3 kb is typically low. However, a recent modification that combines two thermostable DNA polymerases allows amplification of DNA sequences up to 35 kb (Barnes 1994).

Taq polymerase lacks the $3' \rightarrow 5'$ exonuclease required for proofreading, and thus its error rate is relatively high. Early estimates suggested that the frequency of misincorporation by *Taq* polymerase was approximately 10^{-4} nucleotides per cycle. However, when the reaction conditions were optimized, the error rate decreased to less than 5×10^{-6} nucleotides per cycle. Such rare misincorporations do not cause a problem for many applications that use a population of PCR fragments, but cloning mutant PCR products is a potential problem.

Although *Taq* polymerase is thermally stable, it loses activity after many cycles of heating during PCR. Simply increasing the amount of polymerase used is not a suitable solution to this problem. Typically, 1–5 units of *Taq* polymerase are used in a 100-µl reaction. If the concentration is increased, nonspecific products are increased, and if the concentration is decreased, the amount of the desired product is decreased. Recently, several other thermostable DNA polymerases have been marketed that have the $3' \rightarrow 5'$ proofreading activity and are more thermally stable than *Taq* polymerase, thus eliminating some of the problems of *Taq* polymerase.

PRIMERS

Without knowing the complete DNA sequence of the template in addition to the primers, there is no simple way to predict which primers will work and which will not work for PCR. The following guidelines may help limit failure, but despite carefully adhering to these guidelines, some primers simply will not work. So when all else fails, try a different primer. (1) Good primers should hybridize efficiently to the desired DNA sequence, but poorly or not at all to nonspecific DNA sequences. Usually, oligonucleotides between 20 and 30 nucleotides long have adequate specificity. (2) Secondary structures in the oligonucleotides can prevent proper annealing with the template, resulting in poor priming. (3) When the 3′ end of one primer anneals to the 3′ end of another primer, the polymerase extends each primer to the end of the other, forming "primer-dimers." Since the primer-dimers are short, they are quickly amplified at the expense of the desired PCR product. (4) For many applications, the primers are designed to be exactly complementary to the template, but for some applications, the primers are designed with mismatches (e.g., when using mutagenic primers or primers designed from an amino acid sequence). However, mismatches at the 3′ end of a primer prevent extension, and the closer a mismatch is to the 3′ end of a primer, the more likely it will prevent extension. Therefore, any mismatches should be located in the middle or near the 5′ end of the primer.

Primers are usually added at 0.1–0.5 μM. If the concentration of primers is too high, mispriming may occur, increasing the amount of nonspecific products. If the concentration of primers is too low, the yield of the product may be low.

TEMPLATE DNA

PCR can be done even on very crude samples (e.g., directly from colonies of bacteria [Joshi et al. 1991]). However, certain contaminants in the template DNA (e.g., urea, SDS, sodium acetate, and agarose) can decrease the efficiency of PCR. For this reason, using excess template DNA sometimes decreases the efficiency of PCR by increasing the concentration of contaminants in the reaction. In contrast, when too little template DNA is used, nonspecific products are more common. If nonspecific products are a problem, the desired fragment can often be purified from a gel and reamplified.

dNTPs

Typically, 200 μM of each dNTP is added to the reaction. At this concentration, there is enough dNTP to synthesize 12.5 μg of DNA when half of the dNTPs are incorporated. Increasing the dNTP concentration increases the error rate of *Taq* polymerase. Decreasing the dNTP concentration below the K_m of *Taq* polymerase (10–20 μM) or adding unequal amounts of each dNTP increases the frequency of misincorporation.

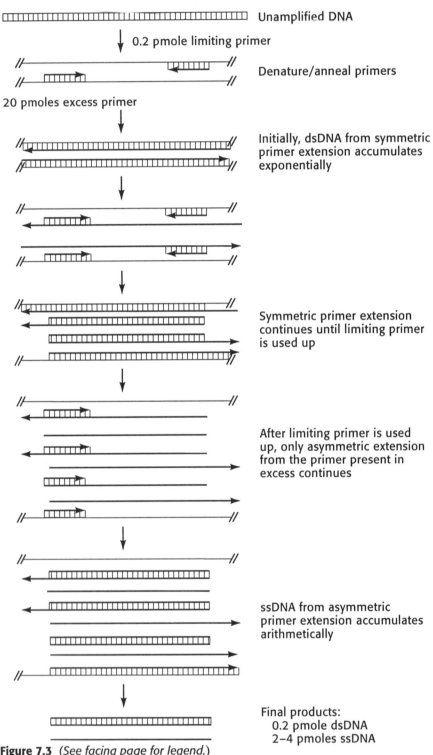

Figure 7.3 (*See facing page for legend.*)

Magnesium

The concentration of Mg^{++} in the reaction is critical. It may affect denaturation, primer annealing, and the activity and fidelity of *Taq* polymerase. Since *Taq* polymerase requires free Mg^{++}, the concentration must be greater than the amount chelated by DNA and dNTPs. However, excess Mg^{++} may promote nonspecific annealing of the primers, decreasing the yield and purity of the desired product.

Number of Cycles

Too few cycles of PCR amplification result in a low yield of product, but too many cycles are not good either. Synthesis of the specific PCR product usually continues until approximately 0.3–1 pmole is accumulated, and then it reaches a plateau. However, synthesis of nonspecific products may continue beyond this time. Therefore, too many cycles may increase the amount and complexity of nonspecific products obtained. Typically, 30 cycles is a good compromise for amplifying a sample with 10^4 to 10^5 DNA molecules.

Asymmetric PCR

When one primer is limiting and the other is added in excess, only one strand of the DNA is amplified (Fig. 7.3) (McCabe 1990). The resulting single-stranded DNA is a useful template for DNA sequencing (Gyllensten 1989). In contrast to symmetric PCR, the asymmetric PCR product accumulates arithmetically with each cycle, not exponentially (Fig. 7.4). Thus, to provide sufficient template DNA, the DNA is often symmetrically amplified before asymmetric amplification. Because the primers are present in excess in the standard PCRs (~50 pmoles of each primer), the limiting primer must be used up in the preceding reaction or must be removed before asymmetric PCR. Although the concentration of primers must be determined empirically for each reaction, the limiting primer is typically added at a concentration of about 0.2 pmole and the excess primer is added to about 20 pmoles.

PCR Mutagenesis

PCR can also be used to isolate random or directed mutations. Numerous approaches have been developed for construction of a variety of types of mutations in vitro using PCR. A few examples are shown in Figure 7.5.

Figure 7.3 Asymmetric PCR amplification of DNA. When one primer is limiting and the other primer is added in excess, only one strand of the DNA is amplified during PCR.

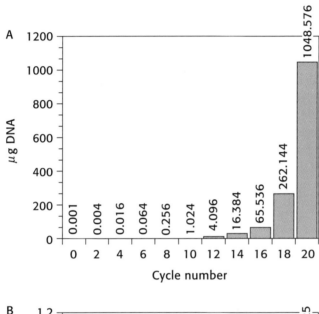

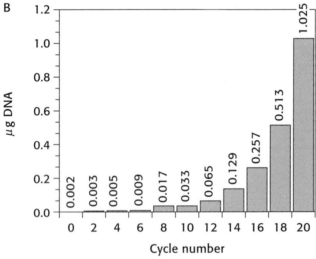

Figure 7.4 Accumulation of amplified DNA during symmetric (*A*) and asymmetric (*B*) PCR. Note the difference in the concentration of DNA shown in the two figures.

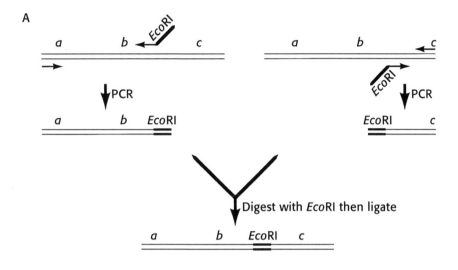

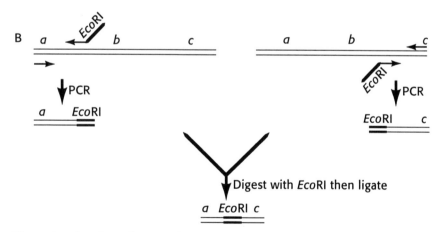

Figure 7.5 A variety of approaches can be used for construction of mutations in vitro using PCR. (*A*) Addition of a specified restriction enzyme cleavage site within a DNA fragment. (*B*) Construction of a deletion mutation with a specified restriction enzyme cleavage site at the join point of the deletion. (*Figure/legend continued on following page.*)

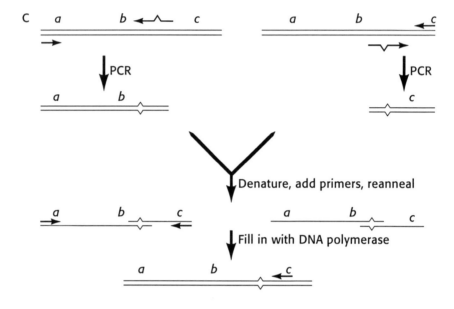

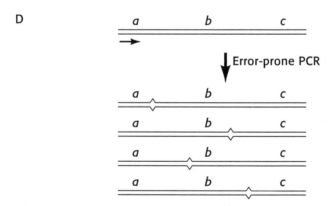

Figure 7.5 (*Continued from preceding page.*) (*C*) Construction of a site-directed nucleotide substitution within a DNA fragment. (*D*) Random mutagenesis of a DNA fragment by error-prone PCR. Two methods for increasing the error frequency during PCR are to alter the dNTP concentration or to alter the divalent cation composition.

Experiments

"[Lou] Baron would often say that if an experiment involved more than four plates and two pipettes, it was probably overdesigned. Baron was a master experimentalist. He did not agonize about all the reasons an experiment would not work; he did not wait until he had everything ready for the perfect well-controlled experiment. Baron did simple genetic experiments that asked a single, relatively straightforward question. He depended on genetic selection and the resolving power of recombination to provide the answer."

Stanley Falkow
Reprinted, with permission, from A Look Through the Retrospectoscope.
In *Molecular Genetics of Bacterial Pathogenesis: A Tribute to Stanley Falkow*
(ed. V.L. Miller et al.), pp. xxiii–xxxix. ASM Press, Washington, DC. (1994)

Introduction to Experiments 1–6

Experiments 1–6 illustrate some fundamental yet powerful genetic techniques as applied to the study of *Salmonella typhimurium*. For illustrative purposes, we have chosen to examine the genetics of sorbitol catabolism, because this provides several examples of screens and selections and also because the *srl* genes and their regulation are not well-characterized in *S. typhimurium*. Thus, these experiments illustrate approaches to examine a previously unstudied genetic system.

The first three experiments employ three different types of transposon delivery systems and demonstrate three different uses of transposons. Clearly, alternate methods and transposon elements can be employed in different contexts. Experiment 4 presents two methods for finding two classes of conditional mutations. Experiment 5 illustrates a different use of transposons, in molecular cloning, and Experiment 6 provides methods for physical map construction.

Together, these experiments depict the range of approaches available for in-depth genetic analysis of *S. typhimurium*. The considerations and overall strategies should prove inspirational for similar analysis of other genetic systems in other organisms.

EXPERIMENT 1

Transposons: Mini-Mu-*lac* Insertions in Structural Genes

Experiment 1 demonstrates one method for isolating transposon insertions within genes of interest: The transposition method is transitory cis-acting complementation, the transposon is MudJ, and the genes of interest are those required for catabolism of sorbitol (*srlABD*). This general procedure for isolating MudJ insertions is easily adapted to a variety of other genetic systems. The approaches used here were adapted from Benson and Goldman (1992), Davis et al. (1980), Hoppe et al. (1979), and Hughes and Roth (1985, 1988).

The donor MudJ insertion is within the *hisD* gene, so the recipient strain carries a large nontransducible deletion of the *his* operon, to prevent inheritance of the donor *hisD*::MudJ insertion by homologous recombination.

The insertions are backcrossed to the wild type to confirm that the MudJ insertion is in the gene of interest. The MudJ insertions are mapped by Mud-P22 transduction and are also used to construct directed deletions with defined endpoints.

The MudJ element forms *lacZ* operon fusions, so that regulation of the gene of interest can be monitored. Three types of regulatory phenotypes have been observed with *srl*::MudJ insertions. The first is Lac+ only in the presence of Srl (Reg+ phenotype) and represents the authentic pattern of *srl* operon regulation. The second type is Lac+ irrespective of Srl (Reg^c [constitutive] phenotype) and may represent insertion-deletion events whereby the *lacZ* gene within the MudJ element is under the control of an adjacent promoter,

251

outside of the *srl* operon. The third type is Lac⁻ irrespective of the presence of sorbitol and represents insertions in which the *lacZ* gene is oriented opposite from the *srl* operon. This diversity of Lac and Reg phenotypes underscores the importance of isolating and characterizing several independent fusions when approaching a gene whose regulation is unknown. Below are listed the bacterial strains and bacteriophage lysates used in this experiment.

BACTERIAL STRAINS

TSM1(prototroph)
TSM101 (Δ*his*)
TSM102 (*hisD*::MudJ *hisA*::Mud1)
TSM105 (*srl*::MudJ)
TSM123 (Δ*trp*)
TSM124 (*zgc*::MudA)
TSM125 (P22ʳ)
TSM140 (Δ[*srl-zgc*]::MudA)
srl::MudJ strains (for deletion mapping)
Δ(*srl-zgc*) strains (for deletion mapping)

BACTERIOPHAGE LYSATES

P22 broth prepared with a P22 HT *int* lysate of strain TSM1
P22 broth prepared with a P22 HT *int* lysate of strain TSM140
P22 HT *int* lysate of strain TSM102 (*hisD*::MudJ *hisA*::Mud1)
P22 HT *int* lysate of strain TSM124 (*zgc*::MudA)
P22 HT *int* lysates of *srl*::MudJ strains (for deletion mapping)
Mud-P22 lysates

Generate Random MudJ "Hops"

MATERIALS

For culture media and supplements, see Appendix B.

EBU plates
EBU-Km plates
LB broth
LB broth-EGTA (10 mM) plates
MacConkey-Km-EGTA-Srl plates
NA-TCC-Km-EGTA-Srl plates
NA-XGal plates
0.85% NaCl
Microcentrifuge tubes
Sterile toothpicks

PROCEDURE

1. Inoculate a single colony of strain TSM101 (Δhis) into 2.5 ml of LB broth. Aerate overnight at 37°C. Store culture on ice.

2. Pipette 0.1 ml of culture into each of nine microcentrifuge tubes.

3. Dilute a P22 HT *int* lysate of strain TSM102 (*hisD*::MudJ *hisA*:: Mud1) to approximately 10^{10} pfu/ml in cold 0.85% NaCl.

4. Add 0.1 ml of diluted lysate as shown below.

Tube	Cells (ml)	Phage (ml)	Comments
A	0.1	0.1	pool A
B	0.1	0.1	pool B
C	0.1	0.1	pool C
D	0.1	0.1	pool D
E	0.1	0.1	pool E
F	0.1	0.1	pool F
G	0.1	0.1	pool G
H	0.1	0.1	pool H
I	0.1	0	cells-only control
J	0	0.1	phage-only control

5. Gently mix the contents of each tube.

6. Incubate for 10–15 minutes at 37°C to allow the phage to adsorb to the cells.

7. Add 1.0 ml of LB broth-EGTA (10 mM) and gently mix (EGTA chelates Ca^{++}, thereby inhibiting further infection).

8. Incubate for 15–20 minutes at 37°C to allow time for expression of Km^r.

9. Plate 0.2 ml of tube A on five NA-TTC-Km-EGTA-Srl plates. Do the same for tubes B–H (40 plates total).

10. Plate 0.2 ml each of tubes I and J on one NA-TTC-Km-EGTA-Srl plate (2 plates total).

11. Incubate all plates overnight at 37°C.

12. Carefully inspect each of the plates. No colonies should be seen on the two control plates, whereas the experimental plates should have 100–500 colonies per plate, the vast majority of which should be white. Rare dark-red colonies are presumptive *srl* mutants.

13. Estimate the total number of white and red colonies in each pool. Record the results. Assign isolation numbers to each of the red colonies chosen for further analysis. Be sure to keep track of which colonies came from which pools.

14. Use sterile toothpicks to pick each of the chosen red colonies, along with one or two randomly chosen white colonies. Streak for single colonies on the following media:

    a. EBU-Km agar (to isolate phage-free clones).

    b. MacConkey-Km-EGTA-Srl (to ensure that the mutants are genuine).

    c. NA-TTC-Km-EGTA-Srl (to ensure that the mutants are genuine).

    d. NA-XGal spread with 0.1 ml of 10% Srl (to identify Lac$^+$ strains).

    e. NA-XGal (to identify Reg$^+$ strains).

15. Incubate all plates overnight at 37°C.

16. Inspect the plates and record the results.

    • The streaks on the EBU-Km plates should be mixtures of white and dark-blue colonies. The blue colonies are those that are releasing phage. Hopefully, several white colonies will be well-isolated.

    • The isolated colonies on the MacConkey plates should be white (Srl$^-$). If all colonies in a given streak are red, then that candidate colony was a fake. If the streak is a mixture of red and white colonies, then that candidate colony was a mixture of mutant and wild type. This can be salvaged by restreaking white colonies.

    • The NA-XGal plates reveal which mutants are Lac$^+$ Reg$^+$ and are the isolates upon which to concentrate.

17. Use sterile toothpicks to pick two to four well-isolated white colonies from the EBU-Km plate for each of the mutants chosen for further analysis. Streak for single colonies on the following media:

    a. EBU agar (to confirm phage-free clones).

    b. MacConkey-Km-EGTA-Srl (to confirm that the mutants are genuine).

    c. NA-XGal spread with 0.1 ml of 10% Srl (to confirm Lac$^+$ strains).

    d. NA-XGal (to confirm Reg$^+$ strains).

18. Streak each colony across a line of P22 H5 on an EBU plate to test for P22 sensitivity. Use strains TSM1 (P22^s) and TSM125 (P22^r) as controls.

19. Incubate all plates overnight at 37°C.

20. Inspect the plates and record the results. Choose one P22-sensitive Srl⁻ Lac⁺ Reg⁺ isolate from each pool for further analysis. Save desired strains for the short term on EBU plates at 4°C.

Backcross MudJ Insertions to the Wild Type

MATERIALS

For culture media and supplements, see Appendix B.

EBU plates
EBU-Km plates
LB broth
MacConkey-Km-EGTA-Srl plates
NA-XGal plates
Sterile toothpicks

PROCEDURE

1. Prepare P22 HT *int* lysates on each strain (see Appendix D, Generalized Transduction with Phage P22).

2. Use 10 µl of each lysate for delayed expression transductions with strain TSM1 (see Appendix D, Generalized Transduction with Phage P22). Plate on MacConkey-Km-EGTA-Srl plates.

3. Incubate all plates overnight at 37°C.

4. Inspect the plates and record the results. If the MudJ insertion is in *srl*, then 100% of the Km^r transductants will be Srl⁻ (white on MacConkey).

5. Estimate the titers of the P22 HT *int* lysates as both pfu/ml and as *trp*⁺ transductants/ml and calculate the approximate frequency of transduction for each cross.

6. Use sterile toothpicks to pick eight Srl⁻ Km^r transductants from each cross. Streak for single colonies on the following media:

   a. EBU-Km agar (to isolate phage-free clones).

   b. MacConkey-Km-EGTA-Srl (to ensure that the transductants are genuine).

   c. NA-XGal spread with 0.1 ml of 10% Srl (to identify Lac$^+$ transductants).

   d. NA-XGal (to identify Reg$^+$ transductants).

7. Incubate all plates overnight at 37°C.

8. Inspect the plates and record the results.

9. Use sterile toothpicks to pick two to four well-isolated white colonies from the EBU-Km plate for each of the mutants chosen for further analysis. Streak for single colonies on the following media:

   a. EBU agar (to confirm phage-free clones).

   b. MacConkey-Km-EGTA-Srl (to confirm that the transductants are genuine).

   c. NA-XGal spread with 0.1 ml of 10% Srl (to confirm Lac$^+$ strains).

   d. NA-XGal (to confirm Reg$^+$ strains).

10. Streak each colony across a line of P22 H5 on an EBU plate to test for P22 sensitivity. Use strains TSM1 (P22^s) and TSM125 (P22^r) as controls.

11. Incubate all plates overnight at 37°C.

12. Inspect the plates and record the results. Choose one P22-sensitive Srl$^-$ Lac$^+$ Reg$^+$ strain from each transduction for further work. Save desired strains for the short term on EBU plates at 4°C.

13. Assign collection and allele numbers to each of the desired strains and culture each in LB broth to saturation.

14. Stock each of the new srl::MudJ strains in the central strain collection (see Appendix C, Storing Strains).

Determine the Chromosomal Map Location of *srl*::MudJ Insertions via Mud-P22 Mapping

MATERIALS

For culture media and supplements, see Appendix B.

LB broth
NCE-Srl plates
Frog (48-pin replica plates; Sigma R 2383)

PROCEDURE

1. Inoculate a single colony of each strain for mapping into 2.5 ml of LB broth. Aerate overnight at 37°C. Include strain TSM105 (*srl*::MudJ) as a control.

2. Store cultures on ice until use. Spread 0.1 ml of each culture on an NCE-Srl plate.

3. Spot on Mud-P22 lysates with a frog (see Appendix D, Rapid Mapping in *S. typhimurium* Mud-P22 Prophages).

4. Allow the spots to dry and incubate the plates at 37°C. Unambiguous results may require 36–48 hours of incubation.

5. Inspect the plates and record the results.

Construct Directed Deletions with *srl*::MudJ Insertions

Each of the Lac+ *srl*::MudJ insertions can be converted to a deletion with defined endpoints. Set one endpoint by the position of the *srl*::MudJ insertion and the other by the position of *zgc-7205*::MudA, which is located downstream from the *srl* locus and in the same orientation. For a discussion of the formation of directed deletions, see Section 3 (Deletion Mapping). P22 HT *int* transduction delivers part of the *zgc-7205*::MudA insertion into an *srl*::MudJ strain. Selection for Ap^r results in a deletion whose endpoints are specified by the Mud insertions (draw this out for yourself). In our case, deletion derivatives (as opposed to simple MudA→MudJ replacements) are recognized as those that express β-galactosidase constitutively (Lac$^+$ Reg^c). This is because the *srlR* gene, encoding the *srl* repressor, lies between the *srlABD* operon and the *zgc-7205*::MudA insertion. Once constructed and characterized, the deletion strains are used to help construct a deletion map of the *srl* locus.

MATERIALS

For culture media and supplements, see Appendix B.

EBU plates
EBU-Ap (30 µg/ml) plates
LB broth
LB broth-EGTA (10 mM)
LB plates
LB-Km plates
MacConkey-Srl plates
NA-Ap (30 µg/ml)-XGal-EGTA plates
NA-XGal plates
Microcentrifuge tubes
Sterile toothpicks

PROCEDURE

1. Choose up to four Lac$^+$ *srl*::MudJ strains. Inoculate a single colony of each into 2.5 ml of LB broth. Aerate overnight at 37ºC.

2. For each transduction, mix 0.1 ml of *srl*::MudJ culture and 10 µl of strain TSM124 (*zgc-7205*::MudA) phage lysate in a microcentrifuge tube.

3. Incubate for 90 minutes at room temperature.

4. Add 0.9 ml of LB broth-EGTA (10 mM) and gently mix.

5. Sediment the cells in a microcentrifuge at 14,000 rpm for 2 minutes.

6. Resuspend the cell pellets each in 1 ml of LB broth-EGTA and sediment for 2 minutes.

7. Resuspend the cell pellets each in 1 ml of LB broth-EGTA and sediment for 2 minutes. This repeated washing removes β-lactamase carried over from the donor lysate and makes for cleaner selection on the Ap plates.

8. Resuspend the cell pellets each in 0.1 ml of LB broth-EGTA and plate each on an NA-Ap (30 µg/ml)-XGal-EGTA plate.

9. Plate 0.1 ml of each of the *srl*::MudJ strains on an NA-Ap (30 µg/ml)-XGal-EGTA plate (cells-only control). Spot 10 µl of the *zgc-7205*::MudA P22 lysate on an LB plate (phage-sterility control).

10. Incubate all plates overnight at 37ºC.

11. Inspect the plates and record the results. Approximately 50–200 Ap^r colonies should result from each of the transductions and approximately half of the colonies should be blue (Lac^+).

12. Use sterile toothpicks to pick eight blue transductants from each cross. Streak for single colonies on each of the following media:

    a. EBU-Ap (30 µg/ml) agar (to isolate phage-free clones).

    b. NA-Ap (30 µg/ml)-XGal-EGTA agar (to confirm Reg^c transductants).

    c. LB-Km agar (to confirm loss of MudJ).

    d. MacConkey-Srl (to confirm Srl^- phenotype).

13. Incubate all plates overnight at 37ºC.

14. Inspect the plates and record the results. Deletion strains should be Reg^c Km^s.

15. Use sterile toothpicks to pick two to four well-isolated white colonies from the EBU-Ap plate for each of the deletion strains chosen for further analysis. Streak for single colonies on the following media:

   a. EBU agar (to confirm phage-free clones).

   b. NA-XGal (to confirm Reg^c strains).

   c. LB-Km agar (to confirm loss of MudJ).

   d. MacConkey-Srl (to confirm Srl⁻ phenotype).

16. Streak each colony across a line of P22 H5 on an EBU plate to test for P22 sensitivity. Use strains TSM1 (P22ˢ) and TSM125 (P22ʳ) as controls.

17. Incubate all plates overnight at 37°C.

18. Inspect the plates and record the results.

19. Assign collection and allele numbers to each of the newly constructed deletion strains and culture each in LB broth to saturation.

20. Stock each of the new Δ(*srl-zgc*) strains in the central strain collection (see Appendix C, Storing Strains).

Construct a Deletion Map of the *srl* Region

Donor *srl*::MudJ strains are crossed with recipient Δ*srl* strains, with selection for Srl⁺. One potential complication for deletion mapping is the extreme efficiency of P22-mediated transduction. If using P22 grown on the wild type to make a lysate of a donor *srl* mutant, a small but detectable number of *srl⁺* transducing particles are carried over that contaminate the newly made lysate. These *srl⁺* transducing particles greatly obscure the deletion mapping data by giving false-positive results. To overcome this problem, use donor lysates made with P22 that has been grown on a Δ*srl* strain (see also Hughes et al. 1991).

MATERIALS

For culture media and supplements, see Appendix B.

LB broth
NCE-Srl plates

PROCEDURE

1. Prepare P22 HT *int* lysates on each backcrossed *srl*::MudJ strain, including previously characterized control strains (Table 1.1) (see Appendix D, Generalized P22 Transduction with Phage P22). Use P22 broth prepared with a P22 lysate grown on strain TSM140.

2. Inoculate a single colony of each deletion strain into 2.5 ml of LB broth. Aerate overnight at 37°C. Include strain TSM123 (Δ*trp*) as a control.

3. Use 10 μl of each donor *srl*::MudJ lysate for plate transductions with each of the deletion recipients, including previously characterized control strains (Table 1.1) (see Appendix D, Generalized Transduction with Phage P22). Plate on NCE-Srl plates.

4. Spread 0.1 ml of each deletion culture on an NCE-Srl plate (cells-only control). Also perform the lysate sterility, lysate titer, and transduction controls.

5. Incubate all plates for 24–48 hours at 37°C.

6. Inspect the plates and record the results. Each of the lysates should have efficiently transduced the Δ*trp* strain. Score *srl*⁺ transduction efficiency for each of the donor lysates. Construct

Table 1.1 Characterized Strains for Deletion Mapping

Δ(*srl-zgc*) Strain	Parental *srl*::MudJ Strain
TSM131 (VJSS337)	TSM1078
TSM134 (VJSS342)	TSM105 (VJSS068)
TSM135 (VJSS343)	TSM126 (VJSS069)
TSM136 (VJSS344)	TSM127 (VJSS070)
TSM139 (VJSS347)	TSM1002
TSM140 (VJSS348)	TSM1003
TSM141 (VJSS349)	TSM1004
None	TSM1005
TSM144 (VJSS352)	TSM1008
TSM1143	TSM1141
TSM1144	TSM1142

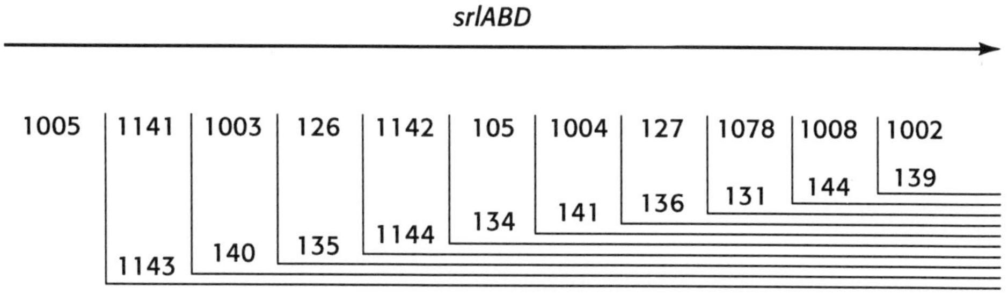

Figure 1.1 Genetic map of the *srl* region deduced from crosses among the strains listed above. Lines indicate deleted material.

a deletion map of the *srl* region. Previously characterized deletion strains used for map construction should be included for reference (for full details, see Appendix A). The deletion map constructed with these strains is shown in Figure 1.1.

EXPERIMENT 2

Transposons: Tn*10*d Insertions Linked to Structural Genes

Experiment 2 demonstrates a method for isolating transposon insertions adjacent to genes of interest. First, a pool of random Tn*10*d(Tc) insertions is generated by introducing the element into a *srl*+ strain that overproduces the Tn*10* transposase. A generalized transducing lysate grown on this pool is used to correct the *srl* defect of a recipient strain with concomitant selection for Tc^r. This general procedure for isolating linked transposon insertions is easily adapted to a variety of other genetic systems. The approaches used here were adapted from Elliott and Roth (1988) and Kleckner et al. (1991).

The recipient *srl*::MudJ insertion confers three phenotypes: Srl$^-$, Lac$^+$, and Km^r. In principle, any of these phenotypes can be used in screening for rare Tc^r *srl*+ transductants from the random pools, although the first two are more convenient. This protocol uses the Lac phenotype (using XGal as the indicator) for screening, as experience has shown that some Tn*10*d(Tc) insertions adjacent to the *srl* locus confer poor growth on MacConkey-Srl medium.

Each Tn*10*d(Tc) insertion is subjected to a three-point cross to determine its genetic map location with respect to nearby markers, and each is also mapped by Mud-P22 transduction.

The linked insertions isolated in this experiment are denoted as residing at locus *zgc* to indicate their map position (centisome 62; see Section 2, Genetic Nomenclature).

265

Below are listed the bacterial strains and bacteriophage lysates used in this experiment.

BACTERIAL STRAINS

TSM1 (prototroph)
TSM103 (pNK2881; *ats-1 ats-2*)
TSM104 (*zzf*::Tn*10*d[Tc])
TSM105 (*srl*::MudJ)
TSM106 (*zgc*::Tn*10*d[Tc])
TSM107 (*srl*::MudJ *zgc*::Tn*10*d[Cm])
TSM118 (*zzf*::Tn*10*d[Km])
TSM119 (*zzf*::Tn*10*d[Cm])
TSM120 (pNK976; *ats*+)
TSM123 (Δ*trp*)
TSM125 (P22^r)

BACTERIOPHAGE LYSATES

P22 broth prepared with a P22 HT *int* lysate of strain TSM1
P22 HT *int* lysate of strain TSM104 (*zzf*::Tn*10*d[Tc])
Mud-P22 lysates

Generate Random Tn*10*d(Tc) "Hops"

MATERIALS

For culture media and supplements, see Appendix B.

LB broth
LB broth-Ap
LB broth-EGTA (10 mM)
LB-Tc plates
0.85% NaCl
5 M Glycerol
Microcentrifuge tubes
Sterile spreader

PROCEDURE

1. Inoculate a single colony each of strains TSM1 (prototroph) into 2.5 ml of LB broth, and TSM103 (pNK2881; *ats* transposase) or TSM120 (pNK976; *ats* transposase) into 2.5 ml of LB broth-Ap. Aerate overnight at 37°C.

2. Store the cultures on ice until use.

3. Pipette 0.1 ml of TSM1 culture into one microcentrifuge tube. Pipette 0.1 ml of TSM103 (or TSM120) culture into each of nine microcentrifuge tubes.

4. Dilute a P22 HT *int* lysate of strain TSM104 (*zzf-1831*:: Tn*10*d[Tc]) to approximately 10^{10} pfu/ml in cold 0.85% NaCl.

5. Add 0.1 ml of diluted lysate as shown below.

Tube	Strain	Cells (ml)	Phage (ml)	Comments
A-1	TSM103	0.1	0.1	pool A
A-2	TSM103	0.1	0.1	pool A
A-3	TSM103	0.1	0.1	pool A
A-4	TSM103	0.1	0.1	pool A
B-1	TSM103	0.1	0.1	pool B
B-2	TSM103	0.1	0.1	pool B
B-3	TSM103	0.1	0.1	pool B
B-4	TSM103	0.1	0.1	pool B
C	TSM103	0.1	0	cells-only control
D	none	0	0.1	phage-only control
E	TSM1	0.1	0.1	transposition control

6. Gently mix the contents of each tube.

7. Incubate for 10–15 minutes at 37°C to allow the phage to adsorb to the cells.

8. Add 1.0 ml of LB broth-EGTA (10 mM) and gently mix (EGTA chelates Ca^{++}, thereby inhibiting further infection).

9. Incubate for 15–20 minutes at 37°C to allow time for expression of Tc^r.

10. Plate 0.2 ml of tube A-1 on five LB-Tc plates. Do the same for tubes A-2, -3, -4 and B-1, -2, -3, -4 (40 plates total).

11. Plate 0.2 ml of tube C on one LB-Tc plate. Do the same for tubes D and E (3 plates total).

12. Incubate all plates overnight at 37°C.

13. Inspect the plates. No colonies should be seen on the three control plates, whereas the experimental plates should have approximately 200 colonies per plate. Record the results.

14. Pipette 2.0 ml of 0.85% NaCl onto the surface of each plate. Use a sterile spreader to resuspend the colonies into the liquid. Make sure that the colonies are well-dispersed. The cell suspension should be quite thick and viscous.

15. Collect the cell suspensions in large sterile tubes. Keep the two pools separate.

16. Pipette 1.5 ml of each pool into each of three microcentrifuge tubes (six tubes total).

17. Sediment the cells in a microcentrifuge at 14,000 rpm for 30 seconds. The supernatant may remain cloudy with cells.

18. Remove the supernatant and resuspend each pellet in 1.5 ml of 0.85% NaCl.

19. Repeat steps 17 and 18 two more times to give a total of three washes.

20. Sediment the cells in a microcentrifuge for 30 seconds.

21. Resuspend each pellet in 1.5 ml of LB broth.

22. Combine the three tubes of each of the washed pools into larger sterile tubes. Keep the two pools separate. The pools should be quite dense and viscous.

23. Mix 0.05 ml of each pool with 0.5 ml of LB broth. This should give a turbid suspension equivalent to a saturated culture. Adjust the volumes if necessary.

24. Add 2.0 ml of P22 broth to each diluted culture.

25. Aerate overnight at 37ºC. The culture will be quite dense, but cell debris and "strings" are often visible when lysis has proceeded well.

26. Meanwhile, mix 0.5 ml of each of the concentrated pools with 0.5 ml of sterile 5 M glycerol in a 2-ml screw-cap tube. Invert several times to mix well and place in a –70ºC freezer. These frozen stocks serve as sources of inocula if the P22 lysates of the pools need to be remade.

Screen Pools for Insertions Adjacent to *srl*

MATERIALS

For culture media and supplements, see Appendix B.

EBU plates
EBU-Tc plates
E-Glc-ACH plates
LB broth
LB broth-EGTA (10 mM)
LB plates
LB-Km plates
MacConkey-Tc-Srl plates
NA-XGal-Srl plates
NA-Tc-EGTA-XGal-Srl (0.1%) plates
TBSA (Tryptone Broth–Soft Agar)
Chloroform
Sterile toothpicks

PROCEDURE

1. Inoculate a single colony each of strains TSM105 (*srl*::MudJ) and TSM123 (Δ*trp*) into 2.5 ml of LB broth. Aerate overnight at 37°C.

2. Store the cultures on ice until use.

3. Harvest the lysates (see Appendix D, Generalized Transduction with Phage P22).

4. Store the lysates on ice until use. Allow enough time for the chloroform and debris to settle.

   Caution: Chloroform is a carcinogen and may damage the liver and kidneys. Do not mouth pipette and avoid contact with skin.

5. Pipette 0.1 ml of TSM105 culture into each of six microcentrifuge tubes.

6. Add 10 µl of undiluted lysate as shown below.

Tube	Cells (ml)	Phage (µl)	Comments
A-1	0.1	10	lysate from pool A
A-2	0.1	10	lysate from pool A
B-1	0.1	10	lysate from pool B
B-2	0.1	10	lysate from pool B
C	0.1	0	cells-only control
D	0	100	phage-only control

7. Incubate for 10–15 minutes at 37°C.

8. Add 1.0 ml of LB broth-EGTA (10 mM) and gently mix.

9. Incubate for 15–20 minutes at 37°C.

10. Plate 0.2 ml of tube A-1 on five NA-Tc-EGTA-XGal-Srl plates. Do the same for tubes A-2, B-1, and B-2 (20 plates total).

11. Plate 0.2 ml of tubes C and D on one NA-Tc-EGTA-XGal-Srl plate (2 plates total).

12. For each lysate used, mix 0.05 ml of TSM105 culture with 2.5 ml of TBSA and pour onto an LB plate. Spot on 6/.02, 7/.02, 8/.02, and 9/.02 dilutions of the lysate (see Section 2, Microbiological Procedures) and allow the spots to dry.

13. For each lysate used, spread 0.1 ml of TSM123 culture on an E-Glc-ACH plate. Spot on 0/.02, 1/.02, 2/.02, and 3/.02 dilutions of the lysate and allow the spots to dry.

14. Incubate all plates overnight at 37°C.

15. Carefully inspect each of the plates. No colonies should be seen on the two control plates, whereas the experimental plates should have 100–500 colonies per plate, the vast majority of which should be blue. Rare white colonies are presumptive srl^+ transductants.

16. Estimate the total number of blue and white colonies in each pool. Record the results. Assign isolation numbers to each of the white colonies chosen for further analysis. Be sure to keep track of which colonies came from which pools.

17. Use sterile toothpicks to pick each of the chosen white colonies, along with one or two randomly chosen blue colonies. Streak for single colonies on the following media:

    a. EBU-Tc agar (to isolate phage-free clones).

    b. NA-Tc-EGTA-XGal-Srl agar (to ensure that the transductants are Lac⁻).

    c. MacConkey-Tc-Srl agar (to ensure that the transductants are Srl^+).

    d. LB-Km agar (to ensure that the transductants are Km^s).

18. Incubate all plates overnight at 37°C.

19. Inspect the plates and record the results.

    • The streaks on the EBU-Tc plates should be mixtures of white and dark blue colonies. The blue colonies are those that are releasing phage. Hopefully, several white colonies will be well-isolated.

    • The isolated colonies on the NA-XGal-Srl plates should be white (Lac⁻). If all colonies in a given streak are blue, then that candidate colony was a fake. If the streak is a mixture of blue and white colonies, then that candidate colony was a mixture of mutant and wild type. This can be salvaged by restreaking white colonies.

20. Use sterile toothpicks to pick two to four well-isolated white colonies from the EBU plate for each of the isolates chosen for further analysis. Streak for single colonies on the following media:

    a. EBU agar.

    b. NA-Tc-EGTA-XGal-Srl agar.

    c. MacConkey-Tc-Srl agar.

    d. LB-Km agar.

21. Streak each colony across a line of P22 H5 on an EBU plate to test for P22 sensitivity. Use strains TSM1 (P22^s) and TSM125 (P22^r) as controls.

22. Incubate all plates overnight at 37°C.

23. Inspect the plates and record the results. Choose up to two P22-sensitive Srl⁺ Km^s isolates from each pool for further analysis. Save desired strains for the short term on EBU plates at 4°C.

Confirm Linkage of Tn*10*d(Tc) Insertions to the *srl* Locus

MATERIALS

For culture media and supplements, see Appendix B.

LB broth
NA-Tc-EGTA-XGal-Srl (0.1%) plates

PROCEDURE

1. Prepare P22 HT *int* lysates on each strain (see Appendix D, Generalized Transduction with Phage P22).

2. Use 10 µl of each lysate for delayed expression transductions with strain TSM1 (see Appendix D, Generalized Transduction with Phage P22). Plate on NA-Tc-EGTA-XGal-Srl plates.

3. Inspect the plates and record the results. Typical linkage values range from approximately 10% to 90%.

4. On the basis of these results, choose up to two of the original P22-sensitive Srl+ isolates from each pool for further work if the two strains show clearly different linkage to *srl*.

5. Assign collection and allele numbers to each of the desired strains and culture each in LB broth to saturation.

6. Stock each of the new *zgc*::Tn*10*d(Tc) strains in the central strain collection (see Appendix C, Storing Strains).

Determine the Map Location of Each *zgc*::Tn*10*d(Tc) Insertion via Mud-P22 Mapping

MATERIALS

For culture media and supplements, see Appendix B.

LB-Tc broth
Bochner-Maloy plates
Frog (48-pin replica plater; Sigma R 2383)

PROCEDURE

1. Inoculate a single colony of of each strain for mapping into 2.5 ml of LB-Tc broth. Aerate overnight at 37°C. Include a culture of strain TSM106 (*zgc*::Tn*10*d[Tc]) as a control.

2. Store cultures on ice until use. Spread 0.1 ml of each culture on a Bochner-Maloy plate.

3. Spot on Mud-P22 lysates with a frog (see Appendix D, Rapid Mapping in *S. typhimurium* with Mud-P22 Prophages).

4. Allow the spots to dry and incubate the plates overnight at 37°C.

5. Inspect the plates and record the results. Unambiguous results may require 36–48 hours of incubation, but check the plates after about 18 and 24 hours as well.

Determine the Relative Map Position of the zgc::Tn10d(Tc) Insertion: Three-point Crosses

MATERIALS

For culture media and supplements, see Appendix B.

LB-Tc plates
LB-Tc-Cm plates
LB-Tc-Km plates
NA-Tc-XGal plates

PROCEDURE

1. Prepare P22 HT *int* lysates on each zgc::Tn10d(Tc) strain to be tested (see Appendix D, Generalized Transduction with Phage P22).

2. Use 10 μl of each lysate for delayed expression transductions with strain TSM107 (see Appendix D, Generalized Transduction with Phage P22). Plate on LB-Tc plates.

3. Incubate all plates overnight at 37°C.

4. Patch up to 200 transductants (50 per plate) on the following plates:

   a. LB-Tc-Cm (to score inheritance of the donor Cm^s phenotype).

   b. LB-Tc-Km (to score inheritance of the donor Km^s phenotype).

   c. NA-Tc-XGal spread with 0.1 ml of 10% Srl (to score inheritance of the donor Lac$^-$ phenotype).

5. Incubate all plates overnight at 37°C.

6. Inspect the plates and record the results. Photocopy the blank three-point cross form (see Fig. 2.1) for recording the data (an example form is also shown in Fig. 2.2). Determine the map order and distances of the Tn10d(Tc) insertion with respect to the srl::MudJ and zgc::Tn10d(Cm) insertions. For assistance, see Section 3 (Genetic Mapping by Generalized Transduction).

Donor:						Recipient:					
Selection:						Date:					

#						#					
1						26					
2						27					
3						28					
4						29					
5						30					
6						31					
7						32					
8						33					
9						34					
10						35					
11						36					
12						37					
13						38					
14						39					
15						40					
16						41					
17						42					
18						43					
19						44					
20						45					
21						46					
22						47					
23						48					
24						49					
25						50					

Figure 2.1 Three-point cross form for recording data.

Three-point mapping of *zgc-7203*::Tn*10*d(Tc) relative to *srl-251*::MudJ and *zgc-1623*::Tn*10*d(Cm)

Donor: TSM106 (*zgc*::Tc)   Recipient: TSM107 (*srl*::MudJ *zgc*::Cm)

Selection: Tc resistance   Date: August 12, 1991

#	Cm	Km	XG		Class		#	Cm	Km	XG		Class	
1	R	S	W		2		26	S	S	W		1	
2	S	S	W		1		27	S	S	W		1	
3	S	S	W		1		28	S	R	B		3	
4	S	S	W		1		29	S	R	B		3	
5	S	S	W		1		30	S	R	B		3	
6	S	S	W		1		31	S	S	W		1	
7	S	S	W		1		32	S	S	W		1	
8	S	S	W		1		33	R	R	B		4	!
9	S	R	B		3		34	S	S	W		1	
10	S	S	W		1		35	S	S	W		1	
11	S	R	B		3		36	R	S	W		2	
12	S	S	W		1		37	S	R	B		3	
13	S	S	W		1		38	R	S	W		2	
14	S	S	W		1		39	R	S	W		2	
15	S	S	W		1		40	S	S	W		1	
16	R	S	W		2		41	R	S	W		2	
17	S	R	B		3		42	R	S	W		2	
18	S	S	W		1		43	S	S	W		1	
19	S	S	W		1		44	S	S	W		1	
20	S	S	W		1		45	S	S	W		1	
21	S	R	B		3		46	R	S	W		2	
22	S	S	W		1		47	S	S	W		1	
23	S	R	B		3		48	S	R	B		3	
24	S	R	B		3		49	R	S	W		2	
25	S	R	B		3		50	S	R	B		3	

Figure 2.2 Example of three-point cross form.

EXPERIMENT 3
Transposons: Tn*10*d Insertions in Regulatory Genes

Experiment 3 demonstrates a method for isolating regulatory mutations following transposon mutagenesis. In this case, the transposition method is regulated transposase expression, and the transposon is Tn*10*d(Cm). A Φ(*srl-lacZ*) operon fusion serves as an indicator for gene regulation. Individual colonies are screened for both constitutive ("up"; Reg^c phenotype) and uninducible ("down"; Reg⁻ phenotype) phenotypes. The approaches used here were adapted from Hughes and Roth (1985) and Kleckner et al. (1991).

The donor Tn*10*d(Cm) insertion is within a ColE1-type plasmid; transposase function is provided by an adjacent *tnpA* gene under control of the IPTG-regulated *tac* promoter. A second compatible plasmid expresses the LacI protein. Chromosomal Tn*10*d(Cm) insertions are accumulated in strains that are cultured in the presence of IPTG. Individual colonies are then directly screened for mutant phenotypes. This procedure is broadly adaptable to a variety of genetic systems and is particularly useful for systems in which genetic exchange (and thereby introduction of suicide vectors) is inefficient.

Mutant individuals retain the donor plasmid, a complication for subsequent genetic analysis. P22 transduction is used to backcross the chromosomal mutant allele into a plasmid-free Φ(*srl-lacZ*) operon fusion strain. Many such Cm^r transductants represent individuals that inherited the donor plasmid by transduction; these colonies will be Ap^r due to the *bla* gene which is also on the donor plasmid. Ap^s transductants with the appropriate regulatory phenotypes represent genuine backcross events.

Insertions are then tested for linkage to the *srlACD* operon, and unlinked insertions are tested for linkage to a marker that is near *srlU*, a gene of unknown function required for normal regulation of Φ(*srl-lacZ*) operon fusion expression. Linked insertions might lie within a regulatory gene (*srlR*), or they might affect the promoter/control region for the *srlACD* operon. These possibilities can be distinguished by complementation analysis.

A tandem duplication strain merodiploid for the *srl* operon is constructed according to the method of Hughes and Roth (1985). This general procedure allows construction of a merodiploid strain for virtually any region of the *Salmonella* chromosome. The mutant insertion, along with a Φ(*srl-lacZ*) operon fusion, is transduced into one of the duplicated *srl* loci, and the resultant strain is examined for its regulatory phenotype. Restoration of the wild-type phenotype (Reg+ phenotype) will indicate that the insertion lies within a protein-encoding regulatory gene (recessive). Retention of the original phenotype (Reg^c [constitutive] or Reg− phenotype) will indicate that the insertion lies within a cis-acting control region. (cis-acting Reg− insertions might also lie within the *srl* operon and block *lacZ* expression due to polarity or they might lie within the *lacZ* gene itself.) Below are listed the bacterial strains and bacteriophage lysates used in this experiment.

BACTERIAL STRAINS

TSM1 (prototroph)
TSM105 (*srl*::MudJ)
TSM108 (*srl*::MudJ/pMS421/pNK2884)
TSM113 (*nadB-cysH* duplication) (30 µg/ml Ap)
TSM121 *nadB*::MudA (Lac−; orientation B)
TSM122 *cys[HDC]*::MudA (Lac+; orientation B)
TSM123 (Δ*trp*)
TSM125 (P22^r)

BACTERIOPHAGE LYSATES

P22 broth prepared with a P22 HT *int* lysate of strain TSM1
P22 HT *int* lysate of strain TSM111 (*zgx-7204*::Tn*10*d[Tc]; insertion near *srlU*)
P22 HT *int* lysate of strain TSM121 (*nadB*::MudA)
P22 HT *int* lysate of strain TSM122 (*cys[HDC]*::MudA)

Generate Random Tn*10*d(Cm) "Hops"

MATERIALS

For culture media and supplements, see Appendix B.

E-Glc plates
LB broth-Ap (200 μg/ml)-Sp (50 μg/ml)-IPTG (1 mM)
NA-XGal plates
10% Sorbitol

PROCEDURE

1. Inoculate four different colonies of strain TSM108 (*srl*::MudJ/ pMS421/pNK2884) each into 5 ml of LB broth-Ap-Sp-IPTG (1 mM). Aerate for approximately 24 hours at 37°C.

2. Plate 6/0.2 dilutions (see Section 2, Microbiological Proce-dures) of each culture on each of ten NA-XGal plates. Be sure to mark each plate to designate which culture is on which plate (40 plates total).

3. Incubate all plates overnight at 37°C.

4. Carefully inspect each of the plates. Each should have approxi-mately 200 colonies, the vast majority of which should be white. Search for rare blue (constitutive; Reg^c phenotype) colonies and mark the location of each on the back of the plate. Record the results.

5. Replica print each plate both to an E-Glc plate and to an NA-XGal plate previously spread with 0.1 ml of 10% sorbitol. Be sure to spread the sorbitol over the *entire* surface of the agar.

6. Incubate all plates overnight at room temperature (to avoid overgrowth).

7. Carefully inspect each of the plates. Most of the colonies should have grown on the minimal-glucose plates; search care-fully for auxotrophs, which provide an indication for the level

of mutagenesis in each pool. Most of the colonies should have turned blue on the XGal plus inducer plates. Search for rare white (uninducible; Reg⁻ phenotype) colonies. Record the results.

8. Streak Reg^c and Reg⁻ colonies on NA-XGal and on NA-XGal previously spread with 0.1 ml of 10% Srl.

9. Incubate overnight at 37ºC.

10. Inspect the plates and record the results. Choose up to two Reg^c and Reg⁻ isolates from each pool for further analysis. Assign isolation numbers to each of the colonies chosen for further analysis. Be sure to record which colonies came from which pools.

Backcross Tn*10*d(Cm) Insertions to the Wild Type

MATERIALS

For culture media and supplements, see Appendix B.

EBU plates
EBU-Cm plates
LB broth
LB-Ap (200 μg/ml) plates
NA-XGal plates
NA-XGal-Cm-EGTA plates
10% Sorbitol
Sterile toothpicks

PROCEDURE

1. Prepare P22 HT *int* lysates on each strain (see Appendix D, Generalized Transduction with Phage P22).

2. Use 10 μl of each lysate for delayed expression transductions with strain TSM105 (*srl*::MudJ) (see Appendix D, Generalized

Transduction with Phage P22). Plate on NA-XGal-Cm-EGTA plates. Previously spread the plates with 0.1 ml of 10% sorbitol for transductions involving Reg⁻ donors. Be sure to spread the sorbitol over the *entire* surface of the agar.

3. Incubate all plates overnight at 37°C.

4. Inspect the plates and record the results. Most of the Cmʳ transductants will have the Reg⁺ phenotype, and only a few will have the Regᶜ or Reg⁻ phenotype of the parent, because the donor strain carries pNK2884 (Apʳ Cmʳ), which is efficiently transduced by P22 HT *int*.

5. Use sterile toothpicks to pick two Regᶜ/Reg⁻ colonies from each transduction. Streak for single colonies on the following media:

   a. EBU-Cm agar (to isolate phage-free clones).

   b. LB-Ap (200 µg/ml agar (to ensure that the colonies do not contain pNK2884).

   c. NA-XGal-Cm-EGTA agar (to confirm Regᶜ phenotypes).

   d. NA-XGal-Cm-EGTA agar spread with 0.1 ml of 10% Srl (to confirm Reg⁻ phenotypes). Be sure to spread the sorbitol over the *entire* surface of the agar.

6. Incubate all plates overnight at 37°C.

7. Inspect the plates, and record the results.

   • The streaks on the EBU-Cm plates should be mixtures of white and dark-blue colonies. The blue colonies are those that are releasing phage. Hopefully, several white colonies will be well-isolated.

   • The isolated colonies on the LB-XGal-Cm-EGTA (±Srl) plates should exhibit the appropriate Reg phenotype. All strains should be Apˢ.

8. Use sterile toothpicks to pick two to four well-isolated white colonies from the EBU-Cm plate for each of the transductants chosen for further analysis. Streak for single colonies on the following media:

   a. EBU agar.

   b. NA-XGal.

   c. NA-XGal previously spread with 0.1 ml of 10% sorbitol. Be sure to spread the sorbitol over the *entire* surface of the agar.

9. Streak each colony across a line of P22 H5 on an EBU plate to test for P22 sensitivity. Use strains TSM1 (P22^s) and TSM125 (P22^r) as controls.

10. Incubate all plates overnight at 37°C.

11. Inspect the plates and record the results. Choose one P22-sensitive Reg^c or Reg$^-$ isolate from each cross for further analysis.

12. Assign collection and allele numbers to each of these strains and culture each in LB broth to saturation.

13. Stock each of the new *reg*::Tn*10*d(Cm) strains in the central strain collection (see Appendix C, Storing Strains).

Test Linkage of *reg*::Tn*10*d(Cm) Insertions to the *srl* Operon

MATERIALS

For culture media and supplements, see Appendix B.

LB-Km plates
NA-XGal-Cm-EGTA plates
10% Sorbitol

PROCEDURE

1. Prepare P22 HT *int* lysates on each strain (see Appendix D, Generalized Transduction with Phage P22).

2. Use 10 μl of each lysate for delayed expression transductions with strains TSM1 and TSM105 (*srl*::MudJ) (see Appendix D, Generalized Transduction with Phage P22). Plate on NA-XGal-Cm-EGTA plates. Previously spread the plates with 0.1 ml of 10% sorbitol for transductions involving Reg⁻ donors. Be sure to spread the sorbitol over the *entire* surface of the agar.

3. Incubate all plates overnight at 37°C.

4. Inspect the plates and record the results.

   • The crosses into strain TSM105 (*srl*::MudJ) simply serve to reconfirm that the Tn*10*d(Cm) insertion is linked to the Reg phenotype. These colonies are not analyzed further.
   • The crosses into strain TSM1 distinguish those Tn*10*d(Cm) insertions that are linked to the *srl* operon (coinheritance of *srl*::MudJ from the donor) from those that are not (no inheritance of *srl*::MudJ from the donor). Colonies from this cross are tested for the Km phenotype to confirm the results.

5. Patch up to 50 transductants from the crosses involving strain TSM1 on the following media:

   a. LB-Km (to test for inheritance of MudJ).

   b. NA-XGal-Cm-EGTA (for Reg^c strains).

   c. NA-XGal-Cm-EGTA previously spread with 0.1 ml of 10% sorbitol (for Reg⁻ strains). Be sure to spread the sorbitol over the *entire* surface of the agar.

6. Incubate all plates overnight at 37°C.

7. Inspect the plates and record the results.

Determine Genetic Map Location of Unlinked Reg^c Tn*10*d(Cm) Insertions

To date, all Reg^c *srl* regulatory mutations that are unlinked to the *srl* operon have mapped to a previously unrecognized locus, *srlU*, located near the 68-centisome region of the *S. typhimurium* genetic map. This was determined by mapping with the Mud-P22 collection. For characterization of additional mutants, we use a Tn*10*d(Tc) insertion, *zgi-7204*::Tn*10*d(Tc), that is linked to *srlU*.

MATERIALS

For culture media and supplements, see Appendix B.

LB-Cm-Tc plates
LB-Tc plates
NA-XGal-Tc-EGTA plates

PROCEDURE

1. Use 10 μl of TSM111 (*zgi-7204*::Tn*10*d[Tc]) lysate for delayed expression transductions with each of the strains carrying unlinked *reg*::Tn*10*d(Cm) insertions (see Appendix D, Generalized Transduction with Phage P22). Plate on NA-XGal-Tc-EGTA plates. Do not add sorbitol.

2. Incubate all plates overnight at 37ºC.

3. Inspect the plates and record the results.

4. Patch up to 50 transductants from the crosses on the following media:

   a. LB-Cm-Tc (to test for loss of Tn*10*d[Cm]).

   b. LB-Tc.

5. Incubate all plates overnight at 37ºC.

6. Inspect the plates and record the results. Determine whether the Tn*10*d(Cm) regulatory insertion mutations map to *srlU*.

Construction of *srl* Tandem Duplication Strains

Salmonella geneticists are blessed by having a general method to construct a directed tandem duplication of any defined chromosomal region (Hughes and Roth 1985; see Section 4, Genetic Complementation). The protocol used to construct the duplication strain TSM113 is reproduced here and may be pursued if time permits. Alternatively, strain TSM113 may be used for the complementation analysis described in the subsequent section.

MATERIALS

For culture media and supplements, see Appendix B.

EBU-Ap (30 µg/ml) plates
E-Glc-Ap (15 µg/ml) plates
LB broth
LB broth-EGTA (10 mM)
LB plates
LB-Ap (30 µg/ml) plates
LB-Ap (30 µg/ml)-EGTA plates
Microcentrifuge tubes
Inoculating sticks
Sterile toothpicks

PROCEDURE

1. Prepare high-titer (>10^{11} pfu/ml) P22 HT *int* lysates of strains TSM121 (*nadB*::MudA) and TSM122 (*cys[HDC]*::MudA) (see Appendix D, Generalized Transduction with Phage P22). Add cysteine to cultures of *cys* auxotrophs, as LB medium is deficient in cysteine. Use P22 broth prepared with a P22 lysate grown on strain TSM1.

2. Inoculate a single colony of strain TSM1 into 2.5 ml of LB broth. Aerate overnight at 37ºC.

3. Store the culture on ice until use.

4. Pipette 0.1 ml into each of eight microcentrifuge tubes.

5. Add approximately 2×10^{10} pfu/ml of each lysate as shown. For illustrative purposes, this chart assumes that the TSM121 lysate has a titer of 2×10^{11} pfu/ml and that the TSM122 lysate has a titer of 5×10^{11} pfu/ml. Adjust volumes accordingly.

Tube	TSM121 (μl)	TSM122 (μl)	Comments
A	100	40	pool A
B	100	40	pool B
C	100	40	pool C
D	100	40	pool D
E	100	40	pool E
F	100	0	TSM121-only control
G	0	40	TSM122-only control
H	0	0	cells-only control

6. Gently mix the contents of each tube.

7. Incubate for 90 minutes at room temperature.

8. Add 0.9 ml of LB broth-EGTA (10 mM) and gently mix.

9. Sediment the cells in a microcentrifuge at 14,000 rpm for 2 minutes.

10. Resuspend the cell pellets each in 1 ml of LB broth-EGTA and sediment at 14,000 rpm for 2 minutes.

11. Resuspend the cell pellets each in 1 ml of LB broth-EGTA and sediment at 14,000 rpm for 2 minutes. This repeated washing removes β-lactamase carried over from the donor lysate and makes for cleaner selection on the Ap plates.

12. Resuspend the cell pellets each in 0.1 ml of LB broth-EGTA and plate each on an LB-Ap (30 μg/ml)-EGTA plate.

13. Spot 10 μl of the donor P22 lysates on an LB plate (phage-sterility control).

14. Incubate all plates overnight at 37°C.

15. Inspect the plates and record the results. Approximately 50–200 Ap^r colonies should result from each of the transductions.

16. Replica print each plate to the following media:

    a. E-Glc-Ap (15 μg/ml) agar.

    b. LB-Ap (30 μg/ml) agar.

17. Incubate all plates overnight at room temperature (to avoid overgrowth).

18. Inspect the plates and record the results. Rare prototrophic colonies represent potential tandem duplication strains; auxotrophs represent strains that inherited the donor *nadB* or *cys* insertion. Some prototrophs may arise from (inefficient) transposition of the MudA element.

19. Use sterile toothpicks to pick up to four colonies from each E-Glc-Ap plate. Streak for single colonies on the following media:

    a. E-Glc-Ap (15 μg/ml) agar (to confirm that the strains are prototrophs).

    b. EBU-Ap (30 μg/ml) agar (to isolate phage-free clones).

20. Incubate all plates overnight at 37ºC.

21. Inspect the plates and record the results. The streaks on the EBU-Ap plates should be mixtures of white and dark-blue colonies. The blue colonies are those that are releasing phage. Hopefully, several white colonies will be well-isolated.

22. Use sterile toothpicks to pick two to four well-isolated white colonies from the EBU-Ap plate for at least two each prototrophic transductants from each cross. Streak for single colonies on the following media:

    a. E-Glc-Ap (15 μg/ml) agar.

    b. EBU-Ap (30 μg/ml) agar.

23. Streak each colony across a line of P22 H5 on an EBU plate to test for P22 sensitivity. Use strains TSM1 (P22^s) and TSM125 (P22^r) as controls.

24. Incubate all plates overnight at 37°C.

25. Inspect the plates and record the results. To identify diploid strains, perform segregation analysis. Culture the strains for several generations in the absence of selection (Ap) and test single colonies for loss of the MudA element (Ap^r) located at the duplication join-point.

26. Inoculate two colonies from each cross into 2.5 ml of LB broth with no Ap.

27. Aerate at 37°C until saturated (overnight if necessary).

28. Subculture each with an inoculating stick into 2.5 ml of fresh LB broth with no Ap.

29. Aerate at 37°C until saturated (overnight if necessary).

30. Plate 6/0.2 dilutions (see Section 2, Microbiological Procedures) from each culture onto LB agar.

31. Incubate all plates overnight at 37°C.

32. Replica print each plate to the following media:

    a. LB-Ap (30 μg/ml) agar. (Which ones retain the duplication?)

    b. LB agar.

33. Incubate all plates overnight at room temperature (to avoid overgrowth).

34. Inspect the plates and record the results. Duplication strains should yield 5% segregants (Ap^s).

Complementation Analysis of Linked Reg^c Tn*10*d(Cm) Insertions

Do the linked Reg^c Tn*10*d(Cm) insertions define the gene for an *srl* operon repressor? If so, they should exhibit characteristic behavior in a complementation test.

MATERIALS

For culture media and supplements, see Appendix B.

EBU-Cm-Ap (30 µg/ml) plates
LB-Ap (30 µg/ml) plates
LB broth
LB broth-Ap (30 µg/ml)
LB-Cm plates
LB-Cm-Ap (30 µg/ml)-EGTA plates
LB-Km plates
LB-Km-Ap (30 µg/ml) plates
LB plates
MacConkey-Srl plates
MacConkey-Srl-Ap (30 µg/ml) plates
NA-XGal-Ap (30 µg/ml) plates
10% Sorbitol
Sterile toothpicks
Inoculating sticks

PROCEDURE

1. Inoculate a single colony of the *nadB-srl-cysH* duplication strain (or strain TSM113) into 2.5 ml of LB broth-Ap (30 µg/ml). Aerate overnight at 37ºC.

2. Use 10 µl of each lysate grown on the *srl*-linked Reg^c Tn*10*d(Cm) insertion strains for delayed expression transductions with the *nadB-srl-cysH* duplication strain (see Appendix D, Generalized Transduction with Phage P22). Plate on LB-Cm-Ap (30 µg/ml)-EGTA plates.

3. Incubate all plates overnight at 37ºC.

4. Use sterile toothpicks to pick four colonies from each transduction. Streak for single colonies on the following media:

   a. EBU-Cm-Ap (30 µg/ml) agar (to isolate phage-free
      clones).

   b. LB-Km-Ap (30 µg/ml) agar (to ensure that srl::MudJ was
      coinherited).

5. Incubate all plates overnight at 37°C.

6. Inspect the plates and record the results. The streaks on the
   EBU-Cm plates should be mixtures of white and dark blue
   colonies. The blue colonies are those that are releasing phage.
   Hopefully, several white colonies will be well-isolated. Virtually all of the transductants should be Km^r.

7. Use sterile toothpicks to pick two to four well-isolated white
   colonies from the EBU-Cm-Ap plate for at least two each Km^r
   transductants from each cross. Streak for single colonies on the
   following media:

   a. EBU-Cm-Ap (30 µg/ml) agar.

   b. MacConkey-Srl-Ap (30 µg/ml).

   c. NA-XGal-Ap (30 µg/ml).

   d. NA-XGal-Ap (30 µg/ml) previously spread with 0.1 ml of
      10% sorbitol. Be sure to spread the sorbitol over the *entire*
      surface of the agar.

8. Streak each colony across a line of P22 H5 on an EBU plate to
   test for P22 sensitivity. Use strains TSM1 (P22^s) and TSM125
   (P22^r) as controls.

9. Incubate all plates overnight at 37°C.

10. Inspect the plates, and record the results. Diploids will be Srl$^+$
    as revealed by MacConkey-Srl agar. All Km^r diploids will be
    Lac$^+$ as revealed by XGal-Srl agar. The XGal-only (no Srl) agar
    gives the experimental result. What is the Lac phenotype if the
    reg::Tn10d(Cm) mutation is recessive? If it is dominant? Why?

To confirm further that the strains are diploids, perform segregation analysis. Culture the strains for several generations in the absence of selection (Ap) and test single colonies for loss of one of the two duplicated *srl* regions and for loss of the MudA element (Ap^r) located at the duplication join-point.

11. Inoculate two colonies from each cross into 2.5 ml of LB broth with no Ap.

12. Aerate at 37°C until saturated (overnight if necessary).

13. Subculture each with an inoculating stick into 2.5 ml of fresh LB broth with no Ap.

14. Aerate at 37°C until saturated (overnight if necessary).

15. Plate 6/0.2 dilutions (see Section 2, Microbiological Procedures) from each culture onto LB agar.

16. Incubate all plates overnight at 37°C.

17. Replica print each plate to the following media:

    a. LB-Ap (30 µg/ml) agar. (Which ones retain the duplication?)

    b. LB-Cm agar. (Which ones retain *reg*::Tn*10*d[Cm]?)

    c. LB-Km agar. (Which ones retain *srl*::MudJ?)

    d. MacConkey-Srl agar. (Which ones retain *srl*$^+$?)

18. Incubate all plates overnight at room temperature (to avoid overgrowth).

19. Inspect the plates and record the results. There should be ≥5% segregants. All Ap^s strains should be either Srl$^+$ or Km^r. Cm^r and Km^r should cosegregate, as they are tightly linked.

EXPERIMENT 4

Isolation of Conditional (Heat-sensitive and Amber) Mutations

Experiment 4 demonstrates two methods for isolating point mutations in structural genes. The first method involves generalized mutagenesis of cultures with an alkylating agent, diethylsulfate (DES). The second method (localized mutagenesis) uses hydroxylamine (HA) to mutagenize P22 HT *int* phage grown on a strain with a selectable marker linked to the *srl* operon. Both methods screen colonies on tetrazolium plates to recover mutants. Mutants are isolated at 42°C to search for heat-sensitive mutants. The remaining *srl* lesions are then transduced into an amber suppressor strain (*supD*) to determine which are amber mutations.

DES mutagenesis is widely adaptable to a variety of bacterial species. DES is an effective mutagen that acts by alkylation of DNA, causing mutations both directly by mispairing during replication and indirectly by inducing error-prone repair (SOS system). DES induces a variety of single base-pair lesions, including $G \cdot C \rightarrow A \cdot T$ transversions and $G \cdot C \rightarrow C \cdot G$ and $G \cdot C \rightarrow T \cdot A$ transitions. This protocol was adapted from Roof and Roth (1988).

Localized mutagenesis is a highly efficient procedure that is adaptable to any bacterial species for which generalized transduction is available. HA is an effective in vitro mutagen that acts by converting cytosine to N^4-hydroxycytosine, which base pairs with adenine. Thus, HA action in vitro causes $G \cdot C \rightarrow A \cdot T$ transitions.

295

This protocol was adapted from Hong and Ames (1971) and Davis et al. (1980).

Some of the mutations confer a heat-sensitive phenotype. A significant proportion of the other mutations are nonsense mutations, which are easily identified by transduction into any of a variety of suppressor-containing strains. This is illustrated by using a *supD* strain, which suppresses amber (UAG) mutations. The *supD* suppressor is strong and efficient in most codon contexts (for review, see Eggertsson and Söll 1988). Below are listed the bacterial strains and bacteriophage lysates used in this experiment.

BACTERIAL STRAINS

TSM1 (prototroph)
TSM112 (*zgc*::Tn*10*d[Cm])
TSM123 (Δ*trp*)
TSM125 (P22^r)
TSM140 (Δ[*srl-zgc*]::MudA)
TSM145 (*supD* Δ[*srl-zgc*]::MudJ)
TSM146 (*supD srl*^+ *zgc*::Tn*10*d[Cm])
TSM147 (*srl* [Am] *zgc*::Tn*10*d[Cm])
TSM148 (*srl* [Ts] *zgc*::Tn*10*d[Cm])
TSM149 (*supD srl* [Am] *zgc*::Tn*10*d[Cm])

BACTERIOPHAGE LYSATES

P22 broth prepared with a P22 HT *int* lysate of strain TSM1
P22 broth prepared with a P22 HT *int* lysate of strain TSM140
P22 HT *int* lysate of strain TSM112 (*zgc*::Tn*10*d[Cm])

Isolate DES-induced *srl* Mutations

MATERIALS

For culture media and supplements, see Appendix B.

E-Glc liquid medium
E-Glc plates
LB plates
NA-TCC-Cm-Srl plates
NCE-Srl plates
DES (diethylsulfate)
Screw-cap tubes (16 × 150 mm)
Sterile toothpicks

PROCEDURE

1. Inoculate a single colony of strain TSM112 (*zgc-7206*:: Tn*10*d[Cm]) into 5 ml of E-Glc liquid medium. Aerate overnight at 37ºC.

2. Store the culture on ice until use.

3. Subculture 0.1 ml of the TSM112 culture into 5 ml of E-Glc liquid medium in a 16 × 150-mm screw-cap tube.

4. Plate 4/.1 and 5/.1 dilutions on LB plates (see Section 2, Microbiological Procedures) (*t* = 0).

5. Add 20 µl of DES. Secure the screw-cap tightly and vortex thoroughly to dissolve the DES.

   Caution: DES is a mutagen and suspected carcinogen. *It is also volatile.* Wear gloves when handling DES-treated material and work in a chemical fume hood. Use screw-cap tubes for all DES-treated cultures and mechanical pipettors to manipulate DES solutions. Dispose of all DES-treated cultures in bleach.

6. Incubate overnight at 37ºC.

7. At approximately 15, 30, and 45 minutes after adding the DES, plate 4/.1 and 5/.1 dilutions on LB plates ($t = 15$, $t = 30$, $t = 45$).

8. At approximately 45 minutes after adding the DES, subculture 5 μl into each of five screw-cap tubes containing 5 ml of E-Glc liquid medium. Secure the screw-caps tightly. These subcultures provide five outgrowth pools from which to isolate independent mutants, thereby avoiding siblings.

9. Aerate the subcultures overnight at 37°C. Incubate all plates overnight at 37°C.

10. Inspect the LB plates and count the viable colonies. With strain TSM112, this protocol yields about 20% survival at the 45-minute time point. Other strains or species may require different lengths of time. Note that growth is progressively delayed with increasing time of exposure to the mutagen due to SOS-mediated inhibition of cell division.

11. Continue aerating the subcultures to grow until they reach saturation (24 hours).

12. Plate 5/.2 dilutions of each subculture on eight NA-TTC-Cm-Srl plates each.

13. Incubate all plates overnight at 42°C.

14. Carefully inspect each of the plates. Each should have 200–500 colonies per plate, the vast majority of which should be white. Rare dark-red colonies are presumptive *srl* mutants.

15. Estimate the total number of white and red colonies from each subculture. Record the results. Assign isolation numbers to each of the red colonies chosen for further analysis. Be sure to keep track of which colonies came from which subculture.

16. Use sterile toothpicks to pick each of the chosen red colonies, along with one or two randomly chosen white colonies. Streak for single colonies on NA-TTC-Cm-Srl plates.

17. Incubate all plates overnight at 42°C.

18. Inspect the plates and record the results.

19. Use sterile toothpicks to pick a well-isolated colony of each mutant. Streak for single colonies on two plates each of the following media:

    a. NA-TTC-Cm-Srl agar (to confirm the Srl phenotypes).

    b. NCE-Srl agar (to test for growth with sorbitol as carbon source).

    c. E-Glc agar (to test for growth on defined medium).

20. Incubate one set of plates at 30°C and the second set at 42°C both overnight.

21. Inspect the plates and record the results. Choose up to four isolates from each subculture for further analysis.

Localized Mutagenesis: Isolate Hydroxylamine-induced *srl* Mutations

MATERIALS

For culture media and supplements, see Appendix B.

Cold LBSE (LB Broth + NaCl + EDTA)
E-Glc plates
EBU plates
EBU-Cm plates
Hydroxylamine (HA) solution
LB broth
NA-TTC-Cm-EGTA-Srl plates
NA-TTC-Cm-Srl plates
NCE-Srl plates
0.85% NaCl
1 M MgSO$_4$
Phosphate-EDTA buffer
Microcentrifuge tubes
Sterile toothpicks

PROCEDURE

1. Prepare a high-titer ($>10^{11}$ pfu/ml) P22 HT *int* lysate of strain TSM112 (*zgc-7206*::Tn*10*d[Cm]) (see Appendix D, Generalized Transduction with Phage P22).

2. Add the following solutions to two separate microcentrifuge tubes (see Appendix D, Hydroxylamine Mutagenesis of Plasmid DNA):

Addition	Mutagenesis	Control
Phosphate-EDTA buffer	0.2 ml	0.2 ml
Sterile H_2O	0.3 ml	0.7 ml
Hydroxylamine	0.4 ml	–
1 M $MgSO_4$	10 μl	10 μl

Caution: Hydroxylamine is a suspected carcinogen. Wear gloves and *do not* mouth pipette. Dispose of all waste containing HA in appropriate biohazard waste containers.

3. Add 0.1 ml of undiluted TSM112 lysate to each tube.

4. Dilute 10 μl of each sample into 1 ml of ice-cold LBSE in a microcentrifuge tube. Store at 4ºC ($t = 0$).

5. Incubate the mutagenesis and control tubes for 24–48 hours in a 37ºC incubator.

6. Dilute 10 μl of each sample into 1 ml of ice-cold LBSE in a microcentrifuge tube at 4–8-hour intervals. Store at 4ºC ($t = 4$, $t = 8$, etc.)

7. Inoculate a single colony of strain TSM1 into 2.5 ml of LB broth. Aerate overnight at 37ºC.

8. After taking the 16–18-hour time point, further dilute each sample in 0.85% NaCl in a microtiter plate (see Section 2, Microbiological Procedures). Spot titer on a lawn of strain TSM1 (see Appendix D, Determining Phage Titers).

9. Plot pfu/ml versus time on semi-log paper. Extrapolate to a level of 0.1–1% of the initial pfu/ml (generally ~24–36 hours). Note the number of clear plaques, as an estimate for the level of mutagenesis.

10. Continue to dilute 10 µl of each sample into 1 ml of ice-cold LBSE in a microcentrifuge tube at 4–8-hour intervals. Store at 4ºC.

11. At the time predicted to yield 0.1–1% survivors, dilute 10 µl of each sample into 1 ml of ice-cold LBSE in a microcentrifuge tube. Store at 4ºC.

12. Sediment the remainder of the mutagenized phage sample in a microcentrifuge at 14,000 rpm for 30 minutes at 4ºC.

13. Carefully decant the supernatant and overlay the pellet with 0.2 ml of cold LBSE. Store overnight at 4ºC. Occasionally swirl gently; do not vortex.

14. Inoculate a single colony of strain TSM1 into 2.5 ml of LB broth. Aerate overnight at 37ºC.

15. Gently vortex the mutagenized phage suspension. Titer the resuspended phage and the remaining untitered aliquots on a lawn of strain TSM1. Note the proportion of clear plaques.

16. Use 10 µl of the mutagenized lysate for a delayed expression transduction with strain TSM1 (see Appendix D, Generalized Transduction with Phage P22). Plate 0.1 ml each on ten NA-TTC-Cm-EGTA-Srl plates.

17. Incubate all plates overnight at 42ºC.

18. Carefully inspect each of the plates. Each should have 200–500 colonies per plate, the vast majority of which should be white. Rare dark-red colonies are presumptive *srl* mutants.

19. Estimate the total number of white and red colonies from each subculture. Record the results. Assign isolation numbers to

each of the red colonies chosen for further analysis. Assume that each mutant is of independent origin, as long as the delayed expression in step 16 did not extend for more than 30 minutes.

20. Estimate the total drop in titer over the course of the experiment and plot pfu/ml versus time on semi-log paper for future reference. The titer of the mutagenized-resuspended phage should be approximately 10^9 pfu/ml.

21. Use sterile toothpicks to pick each of the chosen red colonies, along with one or two randomly chosen white colonies. Streak for single colonies on each of the following media:

    a. EBU-Cm agar (to isolate phage-free clones).

    b. NA-TTC-Cm-EGTA-Srl agar (to ensure that the mutants are genuine).

22. Incubate the EBU plates at 37°C and the NA-TTC plates at 42°C both overnight.

23. Inspect the plates and record the results.

24. Use sterile toothpicks to pick two to four well-isolated white colonies from the EBU-Cm plate for each of the mutants chosen for further analysis. Streak for single colonies on one EBU plate, and on two plates each of the other test media:

    a. EBU agar (to confirm phage-free clones).

    b. NA-TTC-Cm-Srl agar (to confirm that the transductants are genuine).

    c. NCE-Srl agar (to test for growth with sorbitol as carbon source).

    d. E-Glc agar (to test for growth on defined medium).

25. Streak each colony across a line of P22 H5 on an EBU plate to test for P22 sensitivity. Use strains TSM1 (P22^s) and TSM125 (P22^r) as controls.

26. Incubate the EBU plates overnight at 37°C. Incubate one set of test plates overnight at 30°C and the second set overnight at 42°C.

27. Inspect the plates and record the results. Choose up to four isolates from each subculture for further analysis.

Backcross and Characterize Heat-sensitive Mutations

MATERIALS

For culture media and supplements, see Appendix B.

EBU plates
EBU-Cm plates
E-Glc plates
LB broth
LB-Cm-EGTA plates
NA-TTC-Cm-EGTA-Srl plates
NA-TTC-Cm-Srl plates
NCE-Srl plates
Sterile toothpicks

PROCEDURE

1. Prepare P22 HT *int* lysates on each *srl*(Ts) strain to be tested (see Appendix D, Generalized Transduction with Phage P22).

2. Use 10 μl of each lysate for delayed expression transductions with strain TSM1 (see Appendix D, Generalized Transduction with Phage P22). Plate on LB-Cm-EGTA plates.

3. Incubate all plates overnight at 37°C.

4. Inspect the plates and record the results.

5. Estimate the titers of the P22 HT *int* lysates both as pfu/ml and as *trp*+ transductants/ml and calculate the approximate frequency of transduction for each cross.

6. Use sterile toothpicks to pick eight Cm^r transductants from each cross. Streak for single colonies on the following media:

   a. EBU-Cm agar (to isolate phage-free clones)

   b. NA-TTC-Cm-EGTA-Srl agar (to identify Srl$^-$ transductants).

7. Incubate the EBU-Cm plates overnight at 37°C and the NA-TTC-Cm-EDTA-Srl plates overnight at 42°C.

8. Inspect the plates and record the results.

9. Use sterile toothpicks to pick two to four well-isolated white colonies from the EBU-Cm plate for at least two Srl$^-$ transductants from each donor. Streak for single colonies on one EBU plate and on three plates each of the other test media:

   a. EBU agar (to confirm phage-free clones).

   b. NA-TTC-Cm-Srl agar (to confirm that the transductants are genuine).

   c. NCE-Srl agar (to test for growth with sorbitol as carbon source).

   d. E-Glc agar (to test for growth on defined medium).

10. Streak each colony across a line of P22 H5 on an EBU plate to test for P22 sensitivity. Use strains TSM1 (P22^s) and TSM125 (P22^r) as controls.

11. Incubate the EBU plates overnight at 37°C. Incubate one set of test plates overnight at 30°C, a second set overnight at 37°C, and the third set overnight at 42°C.

12. Inspect the plates and record the results. Choose one P22-sensitive Srl$^-$ (Ts) strain from each transduction for further work. Save desired strains for the short term on EBU plates at 4°C.

   Note: Some mutations confer an Srl$^-$ phenotype only at 42°C, whereas others confer an Srl$^-$ phenotype even at 37°C. The most useful temperature-sensitive mutations are those that also confer a robust Srl$^+$ phenotype at 30°C.

13. Assign collection and allele numbers to each of the desired strains and culture each in LB broth to saturation.

14. Stock each of the new *srl*(Ts) strains in the central strain collection (see Appendix C, Storing Strains).

OPTIONAL EXPERIMENT: Determine the deletion map position for each of the mutations (see Experiment 1).

Identify and Backcross Amber-suppressible Mutations

MATERIALS

For culture media and supplements, see Appendix B.

E-Glc plates
EBU plates
EBU-Cm plates
LB broth
LB-Km plates
NA-TTC-Cm-EGTA-Srl plates
NA-TTC-Cm-Srl plates
NCE-Srl plates
Sterile toothpicks

PROCEDURE

1. Prepare P22 HT *int* lysates on each *srl* mutant to be tested (see Appendix D, Generalized Transduction with Phage P22).

2. Use 10 μl of each lysate for delayed expression transductions with strains TSM1 and TSM145 (*supD Δsrl*) (see Appendix D, Generalized Transduction with Phage P22). Plate on NA-TTC-Cm-EGTA-Srl plates.

3. Incubate all plates overnight at 37ºC.

4. Inspect the plates and record the results.

   • The transductions with strain TSM1 should yield about 50% each of red (Srl⁻) and white (Srl⁺) colonies. These are the backcrossed strains.

   • The transductions with strain TSM145 will yield a significant number of white (Srl⁺ phenotype; i.e., suppressed) colonies if the donor *srl* allele is an amber mutation. A small number of white colonies might signify that the site of the donor mutation is outside of the Δ*srl* deletion in strain TSM145. In this case, the white colonies represent transductants that have repaired the recipient deletion without incorporating the donor allele. These mutations, and other mutations that do not appear to be amber suppressible, are still worth backcrossing and saving, as they might find use for other purposes (e.g., deletion mapping and complementation analysis).

5. Streak two red colonies from each TSM1 transduction onto each of the following media. Also streak two white colonies from each TSM145 transduction as appropriate:

   a. EBU-Cm agar (to isolate phage-free clones).

   b. NA-TTC-Cm-EGTA-Srl agar (to ensure that the transductants are genuine).

6. Incubate all plates overnight at 37°C.

7. Inspect the plates and record the results.

8. Use sterile toothpicks to pick two to four well-isolated white colonies from the EBU-Cm plates for each of the TSM1 backcross strains. Streak for single colonies on the following media:

   a. EBU agar (to confirm phage-free clones).

   b. NA-TTC-Cm-Srl agar (to confirm that the transductants are genuine).

   c. NCE-Srl agar (to test for growth with sorbitol as carbon source).

   d. E-Glc agar (to test for growth on defined medium).

9. Use sterile toothpicks to pick one well-isolated white colony from the EBU-Cm plates for each of the TSM145 backcross strains. Use strains TSM145 (*supD Δsrl*), TSM146 (*supD srl+*), and TSM149 (*supD srl*[Am]) as controls. Streak for single colonies on the following media:

   a. EBU agar (to confirm phage-free clones).

   b. NA-TTC-Cm-Srl agar (to confirm that the transductants are genuine).

   c. NCE-Srl agar (to test for growth with sorbitol as carbon source).

   d. E-Glc agar (to test for growth on defined medium).

   e. LB-Km agar (to identify strains that inherited the donor *srl* locus).

10. Streak each colony across a line of P22 H5 on an EBU plate to test for P22-sensitivity. Use strains TSM1 (P22^s) and TSM125 (P22^r) as controls.

11. Incubate all plates overnight at 37°C.

12. Inspect the plates and record the results. Choose one P22-sensitive Srl⁻ strain from each TSM1 transduction for further work. Save desired strains for the short term on EBU plates at 4°C.

    Note: The TSM145 transductions were done primarily to identify those mutations that are amber-suppressible. However, it is worth saving one P22-sensitive Srl⁺ strain from each TSM145 transduction for the culture collection.

13. Assign collection and allele numbers to each of the desired strains and culture each in LB broth to saturation.

14. Stock each of the new *srl* strains in the central strain collection (see Appendix C, Storing Strains).

OPTIONAL EXPERIMENT: Determine the deletion map position for each of the mutations (see Experiment 1).

EXPERIMENT 5
In Vivo Molecular Cloning

Experiment 5 demonstrates a method for cloning a gene of interest by screening a clone bank for complementing clones. In this case, the clone bank is generated in vivo by using a bacteriophage Mu-based cloning vehicle. Clones carrying srl^+ are identified on indicator media after transducing srl::Tn10 mutants with the clone bank. Chromosomal fragments from chosen clones are then subcloned for further analysis. This in vivo cloning method was developed by Groisman and Casadaban (1986).

Once Srl$^+$ colonies have been isolated, plasmid DNA must be isolated from each colony. This general procedure is known as a "miniprep" or a "quickscreen." The procedure used here was developed by Birnboim and Doly (1979) and modified by Pelham (1985).

Candidate clones are transformed into the srl recipient strains to confirm that the plasmids actually contain srl^+ DNA. The rapid transformation procedure was developed by Chung et al. (1989).

The next steps involve restriction endonuclease digestion and agarose gel electrophoresis of the plasmid DNA samples to determine the size and general restriction map of the insert DNA.

The final procedure involves subcloning fragments from the large srl^+ clones into a general-purpose subcloning vector. This is demonstrated with plasmids pSU18 or pSU19 (Bartolomé et al. 1991), which are moderate-copy number vectors based on pACYC-184. The pSU vectors contain the polylinkers from plasmids pUC18 and pUC19, respectively. Insertional inactivation of the $lacZ$ region,

309

into which the polylinker is embedded, results in white trans-formant colonies on XGal agar (α-complementation; Ullmann 1992). Below are listed the bacterial strains used in this experiment.

BACTERIAL STRAINS

TSM114 (*galE*)
TSM116 (Δ*srl*::MudA) (30°C)
TSM117 (Mu *cts* hP1/pEG5005) (30°C)
TSM192 (Δ[*lacZ*]*M15 srl*::Tn*10*)
TSM193 (pSU18)
TSM194 (pSU19)

Construct Clone Bank

Because Mu lysates are very unstable, it is preferable to make the lysate on the same day it is to be used.

MATERIALS

For culture media and supplements, see Appendix B.

LB broth
LB broth-EGTA (10 mM)
LB broth-Km
LB plates
MacConkey-Km-Srl plates
MacConkey-Km-EGTA-Srl plates
TBSA (Tryptone Broth–Soft Agar)
0.5 M CaCl$_2$
1 M MgSO$_4$
Flasks (125 or 250 ml)
Syringe and filters
Sterile screw-cap tubes
Microcentrifuge tubes
Sterile toothpicks

PROCEDURE

1. Inoculate a single colony of strain TSM117 (pEG5005) into 2.5 ml of LB broth-Km. Aerate overnight at 30°C.

2. Inoculate single colonies of strains TSM114 (*galE*) and TSM116 (*galE* Δ*srl*::MudA/Mu *c*ts hP1) each into 2.5 ml of LB broth. Aerate overnight at 30°C.

3. Store the cultures on ice until use.

4. Inoculate 0.1 ml of TSM117 culture into 20 ml of LB broth-Km in a 125- or 250-ml flask.

5. Incubate with shaking at 30°C until the culture reaches the mid exponential phase (~4–5 hours).

6. Shift the TSM117 culture to 42°C with vigorous shaking. Incubate for 2–2.5 hours. The culture will not clear, but phage particles are released nonetheless.

7. Chill the lysate on ice for a few minutes.

8. Sediment 12 1.5-ml aliquots in a microcentrifuge at 14,000 rpm for 2 minutes.

9. Pool and filter the supernatants into a sterile screw-cap tube.

10. Add $MgSO_4$ to 2 mM and $CaCl_2$ to 0.2 mM. For every 10 ml of lysate, add 20 µl of 1 M $MgSO_4$ and 4 µl of 0.5 M $CaCl_2$. Store the lysate on ice.

11. Mix 2.5 ml of TBSA and 25 µl of 0.5 M $CaCl_2$, add 0.05 ml of TSM114 culture, and pour onto an LB plate.

12. Spot on 0/.02, 1/.02, 2/.02, 3/.02, and 4/.02 dilutions of the lysate (see Section 2, Microbiological Procedures) and allow the spots to dry.

13. Incubate this titer plate overnight at 37°C.

14. Pipette 0.1 ml of TSM116 culture into each of 11 microcentrifuge tubes.

15. Add 0.1 ml of undiluted lysate as shown below:

Tube	Cells (ml)	Phage (ml)	Comments
A–J	0.1	0.1	pools A–J
K	0.1	0	cells-only control
L	0	0.1	phage-only control

16. Gently mix the contents of each tube and incubate for approximately 20–30 minutes at 30°C to allow the phage to adsorb to the cells.

17. Add 1.0 ml of LB broth-EGTA and gently mix. EGTA chelates Ca^{++}, thereby inhibiting further infection.

18. Incubate for 45–60 minutes at 30°C to allow time for expression of Km^r.

19. Sediment the cells in a microcentrifuge at 14,000 rpm for 1 minute.

20. Resuspend each pellet in 0.2 ml of LB broth-EGTA.

21. Plate tubes A–L each on a MacConkey-Km-Srl plate (12 plates total).

22. Incubate all plates overnight at 30°C.

23. Carefully inspect each of the plates. Since these strains grow slowly, allow 36–48 hours of incubation at 30°C before picking colonies. No colonies should be seen on the two control plates, whereas the experimental plates should have approximately 100 colonies per plate, the vast majority of which will be white. Rare dark-red colonies are presumptive Srl^+ clones.

24. Estimate the total number of red and white colonies in each pool. Calculate the approximate frequency of Srl+ clones/pfu. Assign isolation numbers to each of the red colonies chosen for further analysis. Be sure to keep track of which colonies came from which pools.

25. Use sterile toothpicks to pick each of the chosen red colonies, along with one or two randomly chosen white colonies. Streak for single colonies on MacConkey-Km-EGTA-Srl plates.

26. Incubate all plates overnight at 30°C.

Isolate Plasmid DNA and Confirm Srl+ Phenotype

MATERIALS

For culture media and supplements, see Appendix B.

LB broth-Km
MacConkey-Km-Srl plates
Plasmid broth-Cm
Plasmid broth-Km
Sterile toothpicks

PROCEDURE

1. Inoculate single colonies each of up to eight isolates into 2 ml of Plasmid broth-Km. Aerate overnight at 30°C.

2. Inoculate a single colony of strain TSM117 (pEG5005) into 2 ml of Plasmid broth-Km. Aerate overnight at 30°C.

3. Inoculate a single colony of strain TSM193 (pSU18) or TSM194 (pSU19) into 2 ml of Plasmid broth-Cm. Aerate overnight at 37°C. The pSU plasmids will be used for subcloning (see below).

4. Extract plasmid DNA (see Appendix F, Plasmid DNA Minipreps).

5. Transform 5 µl of each plasmid into strain TSM116 (Δ*srl::* MudA) cultured at 30°C (see Appendix D, Preparation and Transformation of Competent Cells). Use plasmid pEG5005 as a control.

6. Plate on MacConkey-Km-Srl plates.

7. Incubate all plates overnight at 30°C.

8. Inspect the plates and record the results. The pEG5005 transformant colonies should be white (Srl⁻). The candidate plasmid transformant colonies will be red (Srl⁺) if the candidate plasmid does indeed carry the *srl⁺* locus.

9. Use sterile toothpicks to pick two well-isolated red colonies for each of the Srl⁺ plasmid transformants. Streak for single colonies on MacConkey-Km-Srl plates.

10. Incubate all plates overnight at 30°C.

11. Assign collection and plasmid numbers to each of the desired strains and culture each in 1 ml of LB broth-Km at 30°C to saturation.

12. Stock each of the new Mud5005-*srl⁺* strains in the central strain collection (see Appendix C, Storing Strains).

Restriction Endonuclease Digestion and Agarose Gel Electrophoresis

Most suppliers of restriction endonucleases also provide the appropriate 10x buffer with each enzyme. The buffer designations used here are for the NE Buffers supplied by New England Biolabs. Appropriate buffers from other manufacturers may also be substituted as needed. Consult the supplier's catalog for information on reaction temperature and buffer composition for each particular enzyme. Alternatively, use general-purpose buffers (see Appendix F, Restriction Endonuclease Buffers).

MATERIALS

For solutions and buffers, see Appendix F.

10x Loading dye solution (with RNase)
10x Restriction buffer
Agarose
*Hind*III
*Pst*I
TBE buffer (Tris-Borate-EDTA)
Gel electrophoresis apparatus
Glass microcapillary tubes
Water bath or heat block
Photography equipment

PROCEDURE

1. Mix together DNA, sterile H_2O, and 10x restriction buffer in a microcentrifuge tube such that the DNA concentration is approximately 100 µg/ml or less. A typical reaction contains 1 µl of 10x buffer, 5–8 µl of DNA sample, and 1–4 µl of H_2O, for a final volume of 10 µl. Set up a series of digests adjusting for the concentration of DNA in each miniprep:

Tube	Sample	Amount (µl)	Enzyme	Amount (µl)	Buffer No.	Amount (µl)	H_2O (µl)
1	pEG5005	3	*Hind*III	0.5	2	1	6
2	1	5	*Hind*III	0.5	2	1	4
3	2	8	*Hind*III	0.5	2	1	1
4	pEG5005	3	*Pst*I	0.5	3	1	6
5	1	5	*Pst*I	0.5	3	1	4
6	2	8	*Pst*I	0.5	3	1	1

For illustrative purposes (see chart above), three different digests are shown, for minipreps of high, average, and low DNA concentration. With these plasmids, 5 µl of DNA is generally a good starting point for the first gel; this amount can then be adjusted as necessary for subsequent gels.

2. Keep the tubes on ice.

3. Add an appropriate amount of restriction enzyme, usually 2–5 units in a volume of 0.5 μl. Use a glass microcapillary tube to dispense the enzyme.

4. Mix the reaction cocktails by gently tapping the bottom of the tubes. Place the tubes in a water bath of appropriate temperature.

5. Incubate the reactions for 60–90 minutes at the appropriate temperature.

6. While the digests are incubating, set up an agarose gel (see Appendix F, Agarose Gels).

7. Add 1 μl of 10x Loading dye solution (with RNase) per 10 μl of DNA sample to be loaded. Incubate for 5 minutes at 37ºC.

8. Run the samples on the gel. Photograph the gel and analyze the results.

9. Continue with additional restriction digests as necessary to identify insert subfragments of suitable size for subcloning. For example, the srlABDMR cluster contains five genes, so at least 5 kb of DNA is necessary to contain the complete cluster. Thus, initial subcloning experiments would focus on insert subfragments of 8–12 kb. Once subcloned, these relatively large inserts can then be subjected to detailed restriction mapping and deletion subcloning, to delimit the srl locus to a smaller region (see below).

Subclone: Restriction, Ligation, and Transformation

MATERIALS

For culture media and supplements, see Appendix B and for solutions and buffers, see Appendix F.

LB broth
LB broth-Cm
LB broth-Km
MacConkey-Cm-Srl plates
Na-Cm-XGal-IPTG plates
2x TSS (Tranformation and Storage Solution; see Appendix D)
10x Ligation buffer
10x Loading dye solution (with RNase)
10x Restriction buffer
TE buffer (Tris-EDTA)
*Hind*III
T4 DNA ligase and buffer

Note: Commercial T4 DNA ligase is generally supplied at very high concentration (400 units/μl or more). To prevent waste, use diluted ligase (10–20 units/μl in ligase dilution buffer).

Agarose
Glass microcapillary tubes
Microcentrifuge tubes
Water bath or heat block
Sterile toothpicks
Photography equipment

PROCEDURE

1. Choose an enzyme for subcloning, based on the restriction patterns observed for the original large *srl*+ clones analyzed above and on compatibility with the polylinker sites in the pSU plasmids (see Appendix G, Plasmid and Transposon Restriction Maps). For this example, we use the enzyme *Hind*III.

2. Mix together DNA, sterile H_2O, and 10x restriction buffer in a microcentrifuge tube such that the DNA concentration is approximately 100 μg/ml or less. Keep the tube on ice. A typical

reaction contains 1 μl of 10X buffer, 5–8 μl of DNA sample, and 1–4 μl of H_2O, for a final volume of 10 μl. Set up a series of digests adjusting for the concentration of DNA in each miniprep:

Tube	Sample	Amount (μl)	Enzyme	Amount (μl)	Buffer No.	Amount (μl)	H_2O (μl)
A	pSU	20	HindIII	0.5	2	2.5	2
B	pEG-srl+	20	HindIII	0.5	2	2.5	2

3. Add an appropriate amount of restriction enzyme, usually 2–5 units in a volume of 0.5 μl. Use a glass microcapillary tube to dispense the enzyme.

4. Mix the reaction cocktail by gently tapping the bottom of the tubes. Place the tubes in a water bath of appropriate temperature.

5. Incubate the reactions for 60–90 minutes at the appropriate temperature.

6. While the digests are incubating, set up an agarose gel (see Appendix F, Agarose Gels).

7. Mix 5 μl from each reaction with 5 μl of TE each in a clean microcentrifuge tube and then add 1 μl of 10X Loading dye solution (with RNase). Incubate for 5 minutes at 37°C. Store the remaining samples at 4°C.

8. Run the samples on the gel. Photograph the gel and analyze the results.

9. If the digests are complete, mix together DNA, sterile H_2O, and 10X ligation buffer in a microcentrifuge tube. Add T4 DNA ligase as appropriate. Samples A and B are the HindIII-digested plasmids from step 2 above. Sample pSU is undigested pSU18 or pSU19. Tubes 1–3 are controls for transformation, restriction, and ligation, respectively. Tubes 4–6

represent the actual subcloning experiment. Save the remaining DNA from tubes A and B at 4°C.

Tube	Sample	Amount (μl)	Sample	Amount (μl)	Ligase (μl)	Buffer (μl)	H₂O (μl)
1	pSU	1.2	–	–	–	2	16.8
2	A	1.5	–	–	–	2	16.5
3	A	1.5	–	–	1	2	15.5
4	A	1.5	B	2	1	2	13.5
5	A	1.5	B	4	1	2	11.5
6	A	1.5	B	8	1	2	7.5

10. Incubate the reactions overnight at 15°C.

11. Inoculate single colonies of strain TSM192 (Δ(*lacZ*)M15 *srl*::Tn*10*) into two tubes each with 4 ml of LB broth.

12. Aerate at 37°C until the cultures reach the early exponential phase.

13. Aliquot 1 ml of culture into each of seven microcentrifuge tubes.

14. Prepare competent cells for transformation (see Appendix D, Preparation and Transformation of Competent Cells).

15. Add 10 μl of DNA from steps 9 and 10 to each of six tubes as shown. Save the remainder at –20°C.

Tube	Cells (ml)	DNA	Comments
A	0.1	ligation 1	transformation control
B	0.1	ligation 2	restriction control
C	0.1	ligation 3	ligation control
D	0.1	ligation 4	subcloning
E	0.1	ligation 5	subcloning
F	0.1	ligation 6	subcloning
G	0.1	none	cells-only control

16. Continue with the transformation protocol.

17. Plate on NA-Cm-XGal-IPTG plates.

18. Incubate all plates overnight at 37°C.

19. Inspect the plates and record the results. Plate 1 should have about 100–200 colonies, all of which should be blue (Lac+). Plate 2 should have substantially fewer colonies. Plate 3 should have at least one half as many colonies as Plate 1. A few colonies on plates 2 and 3 might be white (why?). Plates 4–6 should have varying proportions of white (Lac−) and blue colonies. Plate 7 should have no colonies. White colonies from plates 4–6 represent candidate subclones.

20. Use sterile toothpicks to patch up to 48 white colonies from plates 4–6 on the following media. Patch one blue colony from plate 1 (Srl− Lac+) and one colony of strain TSM193 or TSM194 (Srl+ Lac+) as controls.

    a. NA-Cm-XGal-IPTG (to confirm Lac− phenotype).

    b. MacConkey-Cm-Srl (to identify Srl+ subclones).

21. Incubate all plates overnight at 37°C.

22. Inspect the plates and record the results. If the right restriction enzyme was found serendipitously, there should be some Srl+ colonies. Even if all of the colonies are Srl−, however, continue with the analysis to detect any inserts.

23. Use sterile toothpicks to streak up to six Srl+ (or Srl−, or four Srl+ and two Srl−) isolates on the following media. Also streak the Srl− Lac+ and Srl+ Lac+ controls.

    a. NA-Cm-XGal-IPTG (to reconfirm Lac− phenotype).

    b. MacConkey-Cm-Srl (to confirm Srl+ subclones).

24. Incubate all plates overnight at 37°C.

25. Inoculate the strain carrying the pEG5005-*srl*+plasmid used for the subcloning into 2 ml of LB broth-Km. Aerate overnight at 30ºC. Inspect the plates and record the results. Store the culture of the pEG5005-*srl*+strain on ice until use.

26. Inoculate single colonies of up to six isolates in 2 ml of LB broth-Cm. Aerate to saturation (overnight if necessary) at 37ºC. Store the cultures on ice until use.

27. Extract plasmid DNA from each of the cultures (see Appendix F, Plasmid DNA Minipreps).

28. Use the restriction enzyme used for subcloning to digest each of the six candidate subclones, the original pEG5005-*srl*+ large clone, and the pSU18 or pSU19 vector.

29. Run the digested samples on an agarose minigel.

30. Photograph the gel and analyze the results. Determine whether the subclones all contain the same fragment from the original pEG5005-*srl*+ large clone.

OPTIONAL EXPERIMENT: Further optional experiments involve constructing a detailed restriction map of the subcloned fragment. Restriction mapping is an exercise in logic and organization. It is best to start with a generous supply of high-quality DNA (one can never have too much DNA) in order to run the reactions to completion and to have enough material to run several gels. Once basic patterns of single digests have been obtained, it is necessary to run a series of double and triple digests to resolve the final map.

One potential problem with interpreting multiple digests is the patterns that result from incomplete digestion. To circumvent this problem, restrict a relatively large amount of DNA (ten or more gels' worth) with a given enzyme, run an aliquot to ensure that the digestion has gone to completion, and save the remainder for restriction with other enzymes. Thus, it will be known that at least the first reaction has gone to completion. Aliquots of this sample can then be subjected to further digestion with a battery of enzymes.

Devise and execute a restriction digest strategy to determine if subclones with the fragment of interest in both of the possible orientations have been obtained (this is one reason for choosing more than one candidate subclone for analysis). Having both orientations available is quite useful for subsequent restriction mapping and for deletion subcloning. The polylinker sequences in the vector plasmids are invaluable in this regard. For example, the cohesive ends released by restriction with the enzymes *Bam*HI (G↓GATCC) and *Bgl*II (A↓GATCT) are compatible. If the insert contains a unique *Bgl*II site, then a double *Bam*HI-*Bgl*II digestion followed by self-ligation will result in a subclone with the intervening fragment deleted. Having the insert in both orientations with respect to the polylinker sequences makes it easy to delete the DNA on both sides of the *Bgl*II site.

Physical Mapping of Bacterial Chromosomes by Pulsed-field Gel Electrophoresis

Experiment 6 demonstrates the use of pulsed-field gel electrophoresis (PFGE) for physical mapping of the bacterial chromosome. We will physically map the Tn*10* insertions that were isolated in Experiment 2. All of the steps are performed on cells embedded in agarose blocks to prevent shearing of the 4800-kb chromosomal DNA. The chromosome is digested with restriction enzymes that cut at rare sites, and the resulting DNA fragments are separated using TAFE. Two restriction enzymes are used: *Xba*I, which cuts the *Salmonella typhimurium* chromosome into 24 fragments, and *Bln*I, which cuts the *S. typhimurium* chromosome into 12 fragments (Liu et al. 1993a,b; Sanderson et al. 1995). Tn*10*d(Tc) contains one *Xba*I site and one *Bln*I site, and thus the position of a Tn*10*d(Tc) insertion on the *S. typhimurium* chromosome can be determined by comparing the size of the resulting DNA fragments from cells with the Tn*10*d(Tc) insertion with the size of the fragments from cells without the Tn*10*d(Tc) insertion.

For an example of a pulsed-field gel electrophoresis mapping of insertion mutants on the *Salmonella typhi* chromosome, see Figure 6.1, and for the sizes of DNA fragments obtained from *S. typhimurium* L12 after *Xba*I and *Bln*I digestion, see Table 6.1, both at the end of this procedure.

This experiment uses the GeneLine® System (Beckman) which separates DNA molecules by TAFE. Alternatively, a variety of other commercially available or homemade pulsed-field gel electrophoresis set-ups could be used (see, e.g., Birren and Lai 1994).

BACTERIAL STRAINS

TSM1 (prototroph)

Tn*10*d(Tc) insertion mutants

MATERIALS

LB broth (see Appendix B)

Sterile deionized H_2O

1.2% SeaKem agarose (FMC Corp.) in deionized H_2O

1.0% SeaKem agarose in 1x TAFE running buffer

1 mM Phenylmethylsulfonyl fluoride (PMSF; dissolved in isopropanol)

70% Agarose

Ethidium bromide

Proteinase K

100 mg/ml Bovine serum albumin (BSA)

*Xba*I and buffer

*Bln*I and buffer

TAFE Solution 1: 10 mM Tris-HCl (pH 7.2), 100 mM EDTA, 20 mM NaCl

TAFE Solution 2: 10 mM Tris-HCl (pH 7.2), 100 mM EDTA, 50 mM NaCl, 0.2% SDS, 0.15% N-lauryl sarcosine

TAFE Solution 3: 20 mM Tris-HCl (pH 8.0), 50 mM EDTA

TAFE Solution 4: 1 mg/ml proteinase K, 100 mM EDTA, 0.2% SDS, 1% N-lauryl sarcosine

TAFE Solution 5: 10 mM Tris-HCl (pH 7.2), 0.1 mM EDTA

20x TAFE running buffer: 24.2 g of Tris, 2.9 g of EDTA (free acid), 5 ml of glacial acetic acid, brought to a total volume of 1000 ml with deionized H_2O and then autoclaved

Screw-cap vials (5 ml)

Sterile 1.5-ml microcentrifuge tubes

Well plugs (Bio-Rad 1703706)

Spatula (ethanol-flamed)

Scapel (ethanol-flamed)

PFGE apparatus and power supply

UV Transilluminator

Polaroid camera and film

PROCEDURE

1. Start fresh cultures of TSM1 and each Tn*10*d(Tc) insertion mutant in 3 ml of LB broth. Incubate with aeration overnight at 37°C.

2. Pour the cultures into microcentrifuge tubes. Centrifuge the culture in a microcentrifuge for 1 minute to pellet the cells.

   Caution: This procedure is *very* sensitive to contaminating nucleases. Wear gloves throughout the procedure.

3. Pour off the supernatant. Resuspend the pellet from each tube in 0.25 ml of TAFE Solution 1. Vortex for 30 seconds. Combine the cell suspensions in one tube for each strain.

4. Warm the cell suspensions to 70°C.

5. Boil 1.2% SeaKem agarose in deionized H_2O and hold at 70°C.

6. Add 0.5 ml of the 70°C agarose to the cell suspension and thoroughly mix by inverting the tubes.

7. Immediately add aliquots of the cell suspension–agarose mix into well plugs that have been taped on one side of the plug mold.

8. Allow the plugs to harden at room temperature for approximately 20 minutes and then use a clean, ethanol-flamed spatula to poke the plugs into 5-ml screw-capped vials.

9. Lyse the cells by adding 3 ml of TAFE Solution 2 to the plugs and gently shake for 90 minutes at 70°C.

10. Thoroughly remove the fluid from the vials with a pipette. Wash the cells by adding 3 ml of TAFE Solution 3 and gently shake for 15 minutes at room temperature.

11. Thoroughly remove the fluid from the vials with a pipette. Add 3 ml of TAFE Solution 4 (proteinase K) and gently shake for 18 hours at 42°C.

12. Thoroughly remove the fluid from the vials with a pipette. Wash the cells by adding 3 ml of TAFE Solution 3 and gently shake for 15 minutes at room temperature.

13. Add 3 ml of PMSF and gently shake for 1 hour at room temperature to inactivate any remaining proteinase K.

    Caution: PMSF is extremely destructive to the mucous membranes of the respiratory tract, the eyes, and the skin. It is a highly toxic cholinesterase inhibitor. It may be fatal if inhaled, swallowed, or absorbed through the skin. Wear gloves and safety glasses and work in a chemical fume hood. In case of contact, immediately flush eyes or skin with copious amounts of water. Discard contaminated clothing.

14. Thoroughly remove the fluid from the vials with a pipette. Wash the cells by adding 3 ml of TAFE Solution 3 and gently shake for 15 minutes at room temperature.

15. Repeat the wash step as described in step 14.

16. Thoroughly remove the fluid from the vials with a pipette. Add 3 ml of TAFE Solution 3 diluted 1/10 with sterile deionized H_2O and gently shake for 15 minutes at room temperature.

17. Thoroughly remove the fluid from the vials with a pipette. Add 3 ml of TAFE Solution 3 diluted 1/10 with sterile deionized H_2O. Store the plugs at 4ºC until use.

18. Use a clean, ethanol-flamed spatula to remove each plug and place in a separate sterile vial.

19. Add 1.5 ml of TAFE Solution 5 and incubate with gentle shaking for 30 minutes at room temperature.

20. Thoroughly remove the fluid from the vials with a pipette. Add 0.4 ml of TAFE Solution 5 and incubate with gentle shaking for 30 minutes at room temperature.

21. Repeat step 20.

22. Thoroughly remove the fluid from the vials with a pipette. Add 1.5 ml of 1x restriction enzyme buffer and incubate with gentle shaking for 30 minutes at room temperature.

23. Thoroughly remove the fluid from the vials with a pipette. Add the following:

10x Restriction enzyme buffer (*Xba*I or *Bln*I)	40 μl
BSA (100 mg/ml)	2 μl
Restriction enzyme (*Xba*I or *Bln*I)	4 μl
Sterile deionized H_2O	354 μl

24. Place on ice for 30 minutes to allow the solution to equilibrate with the plug and then incubate with gentle shaking overnight at 37ºC.

25. Thoroughly remove the fluid from the vials with a pipette. Add 1.5 ml of TAFE Solution 5 and incubate with gentle shaking for 1 hour at room temperature.

26. Shake the plug and solution into a sterile, empty petri dish. Use an ethanol-flamed scalpel to cut each plug into three pieces.

27. Prepare a 1% SeaKem agarose gel in 1x TAFE running buffer. Save 2–3 ml of the agarose at 70ºC for sealing the wells.

28. Use an ethanol-flamed spatula to load each plug into wells of a gel so that the plug is against the side of well that will face the resolving portion of the gel. Also include prepared DNA size standards.

29. Seal the wells with 1% SeaKem agarose in 1x TAFE running buffer that has been melted and held at 70ºC.

30. Carefully load the gel into the electrophoresis chamber and allow the gel to equilibrate with the running buffer for 30 minutes.

31. Run the gel at 12–14ºC as follows:

Stage	Total time	Constant current	Terminal switch time
1	30 minutes	170 mA	4 seconds
2	30 hours	150 mA	*

*The switch time used depends on the size of the DNA fragments to be separated. A 40-second switch time should be adequate to separate the large DNA fragments expected from the *srl* region of *S. typhimurium*.

32. Following electrophoresis, remove gel, stain with ethidium bromide, and then visualize the DNA bands on a UV transilluminator (see Appendix F).

Caution: Ethidium bromide is a powerful mutagen and moderately toxic. Always wear gloves when working with solutions that contain this dye. After use, decontaminate and dispose of these solutions in accordance with the safety practices established by your institution's Safety Office.

UV is a mutagen and carcinogen, as well as an eye-damaging agent. Always wear safety glasses and gloves and work in a glass-enclosed hood if possible.

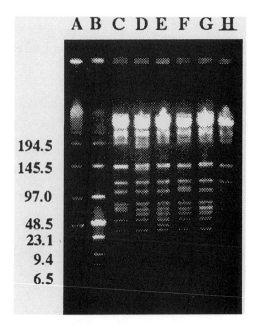

A B C D E F G H

194.5
145.5
97.0
48.5
23.1
9.4
6.5

Figure 6.1 TAFE mapping of Tn*10* insertion mutations in the *Salmonella typhi* chromosome. (Lanes *A* and *B*) DNA size standards; (Lane *A*) concatemers of λ DNA; (lane *B*) mixture of λ concatemers and λ *Hind*III fragments. Size standards in kilobases are indicated at the left side of the gel. (Lanes *C–G*) Different Tn*10* insertion mutants; (lane *H*) wild-type *S. typhi* Ty2. The chromosomal DNA in lanes *C–H* was digested with *Xba*I, and the DNA fragments were separated on a 1% SeaKem agarose gel in 0.5% TAFE buffer at 12–14ºC. The gel was run at 170 mA with 4 seconds switch times for 30 minutes and then at 150 mA with 8 seconds switch times for 18 hours.

Table 6.1 Sizes and Map Positions of DNA Fragments Obtained by Digestion of *S. typhimurium* Chromosomal DNA with *Xba*I and *Bln*I

Restriction enzyme (RE)	Size (kb)	Map position	
		centisome	minute
*Xba*I	708	98.6–13.0	98–12
	800	13.0–29.7	13–25
	49	29.7–30.7	
	243	30.7–35.8	26–30
	457	35.8–45.3	30–42
	224	45.3–50.0	44–46
	104	50.0–52.1	46–47
	225	52.1–56.8	49–54
	35	56.8–57.5	55
	20	57.5–57.9	
	76	57.9–59.5	55–56
	6.6	59.5–59.6	
	<1	59.6–59.7	
	675	59.7–73.8	57–68
	233	73.8–78.6	72–76
	72	78.6–80.1	78
	70	80.1–81.6	80–82
	48	81.6–82.6	82
	65	82.6–84.0	
	32	84.0–84.6	
	18	84.6–85.0	
	6.4	85.0–85.1	
	275	85.1–91.0	83–89
	365	91.0–98.6	91–97
	90		
*Bln*I	1580	91.0–23.9	91–23
	790	23.9–40.3	23–43
	830	40.3–57.5	43–58
	180	57.5–61.3	58–62
	590	61.3–73.7	62–73
	4.1	73.7–73.8	
	1.8	73.8–73.8	
	543	73.8–85.1	75–84
	1.8	85.1–85.1	
	90	85.1–87.0	85–87
	145	87.0–90.0	87–90
	46	90.0–91.0	90–91

Data from Liu et al. (1993a).

Introduction to Experiments 7–11

The genetic approach for characterizing DNA-protein interactions that regulate gene expression requires the isolation of mutants that alter the regulation. Challenge phage provide a powerful approach for isolating regulatory mutations by placing the expression of the phage P22 *ant* gene under the control of a specific DNA-protein interaction. Under appropriate conditions, expression of the *ant* gene determines the lysis-lysogeny decision of P22. Therefore, placing *ant* gene expression under the control of a specific DNA-protein interaction provides very strong genetic selections for regulatory mutations in the DNA-binding protein and DNA-binding site that either increase or decrease the strength of a DNA-protein interaction in vivo.

Experiments 7-11 use challenge phage to dissect the DNA-protein interactions required for regulation by the Nac protein from *Klebsiella aerogenes*. Experiment 7 demonstrates how to construct challenge phage regulated by any specific DNA-protein interaction. Experiment 8 uses challenge phage assays to quantitate the relative affinity of a DNA-protein interaction in vivo, and Experiment 9 uses challenge phage to select rapidly for a variety of mutations in a DNA-binding site that weaken the DNA-protein interaction. Two quick and simple methods are presented in Experiment 10 to determine rapidly the DNA sequence changes in the DNA-binding site mutants. Finally, Experiment 11 demonstrates how challenge phage can be used to isolate mutations in a DNA-binding protein that identify amino acids involved in the DNA-protein interaction. These experiments therefore show how challenge phage provide multiple selections that can be used to characterize quickly the details of DNA-protein interactions.

The methods described in these experiments can be used to study a variety of important regulatory mechanisms that have critical roles in bacterial virulence. For example, challenge phage have recently been used to identify important regulatory sites for virulence genes in *Vibrio cholerae* (Pfau and Taylor 1995).

331

EXPERIMENT 7

Construction of P22 Challenge Phage

Experiment 7 demonstrates the construction of a P22 challenge phage. A DNA-binding site (O_{nac}) cloned on a multicopy plasmid is recombined onto P22 *mnt*::Km9 *arc*(Am) to construct O_{nac} challenge phage.

The O_{nac} site from the *put* operon of *Klebsiella aerogenes* was cloned into the *Sma*I site of pPY190 to produce pPC38. Since pPY190 has homology with P22 on both sides of the *Sma*I site, the plasmid can recombine with P22 and replace the P22 O_{mnt} site with O_{nac}. The homology between the plasmid and P22 is small (~1 kb), but since both the plasmid and the phage are present in multiple copies, the frequency of recombination is quite high. The recombinant O_{nac} challenge phage can be distinguished from the parental phage by their plaque phenotype on a lawn of TSM205 cells. TSM205 contains a P22 prophage that is *sieA* (allowing superinfection with the challenge phage), $c2^+$ (to repress the incoming challenge phage), and mnt^+ (to repress challenge phage that carry O_{mnt}). Two types of mutant phage form clear plaques on this strain: (1) recombinants that replaced O_{mnt} with O_{nac} and (2) mutations in O_{mnt} that prevent Mnt binding. These two classes of mutants are distinguished using both in vitro and in vivo approaches. In vitro, the two classes of mutant phage are distinguished by restriction fragment length polymorphism (RFLP) mapping. In vivo, the two classes of mutant phage are distinguished by infecting cells that express the Nac protein under control of the

P_{tac} promoter: Challenge phage with the O_{nac} substitution will lysogenize cells that were grown in IPTG to induce Nac expression, but they will not lysogenize cells that were grown without IPTG to keep Nac repressed. Since the challenge phage have a *mnt*::Km9 insertion, lysogens can be selected as kanamycin-resistant (Km^r) colonies. Below are listed the bacterial strains and bacteriophage lysate used in this experiment.

BACTERIAL STRAINS

TSM205 (*supE attP*[P22 *sieA*Tpfr])
TSM207 (*supE*/pPC36 P_{tac}-*nac lacI*^{Q} Sp^r Sm^r)
TSM210 (*hsdL* [r$^-$m$^+$] *supE*/pPC38 O_{nac}+ Tc^r)
TSM214 (*supE*)

BACTERIOPHAGE LYSATE

p22 *mnt*::Km9 *arc*(Am)

Cross Plasmids with a Cloned Nac-binding Site (O_{nac}) onto Phage P22 *mnt*::Km9 *arc*(Am)

MATERIALS

For culture media and supplements, see Appendix B.

LB-agar plates
LB broth
LB broth-Tc
LBEDO broth (LB Broth with E Salts and Dextrose) (Appendix D)
TBSA (Tryptone Broth–Soft Agar) (see Appendix D)
P22 *mnt*::Km9 *arc*(Am) phage
Chloroform
0.85% NaCl
Sterile pasteur pipettes
Sterile microtiter dishes
Sterile culture tubes
Heating block (50ºC)
Sterile toothpicks

PROCEDURE

1. Start fresh cultures of TSM205 in 2 ml of LB broth and TSM210 in 2 ml of LB broth-Tc. Incubate with aeration overnight at 37°C.

2. Mix 50 μl of P22 *mnt*::Km9 *arc*(Am) phage (~10^{10} pfu/ml) with 50 μl of TSM210. Store the rest of the phage at 4°C.

3. Let stand for 15–20 minutes at room temperature to allow phage adsorption.

4. Add 2.5 ml of LBEDO broth and aerate for approximately 3 hours at 37°C or until the culture is lysed. At the same time, start a culture with 50 μl of TSM210 in 2.5 ml of LBEDO broth as a no-phage control. Compare the tubes with phage to the no-phage control tube. The no-phage control tube should not show lysis, but cellular debris should be seen in the tubes with phage.

5. Add several drops of chloroform, vortex thoroughly, and allow the chloroform to settle. Pour the supernatant into microcentrifuge tubes avoiding the chloroform. Centrifuge in a microcentrifuge for 2 minutes to pellet the cell debris.

   Caution: Chloroform is a carcinogen and may damage the liver and kidneys. Do not mouth pipette and avoid contact with skin.

6. Transfer the supernatant to a new tube, add a few drops of chloroform, vortex, and allow the chloroform to settle. Store the lysate at 4°C.

7. Dilute the lysate 1/100 by adding 10 μl to 1 ml of LB broth. Vortex. Mix 0.2 ml of the diluted lysate with 0.1 ml of a fresh overnight culture of TSM205 (to give a multiplicity of infection [moi] of ≈10).

8. Let stand for 15–20 minutes at room temperature to allow phage adsorption.

9. Melt TBSA in a microwave. Add 3 ml of the melted top agar to a test tube and place in a 50ºC heating block.

   Caution: Avoid superheating. Always wear gloves when removing solutions from the microwave.

10. Add 3 ml of TBSA to the cells + phage, mix, and quickly pour onto a dry LB-agar plate. Allow the top agar to solidify approximately 10 minutes at room temperature and then incubate overnight at 37ºC. It should be possible to observe plaques after 4–8 hours.

11. Start a fresh culture of TSM205 in 2 ml of LB broth. Incubate with aeration overnight at 37ºC.

12. Carefully check the size and turbidity of the plaques. Pick 8–12 *large clear* plaques with sterile toothpicks and streak for isolation on an LB-agar plate overlayed with 3 ml of TBSA containing 0.1 ml of fresh TSM205. Divide the plate into quarters and streak four plaques per plate beginning from the center of the plate and streaking toward the periphery (to avoid potential cross-contamination from closely spaced plaques). Incubate overnight at 37ºC.

13. Start fresh cultures of TSM205 in 2 ml of LB broth. Incubate with aeration overnight at 37ºC.

14. Pick a large clear plaque from each streak and restreak for isolation on an LB-agar plate overlayed with 3 ml of TBSA containing 0.1 ml of fresh TSM205 (four plaques per plate). Incubate at 37ºC overnight.

15. Start a fresh culture of TSM214 in 2 ml of LB broth. Incubate with aeration overnight at 37ºC.

16. Pick a large clear plaque from each streak and restreak for isolation on an LB-agar plate overlayed with 3 ml of TBSA containing 0.1 ml of fresh TSM214 (four plaques per plate). Also streak P22 *mnt*::Km9 *arc*(Am) phage control. Incubate overnight at 37ºC.

17. Start fresh cultures of TSM214 in 2 ml of LB broth. Incubate with aeration overnight at 37°C.

PCR Amplification and RFLP Mapping from P22 Lysates

MATERIALS

PCR mix (sufficient for 50 reactions)

Sterile deionized H_2O	2.45 ml
dNTPs (0.1 ml of each 10 mM stock solution; pH 7)	0.40 ml
10x PCR amplification buffer	0.50 ml

10x PCR amplification buffer (store at room temperature)

2.5 M KCl	2.0 ml
1 M Tris HCl (pH 8)	1.0 ml
1 M $MgCl_2$	0.150 ml
1% Gelatin	1.0 ml
Sterile deionized H_2O	5.85 ml

P22 *mnt::*Km9 *arc*(Am) phage
Taq polymerase (2.5 units/µl)
O_{mnt} primer (20 pmoles/µl)
Anti-O_{mnt} primer (20 pmoles/µl)
Thermocycler

> *Note:* Some thermocyclers require overlaying the sample with mineral oil or wax to prevent the condensation of water vapor on the upper surfaces of the tube. This can be avoided by using a thermocycler that maintains a higher temperature in the upper portion of the chamber than in the lower part (e.g., the PTC-100™ Thermal Cycler with a Hot Bonnet™ sold by MJ Research).

DNA purification column
*Eco*RI and buffer
Sterile deionized H_2O
Loading dye solution (see Appendix F)
1x TBE (Tris-Borate-EDTA; see Appendix F)
3:1 Agarose
φX174 *Hae*III DNA size standards
Ethidium bromide
Sterile microcentrifuge tubes
UV transilluminator
Polaroid camera and film

The PCR process for amplifying nucleic acids is covered by patents owned by Hoffmann-LaRoche.

PROCEDURE

1. Prepare high-titer P22 lysates from five independent well-isolated, large clear plaques from the potential challenge phage and the P22 *mnt*::Km9 *arc*(Am) phage control as described in Appendix D (Phage P22 Lysates).

2. Phage lysates can be amplified directly without purifying phage DNA. Amplify the potential recombinant phage and the parental P22 *mnt*::Km9 *arc*(Am) phage control. Combine in a small microcentrifuge tube (keep on ice until use):

P22 lysate	10 μl
PCR mix	75 μl
Sterile deionized H_2O	4 μl
O_{mnt} primer (20 pmoles)	5 μl
Anti-O_{mnt} primer (20 pmoles)	5 μl
Taq DNA polymerase (2.5 units/μl)	1 μl

3. Mix by gently pipetting the solution up and down with the pipettor. Amplify in a thermocycler using the following program.

Step	Temperature	Time	Function
1	94ºC	3 minutes	denature
2	45ºC	2 minutes	anneal primers
3	72ºC	2 minutes	DNA synthesis
4	93ºC	1 minutes	denature
5	45ºC	2 minutes	anneal primers/DNA synthesis
6	cycle steps 4–5	repeat 26 times	amplify
7	93ºC	1 minutes	denature
8	45ºC	2 minutes	anneal primers
9	72ºC	5 minutes	complete primer extension
10	4ºC	hold	end

4. Store PCR amplification products at 4ºC until use. If necessary, purify the PCR products as described in Appendix F (Purification and Quantitation of PCR Products).

5. In separate tubes, digest 10 μl of each PCR product with *Eco*RI. For each PCR, add everything except the restriction enzyme to two sterile microcentrifuge tubes. Remove the restriction enzyme from the freezer just before use and immediately place it on ice, add the enzyme, and then immediately return the enzyme to the freezer:

Tube A: *Eco*RI restriction digest

Sterile deionized H_2O	7 μl
DNA	10 μl
10x Restriction enzyme buffer	2 μl
*Eco*RI	1 μl

Tube B: Undigested control

Sterile deionized H_2O	8 μl
DNA	10 μl
10x Restriction enzyme buffer	2 μl

6. Incubate for at least 2 hours at 37°C.

7. Add 4 μl of 5x Loading dye solution to each sample. Run the digested DNA and the undigested control on a 3% agarose (3:1) gel in 1x TBE at 50 mA until the bromophenol blue tracking dye migrates about three-fourths the length of the gel (bromophenol blue dye migrates at ~35 bp). Include a lane with φX174 *Hae*III DNA size standards (see Appendix F, Useful DNA Molecular Weight Standards).

8. Visualize the gel with ethidium bromide and photograph the gel.

   Caution: Ethidium bromide is a powerful mutagen and moderately toxic. Always wear gloves when working with solutions that contain this dye. After use, decontaminate and dispose of these solutions in accordance with the safety practices established by your institution's Safety Office.

   UV is a mutagen and carcinogen, as well as an eye-damaging agent. Always wear safety glasses and gloves and work in a glass-enclosed hood if possible.

9. Compare the restriction digests of the potential recombinant phage with the restriction digests of P22 *mnt*::Km9 *arc*(Am) DNA (see Fig. 7.1).

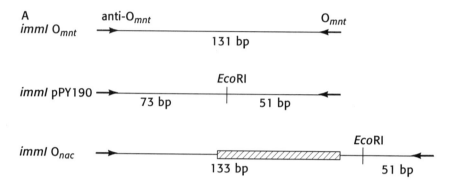

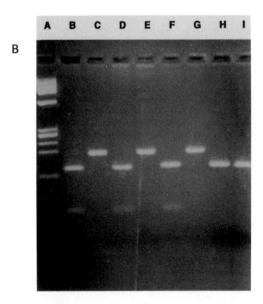

Figure 7.1 RFLP analysis of PCR fragments from O_{nac} challenge phage derivatives. (*A*) Parental phage will only yield a single PCR product that is not digested by *Eco*RI. In contrast, recombinant phage that inherited the O_{nac} site yield a longer PCR product which is digested by *Eco*RI into two fragments. Note that the PCR fragment does not include the *Fnu*4HI site within the *arc* gene. (*B*) A photograph showing a typical RFLP analysis. (Lanes *C, E, G, I*) Undigested PCR products from four different phages. (Lanes *B, D, F, H*) The corresponding *Eco*RI-digested products. The results indicate that the phages analyzed in lanes *B–G* contain the O_{nac} insertion. The phage analyzed in lanes *H–I* is the parental phage control.

RFLP Mapping of P22 DNA

The desired phage recombinants were identified in the previous protocol by RFLP mapping of a PCR fragment from the P22 *immI* region. RFLP can also be identified by restriction digests of purified phage DNA. The following procedure is somewhat more time consuming, but it can be done in 1 day without a thermal cycler.

MATERIALS

*Eco*RI and 10x restriction enzyme buffer
*Fnu*4HI and 10x restriction enzyme buffer
P22 *mnt*::Km9 *arc*(Am) phage
Sterile deionized H_2O
Loading dye solution (see Appendix F)
1x TBE (Tris-Borate-EDTA; see Appendix F)
Agarose
λ *Hind*III size standards
Ethidium bromide
UV transilluminator
Polaroid camera and film

PROCEDURE

1. Prepare a high-titer P22 lysate from each potential challenge phage and purify the phage DNA (see Appendix D, Phage P22 Lysates).

2. For each of the purified phage DNA samples, add everything except the restriction enzyme to three sterile microcentrifuge tubes. Remove the restriction enzyme from the freezer just before use, immediately place it on ice, add the enzyme, and then immediately return the enzyme to the freezer:

   ### Tube A: *Eco*RI restriction digest

Sterile deionized H_2O	6 μl
DNA	2 μl
10x *Eco*RI buffer	1 μl
*Eco*RI	1 μl

Tube B: *Fnu*4HI restriction digest

Sterile deionized H_2O	6 µl
DNA	2 µl
10x *Fnu*4HI buffer	1 µl
*Fnu*4HI	1 µl

Tube C: Undigested control

Sterile deionized H_2O	8 µl
DNA	1 µl
10x Restriction enzyme buffer	1 µl

3. Incubate for at least 2 hours at 37ºC.

4. Add 2 µl of 5x Loading dye solution to each sample.

5. Run the digested DNA and undigested controls on a 1.2% agarose gel in 1x TBE at 50 mA until the bromophenol blue band migrates about three-fourths the length of the gel. Also include a lane with λ *Hin*dIII size standards (Appendix F, Agarose Gels).

6. Stain with ethidium bromide and photograph the gel.

   Caution: Ethidium bromide is a powerful mutagen and moderately toxic. Always wear gloves when working with solutions that contain this dye. After use, decontaminate and dispose of these solutions in accordance with the safety practices established by your institution's Safety Office.
   UV is a mutagen and carcinogen, as well as an eye-damaging agent. Always wear safety glasses and gloves and work in a glass-enclosed hood if possible.

7. Compare the restriction digests of the potential P22 O_{nac} recombinants with that of the P22 *mnt*::Km9 *arc*(Am) DNA (see Fig. 7.2).

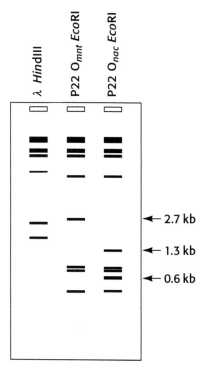

Figure 7.2 *Eco*RI digests of challenge phage derivatives. Recombinant phage that inherited the O$_{nac}$ site will have lost a 2.7-kb *Eco*RI fragment and gained two smaller fragments (1.29 and 0.62 kb).

Identifying O$_{nac}$ Challenge Phage by Selecting Km^r Lysogens In Vivo

MATERIALS

For culture media and supplements, see Appendix B.

LB-agar-Sp-Km plates
LB-agar-Sp-Km-IPTG (1 mM) plates
LB-agar plates
LB broth-Sp
LB broth-Sp + 1 mM IPTG (dioxane-free)
P22 *mnt*::Km9 *arc*(Am) phage
0.85% NaCl
Sterile microtiter dishes

PROCEDURE

1. Titer the phage lysates and the control lysate as described in Appendix D (Phage P22 Lysates).

2. Start a fresh culture of TSM207 in 2 ml of LB broth-Sp. Incubate with aeration overnight at 37°C.

3. Count the number of plaques in each spot. Each viable phage (pfu) will produce one plaque. Calculate the phage titer as follows:

$$\frac{\text{number of plaques}}{5\ \mu l \times \text{dilution factor}} \times \frac{1000\ \mu l}{ml} = \text{pfu}/\mu l$$

4. Subculture 50 μl of TSM207 into 2.5 ml of LB broth-Sp. Grow cells to early exponential phase with aeration at 37°C (~1–2 hours).

5. Subculture 0.5 ml of the early exponential phase culture of TSM207 into two separate sterile test tubes containing 1.5 ml

   LB broth-Sp
   LB broth-Sp-IPTG (1 mM)

6. Grow with aeration for 1–2 hours at 37°C.

7. Calculate the volume of each phage required to give an moi of 10–50 pfu/cell. For example, assuming the cultures are at approximately 5 x 10^8 cells per milliliter, for a multiplicity of infection of 25:

$$\frac{ml}{\text{pfu}} \times \frac{25\ \text{pfu}}{\text{cell}} \times \frac{5 \times 10^8\ \text{cells}}{ml} \times 0.1\ ml \times \frac{1000\ \mu l}{ml} = \mu l\ \text{phage}$$

   Add this volume of the potential challenge phage lysates (φ1–φ5) to columns 1 and 8 of a sterile microtiter dish as shown below. Add 0.1 ml of the TSM207 culture grown without IPTG to the first column (labeled 10^0) of the microtiter dish. Add 0.1 ml of the TSM207 culture grown with 1 mM

IPTG to column 8 (labeled 10^0) of the microtiter dish. Add 45 μl of sterile 0.85% NaCl to columns 2–5 and 9–12 of the microtiter dish.

	No IPTG added							+IPTG			
# 1	#2	#3	#4	#5	#6	#7	#8	#9	#10	#11	#12
$\phi1$	NaCl	NaCl	NaCl	NaCl			$\phi1$	NaCl	NaCl	NaCl	NaCl
$\phi2$	NaCl	NaCl	NaCl	NaCl			$\phi2$	NaCl	NaCl	NaCl	NaCl
$\phi3$	NaCl	NaCl	NaCl	NaCl			$\phi3$	NaCl	NaCl	NaCl	NaCl
$\phi4$	NaCl	NaCl	NaCl	NaCl			$\phi4$	NaCl	NaCl	NaCl	NaCl
$\phi5$	NaCl	NaCl	NaCl	NaCl			$\phi5$	NaCl	NaCl	NaCl	NaCl
ϕC	NaCl	NaCl	NaCl	NaCl			ϕC	NaCl	NaCl	NaCl	NaCl
10^0	10^{-1}	10^{-2}	10^{-3}	10^{-4}			10^0	10^{-1}	10^{-2}	10^{-3}	10^{-4}

8. Include a P22 *mnt*::Km9 *arc*(Am) phage control (ϕC) at an moi of 10–50 in columns 1 and 8 of the last row of the microtiter dish.

9. Mix the cells and phage by gently pipetting the solution up and down and then let stand for approximately 1 hour at 37°C to allow phage adsorption and phenotypic expression.

10. Dilute 5 μl from each column of the microtiter dish into the next column and mix by gently pipetting the solution up and down to achieve sequential tenfold dilutions of the cells + phage (final dilution 10^{-4}). Use new pipette tips for each dilution.

11. Remove 5 μl from each dilution of the cells grown without IPTG and spot onto an LB-agar-Sp-Km plate. Allow the spots to dry at room temperature and then incubate for 1–2 days at 37°C.

12. Remove 5 μl from each dilution of the cells grown with IPTG and spot onto an LB-agar-Sp-Km-IPTG (1 mM) plate. Allow the spots to dry at room temperature and then incubate for 1–2 days at 37°C.

13. Determine the number of viable cells in each culture. Add 45 μl of sterile 0.85% NaCl to the top two rows of a microtiter dish (see below). Add 5 μl of the culture grown without IPTG to duplicate wells in column 1 of the microtiter dish. Add 5 μl of the culture grown with IPTG to duplicate wells in column 8 of the microtiter dish. Mix by gently pipetting the solution up and down with the pipettor. Dilute the cells by sequentially transferring 5 μl into each subsequent column of the microtiter dish. Use new pipette tips for each dilution.

No IPTG added							+IPTG				
# 1	#2	#3	#4	#5	#6	#7	#8	#9	#10	#11	#12
NaCl	NaCl	NaCl	NaCl	NaCl			NaCl	NaCl	NaCl	NaCl	NaCl
NaCl	NaCl	NaCl	NaCl	NaCl			NaCl	NaCl	NaCl	NaCl	NaCl
10^{-1}	10^{-2}	10^{-3}	10^{-4}	10^{-5}			10^{-1}	10^{-2}	10^{-3}	10^{-4}	10^{-5}

14. Spot 5 μl of the 10^{-2}, 10^{-3}, 10^{-4}, and 10^{-5} dilutions onto an LB-agar plate. Allow the spots to dry (usually ~30 minutes) and then incubate the plates upside down overnight at 37°C.

15. Note the number of colonies for each dilution of cells + phage on the LB-agar-Sp-Km and LB-agar-Sp-Km-IPTG plates (if colonies are "too numerous to count," simply indicate TNTC). Also count the number of colonies on each of the plates with cells-only. Calculate the number of Km^r lysogens per cell. For further work, choose a phage lysate that lysogenizes TSM207 at a higher frequency than the control.

EXPERIMENT 8
Challenge Phage Assays

Experiment 8 demonstrates the use of challenge phage assays to quantitate the interaction of a DNA-binding protein with specific DNA-binding sites. The relative affinity of Nac protein for the O_{nac} challenge phage is measured by assaying the percent lysogeny at different IPTG concentrations. This assay can also be used to compare the relative affinity of mutant O_{nac} challenge phage as isolated in Experiment 9 and *nac* suppressor mutants as isolated in Experiment 11. Below is listed the bacterial strain used in this experiment.

BACTERIAL STRAIN

TSM207 (*supE*/pPC36 P_{tac}-*nac lacI*^{Q} Sp^r Sm^r)

MATERIALS

For culture media and supplements, see Appendix B.

LB broth
LB broth-Sp
100 mM IPTG (dioxane-free)
LB broth-Sp-IPTG (10^{-4} mM)
LB broth-Sp-IPTG (10^{-3} mM)
LB broth-Sp-IPTG (10^{-2} mM)
LB broth-Sp-IPTG (10^{-1} mM)
LB broth-Sp-IPTG (10^0 mM)
LB-agar plates
LB-agar-Sp-Km-IPTG (10^{-4} mM) plates

LB-agar-Sp-Km-IPTG (10^{-3} mM) plates
LB-agar-Sp-Km-IPTG (10^{-2} mM) plates
LB-agar-Sp-Km-IPTG (10^{-1} mM) plates
LB-agar-Sp-Km-IPTG (10^{0} mM) plates
$O_{nac}{}^+$ challenge phage (titered)
0.85% NaCl
Sterile microtiter dishes

PROCEDURE

1. Subculture 0.1 ml of TSM207 into 5 ml of LB broth-Sp. Grow cells to early exponential phase with aeration at 37°C (~1–2 hours).

2. Subculture 0.5 ml of the early exponential phase culture of TSM207 into 1.5 ml of

   a. LB broth-Sp-IPTG (10^{-4} mM)

   b. LB broth-Sp-IPTG (10^{-3} mM)

   c. LB broth-Sp-IPTG (10^{-2} mM)

   d. LB broth-Sp-IPTG (10^{-1} mM)

   e. LB broth-Sp-IPTG (10^{0} mM)

3. Grow with aeration for 1–2 hours at 37°C.

4. Calculate the volume of phage required to give a multiplicity of infection (moi) of 10–50 pfu/cell. For example, assuming the cultures are at approximately 5×10^8 cells/ml, for an moi of 25:

$$\frac{ml}{pfu} \times \frac{25 \text{ pfu}}{\text{cell}} \times \frac{5 \times 10^8 \text{ cells}}{ml} \times 0.1 \text{ ml} \times \frac{1000 \text{ μl}}{ml} = \text{μl phage}$$

Add the calculated volume of O_{nac} challenge phage lysates to the wells labeled "ϕ + cells" in Microtiter Plates 1–3 as shown (see pp. 350–351). Also add phage to the final column of Microtiter Plate #3 labeled "ϕ only."

5. Add 0.1 ml of the TSM207 cultures grown with different IPTG concentrations to the wells labeled "φ + cells" in Microtiter Plates 1–3 as shown.

6. Mix the cells and phage by gently pipetting the solution up and down and then let stand for 1 hour at 37°C to allow phage adsorption and phenotypic expression of Km^r.

7. Add 45 µl of sterile 0.85% NaCl to wells labeled "NaCl" as shown. Dilute 5 µl from each column of the microtiter dish into the next column and mix by gently pipetting the solution up and down to achieve the sequential tenfold dilutions shown (final dilution 10^{-4}).

8. Remove 5 µl from each dilution and spot onto an LB-agar-Sp-Km plate with the same concentration of IPTG. Allow the spots to dry at room temperature and then incubate for 1–2 days at 37°C.

9. Prepare the viable cell control dilutions in Microtiter Plate #4 as shown (see p. 351). Remove 5 µl from the 10^{-3}, 10^{-4}, and 10^{-5} dilutions of the cell controls and spot onto an LB-agar plate. Allow the spots to dry at room temperature and then incubate overnight at 37°C.

10. Note the number of colonies for each dilution of cells + phage on each LB-agar-Sp-Km-IPTG plate (if colonies are "too numerous to count," simply indicate TNTC). Also count the number of colonies on each of the LB-agar plates with cells only. Calculate the percent lysogeny at each IPTG concentration:

$$\frac{\text{number of Km}^r \text{ lysogens}}{\text{number of viable cells}} \times 100 = \% \text{ lysogeny}$$

Plot the percent lysogeny versus the [IPTG] on log paper.

MICROTITER PLATE #1

	Cells grown in LB-broth-Sp + 10^{-4} mM IPTG						Cells grown in LB-broth-Sp + 10^{-3} mM IPTG				
Challenge phage 1	φ+ cells	NaCl	NaCl	NaCl	NaCl		φ+ cells	NaCl	NaCl	NaCl	NaCl
Challenge phage 2	φ+ cells	NaCl	NaCl	NaCl	NaCl		φ+ cells	NaCl	NaCl	NaCl	NaCl
Challenge phage 3	φ+ cells	NaCl	NaCl	NaCl	NaCl		φ+ cells	NaCl	NaCl	NaCl	NaCl
Challenge phage 4	φ+ cells	NaCl	NaCl	NaCl	NaCl		φ+ cells	NaCl	NaCl	NaCl	NaCl
Challenge phage 5	φ+ cells	NaCl	NaCl	NaCl	NaCl		φ+ cells	NaCl	NaCl	NaCl	NaCl
Challenge phage 6	φ+ cells	NaCl	NaCl	NaCl	NaCl		φ+ cells	NaCl	NaCl	NaCl	NaCl
Challenge phage 7	φ+ cells	NaCl	NaCl	NaCl	NaCl		φ+ cells	NaCl	NaCl	NaCl	NaCl
Challenge phage 8	φ+ cells	NaCl	NaCl	NaCl	NaCl		φ+ cells	NaCl	NaCl	NaCl	NaCl
	10^{-1}	10^{-2}	10^{-3}	10^{-4}			10^{-1}	10^{-2}	10^{-3}	10^{-4}

Spot 5 μl onto LB-agar-Sp-Km + 10^{-4} mM IPTG   Spot 5 μl onto LB-agar-Sp-Km + 10^{-3} mM IPTG

The first column contains cells + phage. Each subsequent column is a tenfold dilution of the cells + phage in sterile 0.85% NaCl.

MICROTITER PLATE #2

	Cells grown in LB-broth-Sp + 10^{-2} mM IPTG						Cells grown in LB-broth-Sp + 10^{-1} mM IPTG				
Challenge phage 1	φ+ cells	NaCl	NaCl	NaCl	NaCl		φ+ cells	NaCl	NaCl	NaCl	NaCl
Challenge phage 2	φ+ cells	NaCl	NaCl	NaCl	NaCl		φ+ cells	NaCl	NaCl	NaCl	NaCl
Challenge phage 3	φ+ cells	NaCl	NaCl	NaCl	NaCl		φ+ cells	NaCl	NaCl	NaCl	NaCl
Challenge phage 4	φ+ cells	NaCl	NaCl	NaCl	NaCl		φ+ cells	NaCl	NaCl	NaCl	NaCl
Challenge phage 5	φ+ cells	NaCl	NaCl	NaCl	NaCl		φ+ cells	NaCl	NaCl	NaCl	NaCl
Challenge phage 6	φ+ cells	NaCl	NaCl	NaCl	NaCl		φ+ cells	NaCl	NaCl	NaCl	NaCl
Challenge phage 7	φ+ cells	NaCl	NaCl	NaCl	NaCl		φ+ cells	NaCl	NaCl	NaCl	NaCl
Challenge phage 8	φ+ cells	NaCl	NaCl	NaCl	NaCl		φ+ cells	NaCl	NaCl	NaCl	NaCl
	10^{-1}	10^{-2}	10^{-3}	10^{-4}			10^{-1}	10^{-2}	10^{-3}	10^{-4}

Spot 5 μl onto LB-agar-Sp-Km + 10^{-2} mM IPTG   Spot 5 μl onto LB-agar-Sp-Km + 10^{-1} mM IPTG

The first column contains cells + phage. Each subsequent column is a tenfold dilution of the cells + phage in sterile 0.85% NaCl.

Microtiter Plates #1 and #2.

MICROTITER
PLATE #3

Cells grown in LB-broth-Sp
+ 10^0 mM IPTG

Phage controls

Challenge phage 1	φ+ cells	NaCl	NaCl	NaCl	NaCl				φ only	
Challenge phage 2	φ+ cells	NaCl	NaCl	NaCl	NaCl				φ only	
Challenge phage 3	φ+ cells	NaCl	NaCl	NaCl	NaCl				φ only	
Challenge phage 4	φ+ cells	NaCl	NaCl	NaCl	NaCl				φ only	
Challenge phage 5	φ+ cells	NaCl	NaCl	NaCl	NaCl				φ only	
Challenge phage 6	φ+ cells	NaCl	NaCl	NaCl	NaCl				φ only	
Challenge phage 7	φ+ cells	NaCl	NaCl	NaCl	NaCl				φ only	
Challenge phage 8	φ+ cells	NaCl	NaCl	NaCl	NaCl				φ only	

10^{-1} 10^{-2} 10^{-3} 10^{-4}

Spot 5 µl onto
LB-agar-Sp-Km + 10^0 mM IPTG

Spot 5 µl onto
LB-agar

The first column contains cells + phage. Each subsequent column is a tenfold dilution of the cells + phage in sterile 0.85% NaCl. The phage control column (φ) contains the indicated undiluted phage only.

MICROTITER
PLATE #4

Cells grown in
LB-broth-Sp +

Viable cell control

10^{-4} mM IPTG	cells	NaCl	NaCl	NaCl	NaCl	NaCl					
10^{-3} mM IPTG	cells	NaCl	NaCl	NaCl	NaCl	NaCl					
10^{-2} mM IPTG	cells	NaCl	NaCl	NaCl	NaCl	NaCl					
10^{-1} mM IPTG	cells	NaCl	NaCl	NaCl	NaCl	NaCl					
10^0 mM IPTG	cells	NaCl	NaCl	NaCl	NaCl	NaCl					

10^{-1} 10^{-2} 10^{-3} 10^{-4} 10^{-5}

Spot 5 µl onto LB-agar

The first column contains the cells-only control. Each subsequent column is a tenfold dilution of the cells in sterile 0.85% NaCl.

Microtiter Plates #3 and #4.

EXPERIMENT 9
Isolation of Operator Mutations

Experiment 9 demonstrates how challenge phage can be used to isolate mutations in a DNA-binding site that prevent DNA-protein interactions. Following mutagenesis of the O_{nac} challenge phage constructed in Experiment 7, mutations in O_{nac} that are no longer recognized by the Nac protein are selected. The O_{nac} challenge phage are mutagenized by growth in mutator strains and also by chemical mutagenesis.

1. Two types of mutator strains are used: mutants defective for the proofreading subunit of DNA polymerase III (*dnaQ* or "*mutD*") and mutants defective for mismatch repair (*mutS*). The challenge phage are mutated by simply growing the phage lytically in the mutator strain.

2. Hydroxylamine (NH_2OH) is used to mutagenize DNA packaged inside of phage heads in vitro.

Any parental O_{nac} challenge phage will efficiently lysogenize cells producing Nac, but challenge phage with mutations in the O_{nac} site that prevent Nac binding will lyse the cells. Thus, mutations in O_{nac} can be selected as clear-plaque mutants of the challenge phage. The resulting O_{nac} mutant challenge phage is used to determine the sequence specificity of the Nac protein (Experiment 10) and to select for second-site suppressor mutants that identify amino acid residues of the Nac protein that directly interact with the DNA (Experiment 11). Below are listed the bacterial strains used in this experiment.

BACTERIAL STRAINS

TSM214 (*supE*)
TSM215 (*supE mutD*::Tn*10*dTc)
TSM216 (*supE mutS*::Tn*10*)
TSM208 (*supE attP*::[P22 *sieA* Δ(*mnt-a1* Tn*1*) Ap^s Tpfr]/pPC36
P_{tac}-*nac lacI*^Q Sp^r Sm^r)

mutD Mutagenesis

MATERIALS

For culture media and supplements, see Appendix B.

E-medium + 0.2% glucose
LBEDO broth (LB Broth with E Salts and Dextrose)
0.85% NaCl
Chloroform
Microcentrifuge tubes

PROCEDURE

1. Start a fresh overnight culture of TSM215 in 2 ml of E-medium with 0.2% glucose (see comments on *mutD* in Section 4, Mutants and Their Analysis). Incubate with aeration overnight at 37ºC.

2. Add 0.1 ml of TSM215 culture to 2 ml of LBEDO broth. Incubate with aeration for approximately 2–3 hours at 37ºC or until the cells reach early exponential phase.

3. Dilute a high-titer lysate of the challenge phage as follows:

   a. 20 μl of phage lysate into 180 μl of 0.85% NaCl (10^{-1} dilution)

   b. 20 μl of 10^{-1} dilution into 180 μl of 0.85% NaCl (10^{-2} dilution)

   c. 1 ml of 0.85% NaCl (no-phage control)

4. Mix 0.1 ml of each diluted phage lysate with 0.1 ml of exponential phase TSM215 culture. Let stand for 20 minutes at room temperature to allow phage adsorption.

5. Add 2.5 ml of LBEDO broth and incubate with thorough aeration for approximately 3 hours at 37°C or until lysed. Compare the tubes with phage to the no-phage control tube. The no-phage control tube should not show lysis, but cellular debris should be seen in the tubes with phage.

6. Add several drops of chloroform, vortex thoroughly, and allow the chloroform to settle for approximately 5 minutes.

    Caution: Chloroform is a carcinogen and may damage the liver and kidneys. Do not mouth pipette and avoid contact with skin.

7. Transfer the supernatant (avoiding the lower chloroform phase) to microcentrifuge tubes. Centrifuge in a microcentrifuge for 2 minutes to pellet the cell debris.

8. Check the titer of the mutagenized phage (see Appendix D, Phage P22 Lysates) and then store the mutagenized phage lysates at 4°C until use.

mutS Mutagenesis

MATERIALS

For culture media and supplements, see Appendix B.

LB broth
LBEDO broth (LB Broth with E Salts and Dextrose)
0.85% NaCl
Chloroform

PROCEDURE

1. Start a fresh overnight culture of TSM216 in 2 ml of LB broth. Incubate with aeration overnight at 37°C.

2. Add 0.1 ml of TSM216 culture to 2 ml of LBEDO broth. Incubate with aeration for approximately 2–3 hours at 37°C or until the cells reach early exponential phase.

3. Dilute a high-titer lysate of the challenge phage as follows:

    a. 20 μl of the phage lysate into 180 μl of 0.85% NaCl (10^{-1} dilution)

    b. 20 μl of the 10^{-1} dilution into 180 μl of 0.85% NaCl (10^{-2} dilution)

4. Mix 0.1 ml of each diluted phage lysate with 0.1 ml of TSM216. Let stand for 20 minutes at room temperature to allow phage adsorption.

5. Add 2.5 ml of LBEDO broth and incubate with thorough aeration for approximately 3 hours at 37°C or until lysed.

6. Add several drops of chloroform, vortex thoroughly, and allow the chloroform to settle for approximately 5 minutes.

    Caution: Chloroform is a carcinogen and may damage the liver and kidneys. Do not mouth pipette and avoid contact with skin.

7. Transfer the supernatant (avoiding the lower chloroform phase) to microcentrifuge tubes. Centrifuge in a microcentrifuge for 2 minutes to pellet the cell debris.

8. Check the titer of the mutagenized phage (see Appendix D, Phage P22 Lysates) and then store the mutagenized phage lysates at 4°C until use.

Hydroxylamine Mutagenesis

MATERIALS

P22 *mnt*::Km9 *arc*(Am) phage
Sterile deionized H_2O
0.85% NaCl
Hydroxylamine/NaOH (prepare fresh)
   Hydroxylamine (NH_2OH)   350 mg
   4 M NaOH   0.56 ml
   Adjust to 5 ml with sterile deionized H_2O.

Phosphate-EDTA buffer (0.5 M KPO$_4$, pH 6, 5 mM EDTA)

Note: When exposed to oxygen, hydroxylamine solutions form by-products that are toxic to cells (probably peroxides and free radicals). This nonspecific toxicity is decreased by EDTA.

KH$_2$PO$_4$	6.81 g
Deionized H$_2$O	70 ml

Dissolve on a stirrer. Adjust pH to 6 with 1 N KOH. Adjust to 99 ml with deionized H$_2$O. Add 1 ml of 0.5 M EDTA. Autoclave.

1 M MgSO$_4$

LBSE

LB broth	100 ml
0.5 M EDTA	0.2 ml
NaCl	5.85 g

Mix. Autoclave. Store at 4°C.

Sterile microcentrifuge tubes

Sterile microtiter dishes with lids

PROCEDURE

1. Add the following solutions to two separate sterile micro-centrifuge tubes:

Addition	Mutagenesis	Control
Phosphate-EDTA buffer	200 μl	200 μl
Sterile deionized H$_2$O	300 μl	700 μl
Hydroxylamine	400 μl	–
1 M MgSO$_4$	10 μl	10 μl

Caution: Hydroxylamine is a suspected carcinogen. Wear gloves and *do not* mouth pipette. Dispose of all waste containing HA in appropriate biohazard waste containers.

2. Add 0.1 ml of the undiluted high-titer O$_{nac}$ challenge phage to both tubes.

Note: Since phage killing occurs during hydroxylamine mutagenesis, the starting lysates should have a titer of $\geq 10^{11}$ pfu/ml.

3. Incubate for 24–48 hours in a 37°C incubator. Remove samples at 0 time and every 4–8 hours by diluting 10 μl into 1 ml of cold LBSE (10^{-2} dilution).

4. Titer each sample on TSM214 to determine the decrease in viable phage (see Appendix D, Phage P22 Lysates). Dilute the phage by adding 45 μl of sterile 0.85% NaCl to the first 10 columns of a microtiter dish using a multichannel pipettor (see figure below). For each time point (labeled T_0, T_1, etc.), add 5 μl of mutagenized O_{nac} challenge phage (labeled "mut") to an unused well in the first column of the microtiter dish. Add 5 μl of the P22 *mnt*::Km9 *arc*(Am) phage control to the next row of the microtiter dish. Mix by gently pipetting the solution up and down with the pipettor. Dilute the phage by sequentially transferring 5 μl into each subsequent column of the microtiter dish.

	# 1	#2	#3	#4	#5	#6	#7	#8	#9	#10	#11	#12
T_0 mut		NaCl	NaCl	NaCl	NaCl	NaCl	NaCl	NaCl	NaCl			
T_0 control		NaCl	NaCl	NaCl	NaCl	NaCl	NaCl	NaCl	NaCl			
T_1 mut		NaCl	NaCl	NaCl	NaCl	NaCl	NaCl	NaCl	NaCl			
T_1 control		NaCl	NaCl	NaCl	NaCl	NaCl	NaCl	NaCl	NaCl			
T_2 mut		NaCl	NaCl	NaCl	NaCl	NaCl	NaCl	NaCl	NaCl			
T_2 control		NaCl	NaCl	NaCl	NaCl	NaCl	NaCl	NaCl	NaCl			
T_3 mut		NaCl	NaCl	NaCl	NaCl	NaCl	NaCl	NaCl	NaCl			
T_3 control		NaCl	NaCl	NaCl	NaCl	NaCl	NaCl	NaCl	NaCl			
	10^{-1}	10^{-2}	10^{-3}	10^{-4}	10^{-5}	10^{-6}	10^{-7}	10^{-8}	10^{-9}	10^{-10}		

5. Use a multichannel pipettor to spot 5 μl of the 10^{-3}, 10^{-4}, 10^{-5}, 10^{-6}, 10^{-7}, 10^{-8}, and 10^{-9} dilutions onto the TSM214 cell lawn. A wider range of dilutions are plated than usual because the hydroxylamine treatment causes the phage titer to decrease over time. Allow the drops of phage to dry (~30 minutes) and then incubate the plates upside down for 8–16 hours at 37°C.

6. Count the number of plaques in each spot. Each viable phage (pfu) will produce one plaque. Calculate the phage titer as follows:

$$\frac{\text{number of plaques}}{5\ \mu l \times \text{dilution factor}} \times \frac{1000\ \mu l}{ml} = \text{pfu/ml}$$

7. Plot pfu/ml versus time on semi-log paper. Predict when killing will reach 0.1–1.0% survivors (usually 24–36 hours; see Fig. 9.1).

8. At the time point predicted to give 0.1–1.0% survivors (estimated by extrapolating the curve from earlier time points), remove an aliquot for titering and then centrifuge the rest in a microcentrifuge for 30 minutes at 4ºC to pellet the phage.

9. Pour off the supernatant. Overlay the phage pellet with 0.2 ml of cold LBSE. Place overnight at 4ºC, occasionally swirling gently to resuspend the pellet. Do not vortex; handle the hydroxylamine-treated phage gently or the titer will decrease dramatically.

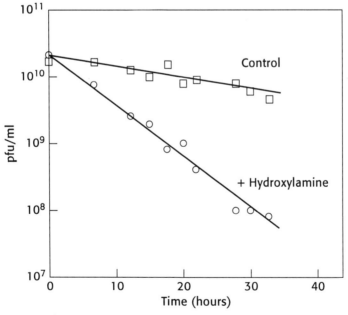

Figure 9.1 An example of a hydroxylamine killing curve for P22. Since small variations in the reagents may result in somewhat slower or faster kinetics of phage killing, a killing curve should be determined empirically each time hydroxylamine mutagenesis is done.

Selection for Mutant Challenge Phage

Clear plaque mutations that inactivate the O_{nac}-binding site are rare (~10^{-6} to 10^{-7}) and the background has numerous faint turbid plaques. It is therefore necessary to plaque-purify the clear plaque mutants several times to avoid confusion if the lysate is contaminated with other phage. Since phage can readily diffuse through the top agar, streak the phage from the center of the plate toward the periphery so that the isolated plaques are well separated.

MATERIALS

For culture media and supplements, see Appendix B.

LB broth
LB broth-Sp
LB broth-Sp-IPTG (1 mM)
LB-agar plates
LB-agar-Sp-IPTG (1 mM) plates
TBSA (Tryptone Broth–Soft Agar)
P22 *mnt*::Km9 *arc*(Am) phage
Sterile microcentrifuge tubes
Sterile test tubes
Sterile pasteur pipettes
Heating block (50ºC)

PROCEDURE

1. Start a fresh overnight culture of TSM208 in 2 ml of LB broth-Sp. Incubate with aeration overnight at 37ºC.

2. Subculture 0.1 ml of the overnight culture of TSM208 into 2 ml of LB broth-Sp. Grow to mid exponential phase with thorough aeration at 37ºC (~2–3 hours).

3. Dilute 0.5 ml of the mid exponential phase TSM208 culture into 1.5 ml of LB broth-Sp-IPTG (1 mM). Incubate the cells with thorough aeration for approximately 2 hours at 37ºC.

4. Mix 200 µl of the induced TSM208 culture with 10^7–10^8 pfu of each mutated phage lysate (100 µl of 10^8–10^9 pfu/ml).

5. Let stand for 20 minutes at room temperature to allow phage adsorption.

6. Melt TBSA in a microwave. For each mixture of cells + phage, add 3 ml of the melted top agar to a test tube and place in a 50°C heating block.

7. Add 3 ml of TBSA to each mixture of cells + phage, mix, and quickly pour onto LB-agar-Sp-IPTG (1 mM) plates. Allow the plates to solidify for 10–15 minutes at room temperature and then incubate upside down at 37°C.

8. Start a fresh overnight culture of TSM208 in 2 ml of LB broth-Sp. Incubate with aeration overnight at 37°C.

9. Subculture 0.1 ml of the overnight culture of TSM208 into 2 ml of LB broth-Sp. Grow to mid exponential phase with thorough aeration at 37°C (~2–3 hours).

10. Dilute 0.5 ml of the mid log phase TSM208 culture into 1.5 ml of LB broth-Sp-IPTG (1 mM). Incubate the cells with thorough aeration for 1–2 hours at 37°C.

11. Melt TBSA in a microwave. Add 3 ml of the melted top agar to three sterile test tubes and place in a 50°C heating block.

12. After the top agar cools to 50°C, add 0.1 ml of TSM208 grown in LB broth-Sp-IPTG (1 mM), mix, and quickly pour onto LB broth-Sp-IPTG (1 mM) plates. Allow the plates to solidify for 10–15 minutes at room temperature.

13. Pick eight large clear plaques and streak on the TSM208 lawn (divide the plate into quadrants and streak four plaques per plate from the center of the plate toward the periphery). Also streak out the unmutagenized O_{nac} challenge phage and P22 *mnt*::Km9 *arc*(Am) as controls. Incubate the plates upside down overnight at 37°C.

14. Start a fresh overnight culture of TSM214 in 2 ml of LB broth. Incubate with aeration overnight at 37°C.

15. Melt TBSA in a microwave. Add 3 ml of the melted top agar to three sterile test tubes and place in a 50°C heating block.

16. After the top agar cools, add 0.1 ml of TSM214, mix, and quickly pour onto LB-agar plates. Allow the plates to solidify for 10–15 minutes at room temperature.

17. Pick large clear plaques from step 13 and restreak on the TSM214 lawn. Incubate the plates upside down overnight at 37°C.

18. Prepare a high-titer lysate of each clear plaque mutant on strain TSM214 as described in Appendix D (Phage P22 Lysates).

19. Titer each phage on TSM214 as described in Appendix D (Phage P22 Lysates). Include a P22 *mnt*::Km9 *arc*(Am) phage control (ϕC) in the last row of the microtiter dish.

20. Count the number of plaques in each spot. Calculate the phage titer as follows:

$$\frac{\text{number of plaques}}{5 \ \mu l \times \text{dilution factor}} \times \frac{1000 \ \mu l}{ml} = \text{pfu/ml}$$

Assign allele numbers to the mutations and add the phage stocks to the phage collection.

DNA Sequence Analysis of Challenge
Phage Mutants

Following isolation of a collection of binding-site mutants on the challenge phage, the DNA sequence changes in the mutants are determined to identify the nucleotides required for the protein-DNA interaction. The DNA sequence of the control region can be determined by cycle sequencing directly from the phage.

Cycle-sequencing reactions are similar to normal dideoxy-sequencing reactions, but a thermostable DNA polymerase is used, allowing multiple cycles of primer annealing and DNA synthesis (Fig. 10.1). The template DNA is first heat-denatured and then a primer is annealed to the template. A thermostable DNA polymerase extends the primer on each template molecule in the population until a ddNTP is incorporated, which terminates further DNA synthesis. The double-stranded DNA is then denatured, allowing another round of primer annealing and DNA synthesis. This process is repeated 20–30 times. The DNA fragments synthesized are separated by electrophoresis on a polyacrylamide/urea gel. Two approaches are described for visualizing the resulting DNA bands: autoradiography of radioactively labeled DNA and silver staining to visualize nonradioactively labeled DNA. The DNA fragments are radioactively labeled by end-labeling the primer with $[\gamma\text{-}^{32}P]$ATP or $[\gamma\text{-}^{33}P]$ATP or by incorporation of $[\alpha\text{-}^{35}S]$dATP into the DNA during synthesis. Using an end-labeled primer has several advantages: (1) spurious bands are avoided due to internal priming caused by partial annealing of template DNA fragments, (2) long and short bands have the same intensity because they all contain a single radioactive label at the 5′ end, and (3) γ-labeled ATP is considerably less expensive than α-labeled dATP. The major advantage of internal labeling is that much more radioactivity is incorporated into the DNA fragments so less time is required for autoradiography.

Alternatively, methods for nonradioactive detection of DNA-sequencing bands have recently been developed, eliminating the need for radioactive labeling. For example, the silver-staining kit sold by Promega is sensitive, easy to use, and does not require special equipment. Below is listed the bacterial strain used for this experiment.

BACTERIAL STRAIN

TSM217 (*E. coli endA1 recA1 hsdR* [r⁻ m⁺])

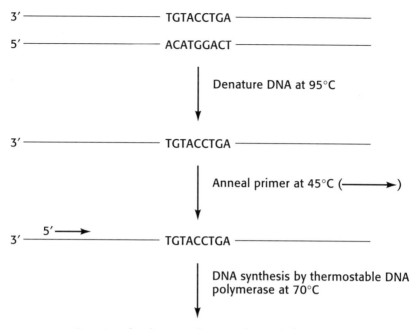

Four termination reactions as shown below:

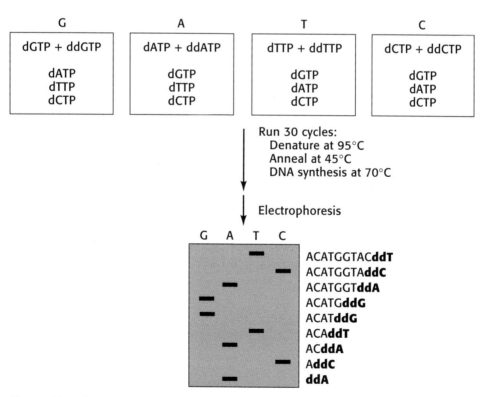

Figure 10.1 DNA sequence analysis by cycle sequencing with a heat-stable DNA polymerase.

PCR Amplification and Cycle Sequencing of Challenge Phage

The DNA sequence of phage lysates is determined directly without purifying phage DNA. However, preamplification of phage DNA usually yields cleaner DNA sequencing results. The sequence of an O_{nac} mutant challenge phage is compared with that of the "wild-type" O_{nac} challenge phage. If the cycle sequencing reactions fail to work properly, the problem may be due to the preparation of the P22 phage lysate or a problem with the sequencing reactions. Therefore, purified λ DNA pretested in the DNA sequencing reactions is included as a control. Reagents for cycle sequencing are available in convenient, reasonably priced kits from several commercial sources.

MATERIALS

PCR mix (sufficient for 50 reactions)
Sterile deionized H_2O	2.45 ml
dNTPs (0.1 ml of each 10 mM stock solution; pH 7)	0.40 ml
10x PCR amplification buffer	0.50 ml

10x PCR amplification buffer (store at room temperature)
2.5 M KCl	2.0 ml
1 M Tris-HCl (pH 8.0)	1.0 ml
1 M $MgCl_2$	0.15 ml
1% Gelatin	1.0 ml
Sterile deionized H_2O	5.85 ml

Taq DNA polymerase (5 units/μl)

Four termination mixes (each with a different ddNTP): G termination mix (ddGTP reaction); A termination mix (ddATP reaction); T termination mix (ddTTP reaction); C termination mix (ddCTP reaction)
7-deaza-dGTP	15 μM
dATP	15 μM
dTTP	15 μM
dCTP	15 μM

ddNTP (0.03 mM ddGTP, 0.45 mM ddATP, 0.9 mM ddTTP, or 0.3 mM ddCTP)

Store at –20ºC, thaw just prior to use, and keep on ice until use.

O_{mnt} primer (20 pmoles)
Anti-O_{mnt} primer (20 pmoles)
O_{nac} mutant challenge phage and wild-type O_{nac} challenge phage
Thermocycler

> *Note:* Some thermocyclers require overlaying the sample with mineral oil or wax to prevent the condensation of water vapor on the upper surfaces of the tube. This can be avoided by using a thermocycler that maintains a higher temperature in the upper portion of the chamber than in the lower part (e.g., the PTC-100™ Thermal Cycler with a Hot Bonnet™ sold by MJ Research).

Sterile H_2O
Sterile 0.5-ml microcentrifuge tubes

Preamplification of Phage DNA

1. Label a sterile 0.5-ml microcentrifuge tube for each phage and place the tubes on ice. Add the following reagents to each tube:

PCR mix	75 µl
O_{mnt} primer (20 pmoles)	1 µl
Anti-O_{mnt} primer (20 pmoles)	1 µl
Sterile deionized H_2O	12 µl
Concentrated O_{nac} challenge phage lysate (from Experiment 8)	10 µl
Taq DNA polymerase (5 units/µl)	1 µl

2. Mix by gently pipetting up and down with a micropipettor.

3. Preheat the thermocycler to 95ºC. Add the tubes to the thermocycler and run the following program.

Step	Temperature	Time	Function
1	95ºC	5 minutes	denature
2	95ºC	0.5 minute	denature
3	45ºC	0.5 minute	anneal primers
4	70ºC	1 minute	DNA synthesis
5	cycle steps 2–4	repeat 29 times	amplify
6	4ºC	hold	end

4. Purify the resulting PCR fragments to remove any partial PCR products and nucleotides as described in Appendix F (Purification and Quantitation of PCR Products).

Cycle Sequencing with ³²P-labeled Primers

MATERIALS

T4 polynucleotide kinase
10x Polynucleotide kinase buffer

Tris-acetate (pH 7.8)	33 mM
Potassium acetate	66 mM
Magnesium acetate	10 mM
Dithiothreitol	0.5 mM
Triton X-100	0.01%

$[\gamma\text{-}^{32}P]$ATP (6000 Ci/mmole, 150 µCi/µl, 25 mM)
Anti-O_{mnt} primer (20 pmoles/µl)
λGT11 control primer (100 ng/µl)
λGT11 control DNA
Four termination mixes (each with a different ddNTP): G termination mix (ddGTP reaction); A termination mix (ddATP reaction); T termination mix (ddTTP reaction); C termination mix (ddCTP reaction)
Amplified O_{nac} challenge phage
5x PCR Sequencing buffer

Tris-HCl (pH 9)	250 mM
MgCl$_2$	10 mM

Dye-stop solution

Formamide	95%
EDTA (pH 7.6)	20 mM
Bromophenol blue	0.1%
Xylene cyanol FF	0.1%

Taq DNA polymerase (sequencing grade; 5 units/µl)
Sterile deionized H$_2$O
Sterile 0.5-ml microcentrifuge tubes

Caution: This experiment uses radioactivity. Obey the following safety precautions for ³²P:

- Wear disposable gloves. Use a Geiger counter to monitor the gloves regularly. Replace gloves frequently.

- Wear a lab coat and safety glasses.

- Always work behind a Plexiglas shield.

- Never mouth pipette.

- Dispose of radioactive materials in the correct radioactive waste container.

- Before leaving the area, use a Geiger counter to check yourself and the work area carefully.

- Clean up spills immediately.

End-labeling Primers with T4 Polynucleotide Kinase

1. Thaw frozen reagents and keep on ice until use. Combine the following in sterile 0.5-ml microcentrifuge tubes.

   a. Anti-O_{mnt} primer

Sterile deionized H_2O	17.5 µl
10x Polynucleotide kinase buffer	2.5 µl
Anti-O_{mnt} primer (20 pmoles/µl)	2.0 µl
T4 polynucleotide kinase	1.0 µl
[γ-^{32}P]ATP (150 µCi/µl, 25 pmoles/µl)	2.0 µl

   b. λGT11 control primer

Sterile deionized H_2O	18.2 µl
10x Polynucleotide kinase buffer	2.5 µl
λGT11 primer (20 pmoles/µl)	1.3 µl
T4 polynucleotide kinase	1.0 µl
[γ-^{32}P]ATP (150 µCi/µl, 25 pmoles/µl)	2.0 µl

2. Incubate for 30 minutes at 37°C.

3. Inactivate the polynucleotide kinase by heating for 5 minutes at 70°C. Allow to cool. If used immediately, keep on ice. Otherwise store at −20°C until use. The end-labeled primers can be used directly for cycle sequencing without removing any unincorporated nucleotides.

Dideoxy Sequencing with Radioactively Labeled Primers

1. For each phage lysate, label four 0.5-ml microcentrifuge tubes G, A, T, or C and place the tubes on ice. Add

G termination mix to each G tube (ddGTP reaction)	2 µl
A termination mix to each A tube (ddATP reaction)	2 µl
T termination mix to each T tube (ddTTP reaction)	2 µl
C termination mix to each C tube (ddCTP reaction)	2 µl

2. Label a sterile 0.5-ml microcentrifuge tube for each phage and place the tubes on ice. Add the following reagents to each tube:

   Challenge phage:

5x PCR sequencing buffer	3 µl
Amplified O_{nac} challenge phage DNA	1 µl
Taq DNA polymerase (sequencing grade, 5 units/µl)	1 µl
^{32}P-labeled anti-O_{mnt} primer	1 µl
Sterile deionized H_2O	9 µl

   λ DNA control:

5x PCR sequencing buffer	3 µl
λGT11 control DNA	1 µl
Taq DNA polymerase (sequencing grade, 5 units/µl)	1 µl
^{32}P-labeled λGT11 primer	1 µl
Sterile deionized H_2O	9 µl

3. Add 3 µl of the template/primer/buffer mix (from step 2) to each tube of nucleotide reaction mix (from step 1), using a new pipette tip for each tube.

4. Mix by gently pipetting with a micropipettor.

5. Preheat the thermocycler to 95ºC. Add the tubes to the thermocycler and run the following program.

Step	Temperature	Time	Function
1	95°C	5 minutes	denature
2	95°C	0.5 minute	denature
3	45°C	0.5 minute	anneal primers
4	70°C	1 minute	DNA synthesis
5	cycle steps 2–4	repeat 29 times	amplify
6	4°C	hold	end

6. Add 3 μl of Dye-stop solution to each tube.

7. Denature the samples for 2 minutes at 70°C and then load 3–7 μl on an 8% polyacrylamide/8 M urea DNA sequencing gel.

Cycle Sequencing Nonradioactive DNA

Reagents for cycle sequencing are available in convenient, reasonably priced kits from several commercial sources.

1. For each DNA template, label four 0.5-ml microcentrifuge tubes G, A, T, C.

2. Add 2 μl of the corresponding dNTP/ddNTP mix to each tube.

3. Prepare the following tubes for each set of sequencing reactions:

PCR-amplified phage DNA	10 μl
5x PCR sequencing buffer	5 μl
Anti-O$_{mnt}$ primer	1 μl
Taq DNA polymerase (sequencing grade, 5 units/μl)	1 μl

Also prepare the following control:

λGT11 control DNA	4 μl
5x PCR sequencing buffer	5 μl
λGT11 primer	4 μl
Sterile deionized H$_2$O	3 μl
Taq DNA polymerase (sequencing grade, 5 units/μl)	1 μl

4. Add 4 µl of template/primer/enzyme mix to each tube containing the dNTP/ddNTP mix. Mix by gently pipetting the solution up and down with the pipettor.

5. Preheat the thermocycler to 95°C. Place the tubes in the preheated thermocycler and run the following program:

Step	Temperature	Time	Function
1	95°C	2 minutes	denature
2	95°C	30 seconds	denature
3	45°C	30 seconds	anneal primers
4	70°C	1 minute	DNA synthesis
5	cycle steps 2–4	repeat 45 times	amplify
6	4°C	hold	end

6. Add 3 µl of Dye-stop solution. Heat the reactions for 2 minutes at 70°C and then immediately load 3–7 µl of each reaction onto an 8% polyacrylamide/8 M urea DNA sequencing gel.

DNA Sequencing Gels

MATERIALS

Acrylamide/bis-acrylamide-urea solution (19:1; store at 4°C)

Note: Commercially available as an ultrapure stock solution containing 8 M urea and 0.05% TEMED in 1X TBE buffer (e.g., GenePage® from Amresco and Acryl-a-Mix® from Promega).

10% Ammonium persulfate: Dissolve 1 g in 9 ml of deionized H_2O. Aliquot and store at –20°C. Thaw just before use.

10x TBE

Tris base	162 g
Boric acid	27.2 g
Na_2EDTA	9.3 g

Dissolve in approximately 600 ml of deionized H_2O on a magnetic stirrer. Adjust to 1 liter with deionized H_2O. Filter through Whatman No. 1 filter paper. Store at room temperature. Discard if a solid precipitate develops.

Sigmacote® (Sigma) or Rainex®
Bind Saline (Promega)
Ethanol (95%)
Glacial acetic acid
Vacuum pump and gel dryer
Sequencing set-up and power supply
Kimwipes (lint-free tissues)
Whatman filter paper sheets
Sequencing plates with 0.8-mm spacers
Gel-sealing tape
Autoradiography cassette
Autoradiography film
Dark room and developing reagents

PROCEDURE

1. Clean the gel plates thoroughly with a nonabrasive detergent. Rinse thoroughly with deionized H_2O and dry with Kimwipes.

2. Thoroughly rinse the plates with ethanol and wipe with Kimwipes. Avoid touching the plates.

3. Prepare the two glass plates as follows:

   Short plate: For radioactive gels, further treatment is not required. For silver-stained gels, treat the short plate as described below.

   a. Prepare fresh binding solution by adding 3 μl of Bind Saline to 1 ml of 95% ethanol + 0.5% glacial acetic acid.

   b. Spread the binding solution onto the clean glass plate using a Kimwipe. Allow to dry for 4–5 minutes.

   c. Remove excess binding solution from the plate with a Kimwipe moistened with 95% ethanol. Use gentle pressure to wipe the plate in one direction and then in a perpendicular direction.

   d. Repeat step c three times using a clean Kimwipe each time.

Long plate: Wipe a clean glass plate with Sigmacote or Rainex and allow the plate to dry for approximately 5 minutes. Remove any excess with a Kimwipe. Repeat this treatment approximately every 10 uses.

4. Assemble the plates with the clean surface facing inside. Insert the side spacers. Tape the sides and bottoms of the plates and clamp the sides.

5. Add 0.4 ml of 10% ammonium persulfate to 75 ml of acrylamide/bis (19:1)–8 M urea–TEMED–TBE solution. Swirl gently to mix.

   Cautions: Acrylamide and bisacrylamide are potent neurotoxins and are absorbed through the skin. Their effects are cumulative. Once polymerized, polyacrylamide is considered to be nontoxic, but it should be treated with care because it may contain small quantities of unpolymerized material. Always wear gloves when handling acrylamide. If solid acrylamide and bis-acrylamide are used, wear gloves and a mask when weighing the powders. Commercially preprepared liquid stock solutions are safer and relatively inexpensive.

   Ammonium persulfate is destructive to tissue of the mucous membranes and upper respiratory tract, eyes, and skin. Inhalation may be fatal. Exposure can cause gastrointestinal disturbances and dermatitis. Wear gloves, safety glasses, and protective clothing. Wash thoroughly after handling.

   TEMED is destructive to tissue of the mucous membranes and upper respiratory tract, eyes, and skin. Inhalation may be fatal. Prolonged contact can cause severe irritation or burns. Wear gloves, safety glasses, and protective clothing. Wash thoroughly after handling. Flammable: Vapor may travel a considerable distance to source of ignition and flash back.

6. Immediately pour the gel. Hold the plate at a 45° angle. Avoid air bubbles by pouring the acrylamide/urea solution down one side of the plate in a single smooth motion. After the plates are full, lay the gel down flat on a table.

7. Insert the flat side of a shark-tooth comb. Clamp the plates together over the comb.

8. Allow the gel to polymerize for at least 1 hour. Save a little of the acrylamide solution in a pasteur pipette to check for polymerization.

9. After polymerization is complete, remove both the clamps and the tape from the bottom. Carefully remove the comb. Rinse the top of the gel with deionized H_2O to remove any unpolymerized acrylamide. Rinse the comb with deionized H_2O to clean off any acrylamide. Invert the comb and insert between the plates so that the teeth just touch the gel surface. Do not puncture the gel surface or move the comb once it has been inserted.

10. Clamp the gel onto the electrophoresis apparatus. Pour 1x TBE buffer into the top and bottom buffer chambers. Flush out the wells with 1x TBE to remove any unpolymerized acrylamide.

11. Pre-run the gel for 30–60 minutes at 60-W constant power.

12. Before loading the sequencing gel, heat the samples at 70°C for 5 minutes, quickly chill on ice, and spin briefly in a microcentrifuge.

13. Carefully load 3–7 µl of each denatured DNA sequencing reaction per lane using a micropipettor. Note the order of the samples loaded on the gel (e.g., GATC).

14. Electrophorese at 60-W constant power until the bromophenol blue dye reaches the bottom of the gel (2–3 hours).

15. Turn off the power and unplug the power supply. Remove the gel from the sequencing setup and lay on the bench with the large plate on the bottom. Carefully separate the plates by gently prying between them with a spatula.

Autoradiography

1. Slowly lay a sheet of Whatman filter paper over the gel. Smooth out the air bubbles so that the paper contacts the entire gel.

2. Gently peel the filter paper off the plate. The gel will stick to the filter paper.

3. Prepare a cold trap between the vacuum pump and the gel dryer. Cover the top of the gel with plastic wrap and dry the gel for at least 1 hour at 80ºC.

4. Remove the plastic wrap and place the dried gel in a film holder. In the dark room, turn off the lights, remove a sheet of autoradiography film, and immediately replace the cover on the box of film. Lay the sheet of autoradiography film over the gel and then close the film holder, making sure it is "light-tight." If no intensifying screen is used, expose the film at room temperature. If an intensifying screen is used, expose the film at –70ºC.

5. Develop the film as listed below:

    a. Remove the film in the dark room with the lights turned off and the safelight turned on.

    b. Submerge the film in developer and agitate intermittently for approximately 2 minutes.

    c. Rinse the film in water for approximately 30 seconds.

    d. Fix the film for 5 minutes in rapid fixer.

    e. Rinse the film in running water for at least 15 minutes.

    f. Hang the film by the edge to air dry.

Silver Staining Protocol

This protocol is based on the Silver Sequencing Kit sold by Promega.

1. Prepare the following solutions:

   a. *Fix/stop solution (10% glacial acetic acid).* Add 200 ml of glacial acetic acid to 1800 ml of "ultrapure" (nanopure or deionized) H_2O.

   b. *Staining solution.* Combine 2 g of silver nitrate and 3 ml of 37% formaldehyde in 2000 ml of ultrapure H_2O.

   c. *Developing solution.* Dissolve 60 g of Na_2CO_3 (ACS grade) in 2000 ml of ultrapure H_2O. Chill on ice. *Immediately* before use, add 3 ml of 37% formaldehyde and 400 µl of sodium thiosulfate (10 mg/ml).

   Cautions: Formaldehyde is toxic and is also a carcinogen. It is readily absorbed through the skin and is irritating or destructive to the skin, eyes, mucous membranes, and upper respiratory tract. Wear gloves and safety glasses.
       Silver nitrate is a strong oxidizing agent and should be handled with care.

2. Separate the gel plates. The gel should stick to the short plate. Place the glass plate with the gel in the fix/stop solution for 20 minutes.

3. Pour off the fix/stop solution and save for use as a stop solution.

4. Rinse the gel for 2 minutes in ultrapure H_2O. Pour off the H_2O and repeat the rinse step two more times.

5. Place the gel in staining solution and agitate well for 30 minutes.

6. Add the formaldehyde and soldium thiosulfate to the developing solution. Very briefly (~2 seconds) rinse the gel in ultrapure H_2O and then place the gel in 1000 ml of developing solution. Agitate for 2–3 minutes until bands begin to appear.

7. Remove the developing solution and replace with the remaining 1000 ml of the fresh developing solution. Allow the gel to develop until the background acquires a light brown color.

8. Add the fix/stop solution to stop further development.

9. Rinse the gel in ultrapure H_2O. Repeat this rinse step.

10. Allow the gel to air-dry on the plate.

Reading DNA Sequencing Gels

Read the DNA sequencing autoradiogram. When runs of bases occur, not all the bands will have the same intensity. Compressions sometimes occur that appear as fatter bands or blank spaces in the autoradiogram. Therefore, always note the distance between the bands when reading an autoradiogram. For trouble-shooting suggestions, see Section 7. Compare the DNA sequence of the O_{nac} mutants with the DNA sequence of the wild-type O_{nac} mutants (see Fig. 10.2).

A   *Klebsiella put* regulatory region:

```
        DraI                          HincII                           SmaI
-----CGTTTTTAAAGGTTGCACCAAACAAAAGTGTTAACTCACGCATACAAATACCCTATGAGCCCCGGGTTAAATT-----
-----GCAAAAATTTCCAACGTGGTTTGTTTTCACAATTGAGTGCGTATGTTTATGGGATACTCGGGGCCCAATTTAA-----
```

B O_{nac} challenge phage sequence (wild type):

```
         . -35 . . . . Pant . . . . . . . . -10 .
    -----ttgacatgatagaagcactctactatattcccGGGCTCATAGGGTATTTGTATGCGTGAGTTggg -----
```

Figure 10.2 O_{nac} challenge phage. (*A*) The DNA sequence of the wild-type *put* regulatory region from *Klebsiella aerogenes* (Chen and Maloy 1991) including the Nac-binding site. (*B*) The position of the Nac-binding site (shown in capital letters) cloned into challenge phage.

Subcloning and dsDNA Sequence Analysis of Challenge Phage Mutants

As an alternative to cycle sequencing, the DNA fragment from the challenge phage can be subcloned onto a plasmid and the DNA sequence directly determined from the plasmid DNA. Both the substituted operator site and the *mnt*::Km9 insertion are on a 1.3-kb *Eco*RI fragment; the DNA fragment can be easily subcloned by selecting for Km^r insertions on the plasmid. The plasmid DNA is purified and the double-stranded DNA is denatured with alkali before dideoxy DNA sequencing.

MATERIALS

pUC18
*Eco*RI and 10x *Eco*RI buffer
λ *Hin*dIII DNA standards
Loading dye solution (Appendix F)
T4 DNA ligase and 10x ligase buffer
Drop dialysis filters (see Appendix F)
Sterile glycerol
Sterile deionized H_2O
Sterile 0.85% NaCl
LB-agar-Ap-Km plates (see Appendix B)
LB-agar-Ap-Km plates (see Appendix B)
Plasmid broth-Ap-Km (see Appendix B)
PEB (see Appendix F)
ALM (see Appendix F)
7.5 M Ammonium acetate (pH 7.6)
2 M Ammonium acetate (pH 7.4)
TE buffer (see Appendix F)
1x TBE (see Appendix F)
Ethidium bromide
Sephadex G-50 spin column (see Appendix F)
2 N NaOH
Isopropanol
Ethanol (70% and 100%)
Anti-O*mnt* primer (20 pmoles/µl)

Sequenase® T7 DNA polymerase

> *Note:* Reagents for dideoxy sequencing using Sequenase® are available as a kit from U.S. Biochemical Corporation.

5x Sequenase buffer

Tris-HCl (pH 7.5)	200 mM
$MgCl_2$	100 mM
NaCl	250 mM

Enzyme dilution buffer. Dilute Sequenase immediately before use—
Do not store the diluted enzyme.
   10 mM Tris-HCl (pH 7.5)
   5 mM DTT (dithiothreitol)
   0.5 mg/ml Bovine serum albumin (nuclease-free)
[α-^{35}S]ATP (1000–1500 mCi/mmole; >10 mCi/ml)
Labeling mix (5x)

dGTP	7.5 μM
dCTP	7.5 μM
dTTP	7.5 μM

Store at –20°C, thaw just prior to use, and then keep on ice.
Four termination mixes (each with a different ddNTP): G termination mix (ddGTP reaction); A termination mix (ddATP reaction); T termination mix (ddTTP reaction); C termination mix (ddCTP reaction)

dGTP	80 μM
dATP	80 μM
dTTP	80 μM
dCTP	80 μM
ddNTP	8 μM
NaCl	50 mM

Store at –20°C, thaw just prior to use, and then keep on ice.
Dye-stop solution

Formamide	95%
EDTA (pH 7.6)	20 mM
Bromophenol blue	0.1%
Xylene cyanol FF	0.1%

Sterile microcentrifuge tubes
Kimwipes (lint-free tissues)

Subcloning Phage DNA onto a Plasmid Vector

1. Prepare high-titer P22 lysates from each mutant O_{nac} challenge phage (see Appendix D, Phage P22 Lysates) and then purify the phage DNA from each lysate (see Appendix D, Phage P22 Lysates).

2. For each of the purified phage DNA samples and for the pUC18 plasmid vector, add everything except the restriction enzyme to a sterile microcentrifuge tube. Remove the restriction enzyme from the freezer just before use and immediately place it on ice, add the enzyme, and then immediately return the enzyme to the freezer.

Addition	Restriction digest	Control
Phage DNA	5 μl	5 μl
pUC18 DNA	2 μl	2 μl
Sterile deionized H_2O	10 μl	11 μl
10x *EcoRI* buffer	2 μl	2 μl
EcoRI	1 μl	–

3. Incubate for approximately 2 hours at 37ºC.

4. Inactivate the *EcoRI* by heating for 20 minutes at 70ºC. Store at 4ºC.

5. Remove 2 μl of each sample. Add 7 μl of TE and 1 μl of 10x Loading dye solution. Run the digested DNA and undigested controls on a 1.2% agarose gel in 1x TBE at 50 mA until the bromophenol blue band migrates approximately three-fourths the length of the gel. Also include a lane with λ *Hind*III size standards (see Appendix F, Agarose Gels).

6. Stain with ethidium bromide and then photograph the gel to estimate the DNA yield and to confirm the restriction digest.

   Caution: Ethidium bromide is a powerful mutagen and moderately toxic. Always wear gloves when working with solutions that contain this dye. After use, decontaminate and dispose of these solutions in accordance with the safety practices established by your institution's Safety Office.

7. Set up DNA ligation reactions for each challenge phage DNA as follows:

Challenge phage + pUC18 DNA cut with	
*Eco*RI (from step 4 above)	20 µl
Sterile deionized H$_2$O	6 µl
Ligase buffer (10×)	3 µl
T4 DNA ligase	1 µl

8. Incubate overnight at 15ºC or let stand on the bench overnight.

9. Remove the excess salt from the ligation mixture by drop dialysis for approximately 1 hour (see Appendix F, Drop Dialysis). Do not touch the filters—use forceps. The volume may increase slightly during this step.

10. Electroporate 2 µl of the dialyzed ligation mix into 40 µl of electrocompetent TSM217 cells (prepared as described in Appendix D, Preparation of Electrocompetent Cells and Electroporation).

11. Centrifuge the broth cultures in a microcentrifuge for 1 minute to pellet the cells.

12. Resuspend the cells in 0.2 ml of 0.85% NaCl. Spot 20 µl on one LB-agar-Ap-Km plate. Allow the spot to dry and then streak for isolated colonies. Incubate both plates for 1–2 days at 37ºC.

13. Divide an LB-agar-Ap-Km plate into four sectors. Streak four colonies of potential Km^r clones from each phage for isolation. Incubate the plate overnight at 37ºC.

Purification of Plasmid DNA for Sequencing

1. Subculture a single Km^r Ap^r colony into 5 ml of Plasmid broth-Ap-Km. Incubate with aeration overnight at 37°C.

2. Pour the culture into microcentrifuge tubes. Centrifuge the culture in a microcentrifuge for 2 minutes to pellet the cells.

3. Pour off the supernatant. Resuspend the pellet from each tube in 200 μl of PEB. Vortex for 30 seconds.

4. Add 400 μl of a freshly prepared ALM. Mix by inverting 10 times and then place the tubes in an ice-water bath for 5 minutes.

5. Add 300 μl of ice-cold 7.5 M ammonium acetate (pH 7.6) to the solution, mix by gently inverting the tube ten times, and then place in an ice-water bath for 5 minutes.

6. Centrifuge in a microcentrifuge for 10 minutes at room temperature. Transfer the supernatant into clean 1.5-ml microcentrifuge tubes. Centrifuge for another 5 minutes.

7. Transfer clear supernatant into new 1.5-ml microcentrifuge tubes. Add 0.6 volume of isopropanol (540 μl), mix by inverting several times, and let stand for 10 minutes at room temperature.

8. Centrifuge in a microcentrifuge for 10 minutes at room temperature.

9. Discard the supernatant. Add 200 μl of 2 M ammonium acetate (pH 7.4) to the pellet. Resuspend the pellet *very thoroughly* with an automatic pipettor and then vortex for 10 seconds. Place in an ice-water bath for 5 minutes.

10. Centrifuge in a microcentrifuge for 5 minutes at room temperature.

11. Remove the supernatant and place in a new microcentrifuge tube. Add 200 μl of isopropanol to each tube and let stand for 10 minutes at room temperature.

12. Centrifuge in a microcentrifuge for 10 minutes at room temperature.

13. Pour off the supernatant and remove excess liquid by inverting on a Kimwipe. Overlay the pellet with 1 ml of ice-cold 70% ethanol and then centrifuge in a microcentrifuge for 5 minutes at room temperature. Dry the pellet under vacuum or by blotting the tube on a Kimwipe.

14. Resuspend the DNA in 20 μl of TE and store at –20ºC until use.

Double-stranded Plasmid Sequencing

1. To denature the plasmid DNA, add 5 μl of 2 N NaOH to each tube with 20 μl of DNA template. Mix and then let stand for 10 minutes at room temperature.

2. Pass each 25-μl sample through a Sephadex G-50 spin column. This step removes the NaOH from the denatured DNA. Keep the denatured DNA on ice or at –20ºC until use.

3. Mix the template and primer as follows:

Denatured DNA	7 μl
5x Sequenase buffer	2 μl
Sequencing primer (2 pmoles/μl)	1 μl

4. Allow the primer to anneal for 30 minutes at 37ºC and then for 10 minutes at room temperature. Keep on ice or at –20ºC until use.

5. Prepare four 0.5-ml microcentrifuge tubes per DNA sample, labeled A, T, G, C. Add 2.5 μl of one of the four termination mixes (ddATP, ddCTP, ddGTP, ddTTP) to the corresponding tube. Keep the tubes on ice.

6. Just before use, dilute the Sequenase T7 DNA polymerase 1:8 in Enzyme dilution buffer and dilute the Labeling mix 1:10 in deionized H_2O.

7. Add the following to each tube in the order shown:

0.1 M Dithiothreitol	1.0 μl
Diluted labeling mix	2.0 μl
Diluted Sequenase	2.0 μl
[α-^{35}S]dATP	0.5 μl

   Let stand for 5 minutes at room temperature for chain elongation.

8. Transfer 3.5 μl of the labeling reaction (from step 7) to each of the four termination tubes (from step 5). Incubate for 5 minutes at 37°C for chain termination.

9. Add 4 μl of Dye-stop solution to each tube. Store at –20°C until use.

10. Analyze the sequencing reactions by electrophoresis and autoradiography as described above in DNA Sequencing Gels.

Isolation of Second-site Suppressor Mutations That Recognize Mutant Operator Sites

Experiment 11 demonstrates the use of challenge phage to select for second-site suppressor mutants of a DNA-binding protein. The O_{nac} mutant challenge phage isolated in Experiment 9 cannot form Km^r lysogens on a nac^+ strain. Figure 11.1 shows some examples of such mutants. The P$_{tac}$-nac plasmid is mutagenized here to isolate nac mutations that alter the sequence specificity of Nac and allow it to bind to a mutant O_{nac} site. Second-site suppressor mutants are selected as Km^r lysogens after infection with the mutant O_{nac} challenge phage. Although the second-site suppressor mutations are expected to be in the nac gene, it is conceivable that a mutation affecting some other protein on the chromosome could also give this phenotype. Thus, identifying potential second-site suppressor mutations in the nac gene requires transferring the P$_{tac}$-nac plasmid from the Km^r lysogens back into the original host and retesting the phenotype. Below are listed the bacterial strains used in this experiment.

BACTERIAL STRAINS

TSM201 ($hsdL$ [r$^-$m$^+$])
TSM204 ($hsdL$ [r$^-$m$^+$]/pGW1700 [$mucA^+B^+$ Tc^r])
TSM207 ($supE$/pPC36 P$_{tac}$-nac $lacI^Q$ Sp^r Sm^r)

Wild-type O_{nac} challenge phage

P_{ant}
tatattcccGGGCTCATAGGGTATTTGTATGCGTGAGTTgggaattcg

O_{nac} challenge phage mutants

5A	tatattcccGGGC**A**CATAGGGTATTTGTATGCGTGAGTTgggaattcg
6T	tatattcccGGGCT**T**ATAGGGTATTTGTATGCGTGAGTTgggaattcg
6G	tatattcccGGGCT**G**ATAGGGTATTTGTATGCGTGAGTTgggaattcg
7T	tatattcccGGGCTC**T**TAGGGTATTTGTATGCGTGAGTTgggaattcg
7G	tatattcccGGGCTC**G**TAGGGTATTTGTATGCGTGAGTTgggaattcg
8A	tatattcccGGGCTCA**A**AGGGTATTTGTATGCGTGAGTTgggaattcg
8C	tatattcccGGGCTCA**C**AGGGTATTTGTATGCGTGAGTTgggaattcg
9G	tatattcccGGGCTCAT**G**GGGTATTTGTATGCGTGAGTTgggaattcg
11A	tatattcccGGGCTCATAG**A**GTATTTGTATGCGTGAGTTgggaattcg
15C	tatattcccGGGCTCATAGGGTA**C**TTGTATGCGTGAGTTgggaattcg
19A	tatattcccGGGCTCATAGGGTATTTG**A**ATGCGTGAGTTgggaattcg
20T	tatattcccGGGCTCATAGGGTATTTGT**T**TGCGTGAGTTgggaattcg
21A	tatattcccGGGCTCATAGGGTATTTGTA**A**GCGTGAGTTgggaattcg
21G	tatattcccGGGCTCATAGGGTATTTGTA**G**GCGTGAGTTgggaattcg
25C	tatattcccGGGCTCATAGGGTATTTGTATGCG**C**GAGTTgggaattcg

Figure 11.1   DNA sequences of the wild-type *Klebsiella aerogenes put* regulatory region and the O_{nac} challenge phage derived from the *put* regulatory region. The parental O_{nac} challenge phage was constructed by subcloning the *Sma*I-*Hin*cII region into the *Sma*I site on pPY190.

MATERIALS

For culture media and supplements, see Appendix B.

LB-agar plates
LB-agar-Sp plates
LB-agar-Sp-Km plates
LB-agar-Sp-Tc plates
LB-agar-Sp-Km-IPTG (1 mM) plates
LB broth
LB broth-Tc
LB broth-Sp
LB broth-Sp-IPTG (1 mM)
LB broth-Sp-Tc-IPTG (1 mM)
Plasmid broth-Sp
Plasmid broth-Km-Sp-IPTG (1 mM)
pPC36 plasmid DNA
Sterile deionized H_2O
100 mM IPTG (dioxane-free)
50% Glycerol
0.85% NaCl

Sterile microtiter plates
Sterile microcentrifuge tubes
Sterile pasteur pipettes
Sterile petri dishes
UV Stratalinker® (Stratagene). This is a reliable, user-friendly instrument for irradiation of samples with a defined dose of 254 nm UV light.

PROCEDURE

1. Start TSM204 recipient cells in 10 ml of LB broth-Tc. Incubate with aeration overnight at 37°C.

2. Place 5 µl of pPC36 DNA in a sterile petri dish.

3. Remove the lid and UV-irradiate to 600×10^2 µJ/cm^2 in a Stratalinker. Save the UV-irradiated plasmid in a sterile microcentrifuge tube at 4°C.

   Caution: UV is a mutagen and carcinogen, as well as an eye-damaging agent. Always wear safety glasses and gloves and work in a glass-enclosed hood if possible.

4. Prepare electrocompetent TSM204 cells from the TSM204 cultures as described in Appendix D (Preparation of Electrocompetent Cells and Electroporation).

5. Mix 2 µl of plasmid DNA with 40 µl of electrocompetent cells:

   a. UV-irradiated pPC36 DNA.

   b. Nonirradiated pPC36 DNA (control for efficiency of UV mutagenesis).

   c. No DNA control.

6. Electroporate the samples as described in Appendix D (Preparation of Electrocompetent Cells and Electroporation). Include a DNA-only control.

7. Pour off the supernatant. Resuspend the cell pellet in 0.1 ml of LB broth and spread the *entire* solution from each culture on an LB-agar-Sp-Tc plate. Incubate the plates overnight at 37°C.

8. Check the electroporation plates and record the number of colonies on each plate (each plate should have a lawn of cells). No growth should be seen on the cells-only and DNA-only controls.

9. Add 1 ml of LB broth to each plate with electroductants (from both the UV-irradiated and nonirradiated plasmids) and suspend the colonies in the liquid with a spreader sterilized by dipping in ethanol then briefly flamed.

10. Remove the cell suspension with a sterile pasteur pipette. To a tube with 10 ml of LB broth-Sp-Tc-IPTG (1 mM), add a sufficient volume of the pooled colonies to obtain a culture that is about as turbid as an early to mid exponential phase culture. Approximately 50 μl is usually sufficient, but the volume may need to be increased depending on density of cell suspension. Freeze the remainder of the cells in 50% glycerol at –70°C.

11. Incubate with aeration for approximately 1 hour at 37°C or until mid exponential phase.

12. For each O_{nac} mutant challenge phage and for the parent O_{nac}^{+} challenge phage control, mix 0.5 ml of cells with phage in a sterile microcentrifuge tube to give a multiplicity of infection (moi) of 10–50 pfu/cell. The volume of each phage required to give an moi of approximately 25 pfu/cell can be calculated as shown in the following equation (assuming the cultures are at ~5 × 10^8 cells/ml).

$$\frac{ml}{pfu} \times \frac{25\ pfu}{cell} \times \frac{5 \times 10^8\ cells}{ml} \times 0.5\ ml \times \frac{1000\ \mu l}{ml} = \mu l\ phage$$

13. Incubate for 1 hour at 37°C to allow phage adsorption and phenotypic expression of Km^r.

14. Plate each tube on an LB-agar-Sp-Km-IPTG (1 mM) plate.

15. Spot 0.1 ml of cells-only (cell control) and 5 μl of phage-only (phage control) on an LB-agar-Sp-Km-IPTG (1 mM) plate, making sure that the spots do not run together. Incubate the plates upside down for 2 days at 37°C.

16. Check the plates. No growth should be seen on the cells-only and phage-only controls. Any Km^r colonies on the challenge plates may be due to second-site suppressor mutations. Determine the efficiency of the UV mutagenesis by comparing the frequency of spontaneous (nonirradiated plasmid) mutants with that of the UV-induced (irradiated plasmid) mutants.

17. Restreak any Km^r colonies on LB-agar-Sp-Km and LB-agar-Sp-Km-IPTG (1 mM) plates (eight colonies per plate). Incubate for 2 days at 37°C.

18. Start cells from single colonies of three potential suppressor mutants in 2 ml of Plasmid broth-Km-Sp-IPTG (1 mM). Also start the parental strain TSM207 in 2 ml of Plasmid broth-Sp. Grow to stationary phase with aeration at 37°C (at least 16 hours, but may take longer for some mutants).

19. Start TSM201 recipient cells in 10 ml of LB broth. Incubate with aeration overnight at 37°C.

20. Purify plasmid DNA from the potential suppressor mutants and the parental control with the *nac*$^+$ plasmid (TSM207) as described in Appendix F (Plasmid DNA Minipreps).

21. Prepare electrocompetent cells from the 10-ml TSM201 cultures as described in Appendix D (Preparation of Electrocompetent Cells and Electroporation).

22. Electroporate each plasmid DNA into TSM201 as described in Appendix D (Preparation of Electrocompetent Cells and Electroporation). Include a control with no plasmid DNA.

23. Resuspend the cell pellets in 0.1 ml of LB broth and spot 50 μl from each culture on an LB-agar-Sp plate. Allow the spots to dry (~30 minutes) and then streak for isolated colonies. Also spot cells-only and DNA-only controls on an LB-agar-Sp plate. Incubate overnight at 37°C.

24. Check the plates. No growth should be seen on the cells-only and DNA-only controls. Pick one colony from each different

plasmid electroporation and restreak on an LB-agar-Sp plate (eight of the electroporants can be streaked on a single plate divided into sectors). Incubate the plates overnight at 37°C.

25. Start single colonies from three of the potential plasmid suppressor mutants and the parental control with the nac^+ plasmid (TSM207) in 2 ml of LB broth-Sp. Incubate with aeration overnight at 37°C.

26. Subculture 0.1 ml of the suppressor mutants and the nac^+ control into 2 ml of LB broth-Sp and grow with aeration at 37°C to mid exponential phase (1–2 hours).

27. Subculture 0.5 ml of the mid exponential phase cultures of the suppressor mutants and the nac^+ control into 1.5 ml of LB broth-Sp and 1.5. ml of LB broth-Sp-IPTG (1 mM).

28. Grow with aeration at 37°C to mid exponential phase (~1–2 hours).

29. To test for allele specificity of the suppressor mutants, assay the efficiency of binding of the three suppressor mutants and the nac^+ control to three different O_{nac} mutant challenge phage and the wild-type O_{nac} challenge phage. Carry out the assays in Microtiter Plates 1 and 2 as shown (see p. 393). The amount of phage added should give an moi of 10–50 when mixed with 10^8 cells. The volume of each phage required to give an moi of approximately 25 pfu/cell can be calculated as shown in the following equation (assuming the cultures are at ~5 × 10^8 cells/ml).

$$\frac{ml}{pfu} \times \frac{25\ pfu}{cell} \times \frac{5 \times 10^8\ cells}{ml} \times 0.1\ ml \times \frac{1000\ \mu l}{ml} = \mu l\ phage$$

For each of the three suppressor mutants and parental control, add 0.1 ml of mid exponential phase cells and calculated volume of phage to columns 1 and 8 of the microtiter dishes.

30. Mix the cells and phage by gently pipetting the solution up and down and then let stand for 1 hour at 37°C to allow phage adsorption and phenotypic expression.

MICROTITER PLATE #1

Cells	Phage	Cells grown in LB-broth-Sp −IPTG						Cells grown in LB-broth-Sp +IPTG				
Suppressor 1	O$_{nac}$ mutant phage 1	1/1	LB	LB	LB	LB		1/1	LB	LB	LB	LB
	O$_{nac}$ mutant phage 2	1/2	LB	LB	LB	LB		1/2	LB	LB	LB	LB
	O$_{nac}$ mutant phage 3	1/3	LB	LB	LB	LB		1/3	LB	LB	LB	LB
	O$_{nac}^{+}$ phage	1/+	LB	LB	LB	LB		1/+	LB	LB	LB	LB
Suppressor 2	O$_{nac}$ mutant phage 1	2/1	LB	LB	LB	LB		2/1	LB	LB	LB	LB
	O$_{nac}$ mutant phage 2	2/2	LB	LB	LB	LB		2/2	LB	LB	LB	LB
	O$_{nac}$ mutant phage 3	2/3	LB	LB	LB	LB		2/3	LB	LB	LB	LB
	O$_{nac}^{+}$ phage	2/+	LB	LB	LB	LB		2/+	LB	LB	LB	LB

10^0 10^{-1} 10^{-2} 10^{-3} 10^{-4} 10^0 10^{-1} 10^{-2} 10^{-3} 10^{-4}

MICROTITER PLATE #2

Cells	Phage	Cells grown in LB-broth-Sp −IPTG						Cells grown in LB-broth-Sp +IPTG				
Suppressor 3	O$_{nac}$ mutant phage 1	3/2	LB	LB	LB	LB		3/2	LB	LB	LB	LB
	O$_{nac}$ mutant phage 2	3/2	LB	LB	LB	LB		3/2	LB	LB	LB	LB
	O$_{nac}$ mutant phage 3	3/3	LB	LB	LB	LB		3/3	LB	LB	LB	LB
	O$_{nac}^{+}$ phage	3/+	LB	LB	LB	LB		3/+	LB	LB	LB	LB
nac^{+} control	O$_{nac}$ mutant phage 1	+/1	LB	LB	LB	LB		+/1	LB	LB	LB	LB
	O$_{nac}$ mutant phage 2	+/2	LB	LB	LB	LB		+/2	LB	LB	LB	LB
	O$_{nac}$ mutant phage 3	+/3	LB	LB	LB	LB		+/3	LB	LB	LB	LB
	O$_{nac}^{+}$ phage	+/+	LB	LB	LB	LB		+/+	LB	LB	LB	LB

10^0 10^{-1} 10^{-2} 10^{-3} 10^{-4} 10^0 10^{-1} 10^{-2} 10^{-3} 10^{-4}

Microtiter Plates #1 and #2.

31. Add 45 µl of LB broth to each well in the microtiter plate marked LB. Dilute 5 µl from each column of the microtiter dish to the next and mix by gently pipetting the solution up and down to achieve sequential tenfold dilutions of the cells + phage. Use new pipette tips for each dilution.

32. Remove 5 µl from each dilution of the cells grown without IPTG and spot onto an LB-agar-Sp-Km plate. Remove 5 µl from each dilution of the cells grown with IPTG and spot onto an LB-agar-Sp-Km-IPTG (1 mM) plate.

33. Spot 5 µl of cells-only (cell control) and 5 µl of phage-only (phage control) on the LB-agar-Sp-Km plate. Allow the spots to dry at room temperature and then incubate upside down for 1–2 days at 37ºC.

34. Determine the number of viable cells used. Dilute *both* the cell culture grown with IPTG and the culture grown without IPTG as follows. Add 45 µl of sterile 0.85% NaCl to the top two rows of a microtiter dish (see below). Add 5 µl of the culture grown without IPTG to duplicate wells in column 1 of the microtiter dish. Add 5 µl of the culture grown with IPTG to duplicate wells in column 8 of the microtiter dish. Mix by gently pipetting the solution up and down with the pipettor. Dilute the cells by sequentially transferring 5 µl into each subsequent column of the microtiter dish.

No IPTG added							+IPTG				
#1	#2	#3	#4	#5	#6	#7	#8	#9	#10	#11	#12
NaCl	NaCl	NaCl	NaCl	NaCl			NaCl	NaCl	NaCl	NaCl	NaCl
NaCl	NaCl	NaCl	NaCl	NaCl			NaCl	NaCl	NaCl	NaCl	NaCl
10^{-1}	10^{-2}	10^{-3}	10^{-4}	10^{-5}			10^{-1}	10^{-2}	10^{-3}	10^{-4}	10^{-5}

35. Spot 5 µl of the 10^{-2}, 10^{-3}, 10^{-4}, and 10^{-5} dilutions onto an LB-agar plate. Allow the spots to dry (~30 minutes) and then incubate the plates upside down overnight at 37°C.

36. Calculate the number of lysogens/viable cell and include the results in the following table. Suppressor mutations in the *nac* gene yield a higher frequency of lysogens/viable cell than the O_{nac}^+ parent plasmid. Suppressor mutations in the *nac* gene are transferred with the plasmid when it is electroporated into a new host. Such suppressor mutations may be due to increased affinity of the Nac protein for O_{nac} or to an altered specificity of recognition of O_{nac}. Suppressor mutants may also be due to mutations elsewhere on the chromosome, but, in contrast to Nac suppressor mutations, suppressor mutations in other genes on the chromosome will not be transferred with the plasmid. Such mutations may define other proteins that interact with the Nac protein or O_{nac}. Assign strain numbers to each of the mutations and include each mutant in the strain collection.

Experiment 11 Results

Phage		TSM207		Suppressor 1		Suppressor 2		Suppressor 3	
		−IPTG	+IPTG	−IPTG	+IPTG	−IPTG	+IPTG	−IPTG	+IPTG
$O_{nac}{}^+$	10^0								
	10^{-1}								
	10^{-2}								
	10^{-3}								
	10^{-4}								
	cells/ml								
	% lysogeny								
O_{nac-1}	10^0								
	10^{-1}								
	10^{-2}								
	10^{-3}								
	10^{-4}								
	cells/ml								
	% lysogeny								
O_{nac-2}	10^0								
	10^{-1}								
	10^{-2}								
	10^{-3}								
	10^{-4}								
	cells/ml								
	% lysogeny								
O_{nac-3}	10^0								
	10^{-1}								
	10^{-2}								
	10^{-3}								
	10^{-4}								
	cells/ml								
	% lysogeny								

Experiment 11 Results

Introduction to Experiments 12–15

Experiments 12 through 15 present various techniques used to approach genetic analysis in diverse bacterial species. They illustrate approaches to examine two commonly studied properties of bacterial pathogens: mechanisms of colonization and regulation of virulence gene expression. Techniques that are broadly applicable for these investigations begin with the identification of relevant genes and then progress to the level of targeted mutational analysis.

The first two experiments involve the isolation of random gene fusions and their initial characterization in bacteria for which genetic systems are not very well developed. Experiment 14 presents broadly applicable methods for allelic exchange into the chromosome to create either genetic disruptions or to study the phenotypes conferred by nonselectable point mutations present within cloned fragments of DNA. The final experiment demonstrates a method for conducting site-directed mutagenesis in vectors that have properties useful for reversion analyses. These mutations are then placed in their normal chromosomal context using the allelic exchange procedure in Experiment 14.

Isolation of *Vibrio cholerae* Tn*phoA* Insertions

Many studies in genetically diverse bacteria address the functions of cell surface or extracellularly secreted products, for which no direct mutant selection has been devised. A method to identify genes that encode protein export signals, and are thus likely to be involved in such processes, is provided by the gene fusion transposon Tn*phoA* (see Section 5, Operon and Gene Fusions). Experiment 12 uses Tn*phoA* mutagenesis to identify genes involved in the elaboration of the *Vibrio cholerae* TCP colonization pilus. To facilitate the use of Tn*phoA* in a wide range of gram-negative bacterial species, the transposon has been inserted in a nonexpressing form (for *phoA* expression) onto a broad host range IncP1 plasmid. Selection for transposition of Tn*phoA* from the pRT291 delivery plasmid into the recipient genome is accomplished using an IncP1 incompatibility method. Active PhoA⁺ fusion strains then provide a pool from which to screen for the lack of surface TCP pili (based on a loss of bacterial autoagglutination in culture, a screen that would be difficult if the number of potential insertion mutants could not first be narrowed down to a manageable size by screening only the insertions in genes that encode secreted products). Insertions are expected to be identified in the pilin structural gene (*tcpA*) or any of a number of genes encoding periplasmic or membrane biogenesis functions required for pilus assembly.

When isolating new mutations, it is essential to backcross them into a clean background to ensure that any mutant phenotype correlates with the mutation being examined and is not due to secondary alterations that may have occurred during the mutagenesis procedure. As demonstrated in Experiments 1–4, backcrossing is greatly facilitated by generalized transducing phages. However, in the case of *V. cholerae*, generalized transduction is not very efficient. In the absence of transduction, several plasmid-mediated allelic exchange systems have been developed (see Section 4, Broad Host Range Allelic Exchange Systems). Below are listed the bacterial strains used in this experiment.

BACTERIAL STRAINS

TSM301   *V. cholerae* O395, *ctxA* Sm^r
TSM302   O395 φ(*tcpA-phoA*)2-1(Hyb) Sm^r
TSM339   (RP4-2 Km^r *tet*::Mu)
          pRT291 (IncP1 Km^r Tc^r Tn*phoA*)
TSM349   pPH1JI (IncP1 Tra$^+$ Gm^r Sp^r)

Introduction of Tn*phoA* into *V. cholerae*

MATERIALS

For culture media and supplements, see Appendix B.

LB-agar plates
LB-agar-Sm plates
LB-agar-Sm-Tc plates
LB-agar-Tc-Km-XP plates
Sterile toothpicks

PROCEDURE

1. Prepare a fresh overnight streak plate of strain TSM339 on LB-agar-Tc-Km-XP to ensure that the donor plasmid is white. Streak recipient strain TSM301 on LB-agar-Sm.

2. Mate donor strain TSM339 with recipient strain TSM301 on LB-agar. Use a toothpick to spread adjacent thick lines of one

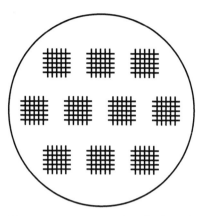

Figure 12.1 Ten crosshatched matings set up on an agar plate.

strain on the plate and then use a second toothpick to apply a thick cross-streak of the other strain (Fig. 12.1). Set up ten cross streaks on one plate for the isolation of independent mutants. Label each cross-streak 1–10. Incubate overnight at 37ºC.

3. Streak mating mix for isolated colonies on LB-agar-Sm-Tc and incubate overnight at 37ºC (these are transconjugants that have received the plasmid from the donor strain). Include a plate with the donor and recipient strains, neither of which should grow (Fig. 12.2).

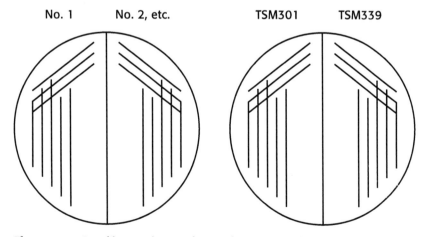

Figure 12.2 Streaking matings and controls, two per selective agar plate.

Vector "Kickout" by Incompatibility and Isolation of Colonies Expressing *phoA* Fusions

MATERIALS

For culture media and supplements, see Appendix B.

LB-agar plates
LB-agar-Sm-Km-Gm-Glu-XP plates
LB-agar-Sm-Km-Glu-XP plates
LB broth
Sterile flat toothpicks and applicator sticks

PROCEDURE

1. Use a flat toothpick or applicator stick to pick up a large number of pRT291 transconjugant cells from a mating (pick a group of colonies together from an area of the plate where the colonies are just becoming isolated). Apply as a fat streak to an LB-agar plate as for the previous mating (Fig. 12.1). Cross-streak with the pPH1JI donor strain TSM349 (Fig. 12.1). Repeat for the rest of the matings. Incubate overnight at 37°C.

2. Use a sterile applicator stick to scrape up a large number of cells from each mating and suspend in separate tubes containing 1 ml of LB broth, labeled 1–10. Spread 0.2-, 0.1-, 0.05-, and 0.02-ml aliquots of the suspended mating mix onto corresponding LB-agar-Sm-Km-Gm-Glu-XP for each mating (total of 40 plates). Incubate the plates at 30°C and examine for blue colonies a couple of times each day during the next 3 days. These colonies are transconjugants that have acquired and are maintaining the pPH1JI plasmid and have thus lost pRT291. Those that retain Tn*phoA* (Km^r) represent transposition events. Incubation at 30°C is used in this case because it is optimal for TCP expression in vitro and thus will promote expression of *phoA* inserts in the chosen genes. Glucose is included in this agar because the acid produced by its fermentation inhibits endogenous *V. cholerae* phosphate activity.

3. During the several days, purify blue colonies onto LB-agar-Sm-Km-Glu-XP. Try to pick various intensities of blue from each mating. (Colonies that appear blue at earlier times generally represent fusions with higher activity than those that appear later.) Incubate at 30°C.

Identification of Insertions That Exhibit a TCP⁻ Phenotype

MATERIALS

Vibrio broth (see Appendix B)

PROCEDURE

1. Test the mutants for TCP production by examining for autoagglutination (bacteria expressing TCP tend to clump together and fall out of suspension when grown in liquid culture medium). This is done by growing 2-ml overnight cultures of insertion mutants in Vibrio broth at 30°C. A light inoculum yields best results.

2. Grow TSM301 and TSM302 as positive and negative autoagglutination controls.

3. Select two autoagglutination-negative mutants for further analysis in Experiment 13. Try to select a dark-blue mutant and a light-blue mutant. Designate the two mutants as EXP12.1 and EXP12.2.

EXPERIMENT 13
Southern DNA Hybridization to Map Tn*phoA* Insertions

Southern DNA hybridization, the use of labeled nucleic acid probes to detect unlabeled DNA fragments separated by electrophoresis and immobilized on a membrane, has an extensive number of applications in molecular genetics. Traditionally, the probe is labeled with a radioactive isotope such as ^{32}P, with the hybridized products being detected by autoradiography. Several newer systems incorporate detection by chemiluminescence, rather than radioactivity. In this experiment, digoxigenin (DIG)-coupled nucleotides are incorporated into the probe. Hybridization of the probe is detected by incubating the membrane with an anti-DIG antibody linked to alkaline phosphatase, which can then be visualized by its reaction with a substrate that yields a luminescent product.

Southern analysis is performed on DNA prepared (Mekalanos 1983) from the mutants isolated in Experiment 12. This analysis is useful for mapping the Tn*phoA* insertion points within a defined region of chromosomal DNA as well as for determining whether a single copy of Tn*phoA* is present. DNA is isolated, digested, separated by agarose electrophoresis, transferred to a membrane, and then hybridized to a labeled probe. Two probes are used. The first is a cosmid clone that carries approximately 30 kb of DNA which covers the complete *tcp* gene cluster. This probe is used to determine whether a particular insertion lies within the previously defined region and whether it is likely to be within the *tcpA* pilin

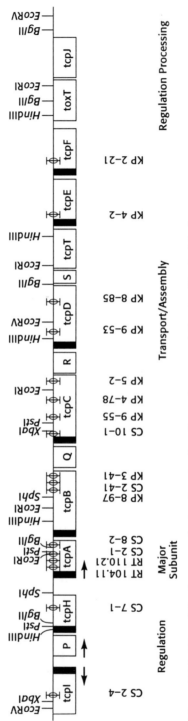

Figure 13.1 Map of the *tcp* gene cluster showing locations of representative *tcp*::TnphoA fusions.

subunit gene by using two different restriction enzyme digestions of the chromosomal DNA and comparing the fusion strain pattern to that of the wild type.

The second probe is an internal fragment of Tn*phoA* which is used to ensure that only one copy of Tn*phoA* is present in the newly isolated insertion mutants and, since it hybridizes asymmetrically, to map a series of unidirectional fusion joints within a gene or along a restriction fragment. When digesting the chromosomal DNA with an enzyme that does not cut within Tn*phoA* (e.g., *Xba*I), only a single band should hybridize. If two enzymes are used and one cuts within Tn*phoA* (e.g., *Xba*I and *Bam*HI), two bands will result, one corresponding to the 5′ side of the internal restriction site and the other to the 3′ side. For the analysis in this experiment, the 5′ side hybridizes more strongly due to the design of the asymmetric probe (see Appendix F, Mapping Fusion Joints by Combining Southern and Western Analyses) (Peterson and Mekalanos 1988). A combination of these digests is used to map positions of the insertions isolated in Experiment 12 relative to a collection of previously isolated *tcp-phoA* fusions (for a map of the *tcp* gene cluster, see Fig. 13.1). The analysis with a Tn*phoA* probe to confirm a single insertion is useful when the chosen bacterial species does not have a genetic method for backcrossing the mutation into a clean, nonmutagenized parent strain. Below are listed the bacterial strains used in this experiment.

BACTERIAL STRAINS

TSM301 O1 Ogawa serotype, classical biotype, *ctxA*
TSM302 φ(*tcpA-phoA*)2-1(Hyb) (*tcpA2-1*::Tn*phoA*)
TSM303 φ(*tcpC-phoA*)5-2(Hyb) (*tcpC5-2*::Tn*phoA*)
TSM304 φ(*tcpH-phoA*)7-1(Hyb) (*tcpH7-1*::Tn*phoA*)
TSM305 φ(*tcpC-phoA*)10-1(Hyb) (*tcpC10-1*::Tn*phoA*)
TSM306 φ(*tcpD-phoA*)8-85(Hyb) (*tcpD8-85*::Tn*phoA*)
TSM307 φ(*tcpF-phoA*)2-21(Hyb) (*tcpF2-21*::Tn*phoA*)
TSM308 φ(*tcpE-phoA*)4-2(Hyb) (*tcpE4-2*::Tn*phoA*)
TSM309 φ(*tcpB-phoA*)3-41(Hyb) (*tcpB3-41*::Tn*phoA*)
TSM310 φ(*tcpA-phoA*)104.11(Hyb) (*tcpA104.11*::Tn*phoA*)
EXP12.1 Putative *tcp*::Tn*phoA* fusion strain #1 (see Experiment 12)
EXP12.2 Putative *tcp*::Tn*phoA* fusion strain #2 (see Experiment 12)

DNA Extraction

MATERIALS

For preparation of materials, see Appendix F (Southern Hybridization Solutions).

LB broth (see Appendix B)
$T_{50}E_{50}$: 50 mM Tris, 50 mM EDTA (pH 8.0)
TE: 10 mM Tris, 1 mM EDTA (pH 7.4)
$T_{50}E_{50}$ containing 20 mg/ml lysozyme
TE containing 500 μg/ml proteinase K
10% SDS (sodium dodecyl sulfate)
3 M Sodium acetate (pH 4.8)
RNase A (DNase-free)
Buffered phenol
Chloroform
Isopropanol (chilled at −20°)
95% Ethanol (chilled at −20°C)
Microcapillary pipettes with flame-sealed, curved tip
Microcentrifuge tubes
Disposable plastic pipettes (1 ml) fitted with a mechanical pipettor
  (blue 3-ml size) or a P1000 Pipetman

PROCEDURE

1. Grow 2-ml overnight cultures of strains TSM301 and TSM302 and two of the Tn*phoA* mutant strains (EXP12.1 and EXP12.2) from Experiment 12 in LB broth at 37°C.

2. Aliquot 1 ml of culture into a microcentrifuge tube and pellet in a microcentrifuge for 1 minute at room temperature. Resuspend the pellet in 0.8 ml of $T_{50}E_{50}$.

3. Add 100 μl of $T_{50}E_{50}$ containing 20 mg/ml of lysozyme (add lysozyme fresh).

4. Incubate for 10 minutes at room temperature.

5. Add 20 μl of 10% SDS and then add 100 μl of TE containing 500 μg/ml proteinase K.

6. Incubate with mixing on a rotator for 1 hour at 37°C (e.g., tape tubes onto front of rotator in 37°C incubator).

7. Add enough buffered phenol to fill the tube almost to the top and mix on rotator in a 37°C incubator for 1 hour.

   Caution: Phenol is highly corrosive and can cause severe burns. Wear gloves, protective clothing, and safety glasses and work in a chemical fume hood. Rinse any areas of skin that come in contact with phenol with a large volume of water or PEG 400 and wash with soap and water; do not use ethanol!

8. Spin in a microcentrifuge for 10 minutes at room temperature. Collect the upper phase into a clean microcentrifuge tube. The viscosity of the DNA makes this manipulation difficult to achieve without contaminating the DNA with the phenol interface. To obtain a high-quality preparation without contamination, collect the top layer by using a slow circular motion with a 1-ml disposable plastic pipette fitted with a mechanical pipettor or use a P1000 Pipetman.

9. Extract with chloroform once by filling the tube to the top with chloroform and then rocking back and forth for 2 minutes.

   Caution: Chloroform is a carcinogen and may damage the liver and kidneys. Do not mouth pipette and avoid contact with skin.

10. Spin in a microcentrifuge for 1 minute at room temperature and collect the upper phase into a clean microcentrifuge tube.

11. Add 3 M sodium acetate to 0.15 M (~35 µl) and mix by shaking back and forth several times. Precipitate with isopropanol (chilled at –20°C) by filling the tube to the top and then inverting it gently back and forth for several minutes. When the DNA precipitate is visible, pull it out on the end of a sealed microcapillary pipette (the end of the microcapillary pipette is sealed in a bunsen burner flame).

12. Resuspend the precipitated DNA by twirling the sealed micro-capillary pipette with the DNA in a microcentrifuge tube already containing 0.5 ml of TE ($T_{10}E_1$).

13. Add 2.5 μg of DNase-free RNase A (e.g., 5 μl of 0.5 mg/ml stock).

14. Digest on a rotator as in step 6 for 1 hour at 37ºC.

15. Add 3 M sodium acetate to 0.15 M (35 μl) and precipitate with chilled isopropanol (fill tube to top) again, as in step 11.

16. Pull out precipitate with a sealed microcapillary pipette and rinse with a few drops of 95% ethanol (use a squirt bottle to apply several drops of ethanol to the DNA precipitate on the microcapillary pipette while holding over a microcentrifuge tube).

17. Redissolve in a microcentrifuge tube containing 0.5 ml of TE ($T_{10}E_1$) as described in step 12. Store at 4ºC. Generally, use 10 μl of this preparation for restriction analysis on gels and blots.

DNA Restriction Digest and Electrophoresis

MATERIALS

*Xba*I
*Bam*HI
*Hind*III
λ *Hind*III standard (preferably digoxigenin-labeled)
10x EndoR buffer (e.g., NEB buffer 2 or see Appendix F)
Agarose
TBE buffer (see Appendix F)
10x Loading dye solution (see Appendix F)
10% Ethidium bromide
Horizontal gel apparatus
Power pack
UV light box and camera
Microcentrifuge tubes
37ºC Water bath or heat block

PROCEDURE

1. Set up each digestion in a microcentrifuge tube and incubate for 2 hours in a 37ºC water bath or heat block. See step 3 loading pattern for appropriate enzymes.

Chromosomal DNA	10 µl
Deionized H_2O	7 µl
10x EndoR buffer (e.g., NEB buffer 2)	2 µl
Enzyme	1 µl

2. Prepare a 100-ml 0.6% agarose gel.

3. Add 2 µl of 10x Loading dye solution to each sample and load gel in the order diagrammed below. Run gel at 100 V until bromophenol blue dye is at the bottom (4–5 hours). Use the following pattern for loading the gel:

lane 1	lane 2	lane 3	lane 4	lane 5	lane 6
TSM301 *Xba*I	TSM302 *Xba*I	Mutant 1 *Xba*I	Mutant 2 *Xba*I	TSM301 *Hind*III	TSM302 *Hind*III

lane 7	lane 8	lane 9	lane 10	lane 11	lane 12
Mutant 1 *Hind*III	Mutant 2 *Hind*III	λStandard DIG-labeled	TSM301 *Xba*I/*Bam*HI	Mutant 1 *Xba*I/*Bam*HI	Mutant 2 *Xba*I/*Bam*HI

   Lanes 1–8 will be probed with the *tcp* cosmid
   Lanes 10–12 will be probed with the internal Tn*phoA* fragment

4. Stain for approximately 30 minutes in deionized H_2O containing 0.5 µg/ml ethidium bromide. Destain for 10–20 minutes in deionized H_2O. Visualize using a UV light box and camera. Photograph for possible later reference regarding quality of DNA and completeness of digestion. Place a fluorescent ruler alongside the gel for deducing sizes of detected bands if DIG-labeled standard is not used.

   Cautions: Ethidium bromide is a powerful mutagen and moderately toxic. Always were gloves when working with solutions that contain this dye. After use, decontaminate and dispose of these solutions in accordance with the safety practices established by your institution's Safety Office.

UV is a mutagen and carcinogen, as well as an eye-damaging agent. Always wear safety glasses and gloves and work in a glass-enclosed hood if possible.

DNA Transfer to Membrane

MATERIALS

0.25 N HCl
Denaturation solution (see Appendix F)
Neutralization solution (see Appendix F)
20x SSC (see Appendix F)
3MM Chromatography paper
Hybridization membrane
Used X-ray film slices
Paper towels
Pyrex dish
Glass plate
UV Stratalinker® (Stratagene). This is a reliable, user-friendly instrument for irradiation of samples with a defined dose of 254 nm UV light.

PROCEDURE

1. Trim the previously stained and photographed gel to include the region to be analyzed. The incubation times in the steps below are optimized for a gel of approximately 20 x 20 cm (cut these times in half for a small minigel). Immerse gel in 200 ml of 0.25 N HCl for 10 minutes or slightly less (no longer than 10 minutes). Rinse briefly in deionized H_2O.

2. Immerse gel in 200 ml of denaturation solution for 30 minutes (occasional swirling). Rinse.

3. Immerse gel in 200 ml of neutralization solution for at least 30 minutes (occasional swirling).

4. While gel is soaking, prepare hybridization membrane (this is a noncharged membrane to prevent nonspecific binding of alkaline phosphatase): Wear dry gloves. Cut a piece of membrane a centimeter or so longer and wider than the gel to be blotted ($\sim$7.5 $\times$ 10 cm). Hydrate first in deionized H_2O and then in 2$\times$ SSC. Cut six pieces of 3MM filter paper to the same size as the membrane. Hydrate one piece in 2$\times$ SSC.

5. Set up 20$\times$ SSC blotting box either with a stack of soaked 3MM filter papers to form a platform underneath the gel or use one sheet of soaked filter paper set on a glass plate support to act as a wick (see Fig. 13.2). Fill with enough 20$\times$ SSC so that it does not run dry during the transfer or to the rim of the filter paper stack.

6. Carefully slide gel onto the filter paper making sure there are no air bubbles between the gel and the paper. Cut off extra piece above comb (mark the positions of the lanes if needed for later cutting of the membrane into sections for different probes).

7. Line the gel with used film strips (these provide a nonporous border to prevent short-circuiting of the flow of transfer buffer around the gel instead of through it) and place the hybridization membrane on top, again removing any air bubbles. Place the wet (with 2$\times$ SSC) filter paper on top of the hybridization membrane and then the five dry pieces of filter paper on top of that.

8. Cut a 3–4-inch stack of paper towels (or 15–20 sheets of gel blot paper) to approximately the same size or slightly larger than the filter paper and stack on top. Place a glass plate and weight ($\sim$500 g) on top (see Fig. 13.2).

9. Let stand overnight or at the very least for 4 hours.

10. Remove the paper towels and filter papers, leaving the membrane. Nick bottom left corner for orientation. Also note which side of the membrane the DNA is on for some of the

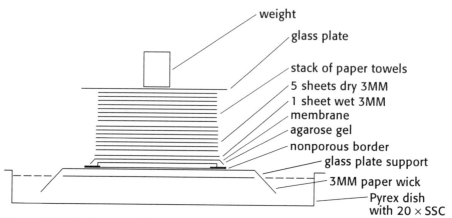

Figure 13.2 Southern transfer setup.

subsequent manipulations. Peel the membrane off smoothly and place in 2x SSC for 20 minutes (preferably on a shaker).

11. Place the membrane, with the DNA side facing up, onto a piece of 2x SSC-saturated 3MM filter paper.

12. UV cross-link the DNA to the membrane. Place the membrane on the 3MM filter paper into the Stratalinker. Push autocross-link. When the display reads 1200 µJ, push start. Make sure that the side of the membrane with the DNA is facing up and the membrane is still moist. Do not allow the paper to cover the sensor device inside the chamber that measures the UV dosage.

    Caution: UV is a mutagen and carcinogen, as well as an eye-damaging agent. Always wear safety glasses and gloves and work in a glass-enclosed hood if possible.

13. Cut the membrane down the center of the λ *Hind*III standard lane.

14. The membrane can be stored at room temperature.

Probe Preparation

Portions of this protocol for preparing a digoxigenin-labeled probe are derived from the Boehringer Mannheim Genius™ System.

MATERIALS

pRT1116 plasmid DNA
pCS9E11 plasmid DNA
*Bam*HI
*Hin*dIII
10x Hexanucleotide mixture
10x dNTP labeling mixture
Klenow enzyme
RNase
10x EndoR for *Bam*HI (see Appendix F)
2x SSC (see Appendix F)
Low-melting temperature (LMT) GTG agarose
Loading dye solution (see Appendix F)
0.2 M EDTA (pH 8.0)
Chemiluminescent labeling reagents
Ethanol/dry ice
Water bath (boiling and 37ºC)
UV light box and camera

PROCEDURE

1. Digest pRT1116 plasmid (source of asymmetric Tn*phoA* probe described in Appendix F, Mapping Fusion Joints by Combining Southern and Western Analyses) as follows for 1–2 hours at 37ºC. Cosmid clone pCS9E11 (*tcp* region probe) is used undigested:

DNA	10 μl
10x EndoR for *Bam*HI	5 μl
Deionized H$_2$O	32 μl
*Hin*dIII	1 μl
*Bam*HI	1 μl
RNase	1 μl

2. Cast a 30-ml 1% minigel with LMT GTG agarose.

3. Add 5 μl of Loading dye solution to digest and run sample, split into four adjacent lanes, stain, and photograph gel. Save gel.

4. Cut out appropriate band (1862 bp) with as little agarose as possible. Wear a face shield and work as quickly as possible to avoid "sunburn" on arms and face. Place gel slice in microcentrifuge tube. Purify the DNA band from the agarose as described in Appendix F (Purification of DNA Bands from Agarose by QIAquick Gel Extraction).

5. Perform the following reactions in duplicate, one set for the gel-purified probe (TnphoA) and one set for the intact plasmid (pCS9E11 tcp cosmid clone).

6. Add 15 μl of DNA to a microcentrifuge tube.

7. Heat-denature the DNA template in a boiling water bath for 10 minutes and then quickly chill on ethanol/dry ice for 30 seconds.

8. Add 2 μl of 10x hexanucleotide mixture and 2 μl of 10x dNTP labeling mixture (contains DIG-11-dUTP) into the tube on ice.

9. Thaw the mixture and add 1 μl (2 units) of Klenow enzyme. Mix.

10. Incubate for 60 minutes or longer (up to 20 hours) at 37ºC.

11. Stop the reaction by adding 2 μl of 0.2 M EDTA (pH 8.0). Store the reactions at 4ºC.

12. Use one half of the reaction volume for hybridization.

Hybridization

MATERIALS

Hybridization solution (see Appendix F, Southern Solutions: Genius System)
2x SSC (see Appendix F)
37–42°C Incubator
Heat-seal bags
UV cross-linked membranes

PROCEDURE

1. Prepare 40 ml of hybridization solution.

2. Briefly rinse the UV cross-linked membrane in 100 ml of 2x SSC. Place the membrane in a sealable bag with 20 ml of hybridization solution. Remove bubbles and seal.

   Caution: UV is a mutagen and carcinogen, as well as an eye-damaging agent. Always wear safety glasses and gloves and work in a glass-enclosed hood if possible.

3. Place in a 37–42°C incubator for 90 minutes or more. This is the prehybridization step.

4. Boil one half of the probe for 5 minutes. Add the probe to 5 ml of hybridization solution prewarmed to 65°C. Pour the pre-hybridization solution out of the bag and then add the hybridization solution containing the probe. Remove air bubbles and reseal.

5. Incubate overnight at 37–42°C.

Developing and Visualizing the Reaction

Portions of this protocol are from the Boehringer Mannheim Genius™ System.

MATERIALS

Washing and detection reagents, buffers (see Appendix F, Southern Solutions: Genius System)

0.2 x SSC (see Appendix F)

0.1% SDS (sodium dodecyl sulfate; see Appendix E)

Lumi-phos 530

Sealable bags

Plastic wrap

P1000 Pipetman

Pyrex dishes

Autoradiographic film

Film cassette

Film developer

PROCEDURE

1. Rescue the probe (it can be used repeatedly) and store at $-20^{\circ}C$.

2. Wash the membrane *twice* with occasional agitation in approximately 100 ml of 0.2x SSC/0.1% SDS for 5 minutes each at room temperature.

3. Wash the membrane *twice* with occasional agitation in approximately 100 ml of 65°C (preheated) 0.5x SSC/0.1% SDS for 15 minutes each at 65°C.

4. Rinse the membrane in approximately 100 ml of Genius Buffer 1 for 2 minutes at room temperature.

5. Incubate in 100 ml of Genius Buffer 2 for 60 minutes at room temperature.

6. Incubate *with rocking* in alkaline-phosphatase-conjugated anti-DIG diluted 1:10,000 (generally use 2 μl of antibody into 20 ml of Genius Buffer 2) for 1 hour at room temperature in a sealable bag.

7. Remove Lumi-phos 530 from refrigerator and allow to warm to room temperature.

8. Wash the membrane *twice* with rocking in approximately 100 ml of Genius Buffer 1 for 15 minutes each at room temperature.

9. Equilibrate the membrane in approximately 100 ml of Genius Buffer 3 for 1 minute at room temperature.

10. Place the membrane on plastic wrap. Use a P1000 Pipetman to dribble approximately 500 μl of Lumi-phos 530 per 100 cm^2 onto the DNA side of the membrane (~400 μl for the large blot and 150 μl for the small blot). Cover the blot with plastic wrap and disperse the solution evenly over the surface.

11. Remove the membrane from first wrap. Place the membrane on top of a fresh sheet of plastic wrap and fold the sheet over to cover the blot. Smooth out any bubbles.

12. Immediately expose the membrane to autoradiographic film for approximately 60 minutes to overnight. For best results, place the film on the side of the membrane to which the DNA is bound.

13. Develop the film and analyze the results.

Mapping Relative Positions of the Fusion Insertion Points

The positions of the new *tcp-phoA* fusion joints are mapped relative to known fusion joint locations.

MATERIALS

Washing and detection reagents, buffers (see Appendix F, Southern Solutions: Genius System)
Loading dye solution (see Appendix F)
10x EndoR for *Bam*HI (see Appendix F)
Agarose
Ethidium bromide
Pyrex dishes
Microcentrifuge tubes
Autoradiographic film
Film cassette
Film developer

Caution: Ethidium bromide is a powerful mutagen and moderately toxic. Always wear gloves when working with solutions that contain this dye. After use, decontaminate and dispose of these solutions in accordance with the safety practices established by your institution's Safety Office.

PROCEDURE

1. Prepare DNA from reference strains (see gel lane order in step 3) using the DNA extraction protocol above.

2. Set up each digestion in a microcentrifuge tube and incubate for 2 hours at 37°C:

Chromosomal DNA	10 μl
Deionized H$_2$O	7 μl
10x EndoR for *Bam*HI	2 μl
*Bam*HI	1 μl
*Xba*I	1 μl

3. Prepare a 0.6% agarose gel. Add 2 μl of Loading dye solution and load gel as below. Use the following pattern for the gel if mapping fusions on the 5-kb *Xba*I fragment:

lane 1	lane 2	lane 3	lane 4	lane 5	lane 6	lane 7
TSM304	TSM310	TSM302	TSM309	TSM305	Mutant	Mutant

Use the following pattern for the gel if mapping fusions on the 20-kb *Xba*I fragment:

lane 1	lane 2	lane 3	lane 4	lane 5	lane 6
TSM303	TSM306	TSM308	TSM307	Mutant	Mutant

All lanes will be probed with the internal Tn*phoA* probe.

4. Run gel until xylene cyanol (blue-green dye) is approximately two thirds of the way toward the bottom. Stain for approximately 30 minutes in deionized H_2O containing 0.5 μg/ml ethidium bromide and then destain 10–20 minutes in deionized H_2O. Photograph for later reference. Follow above protocols for transfer to membrane, hybridization, and development.

5. Perform hybridization with Tn*phoA* probe as described above.

6. Wash, develop, and analyze hybridization pattern (see Appendix F, Mapping Fusion Joints by Combining Southern and Western Analyses).

Allelic Exchange in Gram-negative Bacteria Utilizing Suicide Plasmid Vectors

Experiment 14 demonstrates two general methods to cross plasmid-borne mutations into the chromosome. The first is useful for achieving disruptions in a gene generated by a single crossover that integrates plasmid sequences into the chromosome. The second method is a two-step positive selection for recombination of a variety of mutation types into the chromosome. Both methods utilize vectors that are conditional for their replication such that they replicate in the donor and not in the recipient strain.

The first method (Experiment 14A) uses plasmid pVM55 (Miller and Mekalanos 1988) to inactivate the *toxR* gene of *Vibrio cholerae.* This technique generally utilizes a DNA fragment cloned into the plasmid vector pGP704 (see Appendix G, Plasmid pGP-704), which can only replicate in host cells that provide the π protein, generally supplied from a λ*pir* prophage. For this experiment, the conditional plasmid carries a DNA fragment from within an internal portion of the *toxR*-coding sequence. Because the plasmid cannot replicate in *V. cholerae*, Ap^r transconjugants should contain the mobilized plasmid integrated into the genome by homologous recombination between the *toxR* gene on the chromosome and the cloned sequences present in the plasmid. Since an internal fragment of *toxR* is used, the resulting strains will be merodiploid with respect to the region cloned into the vector plasmid such that they contain two partial copies of *toxR* as diagrammed in Section 4 (Broad Host Range Allelic Exchange Systems). The ToxR protein is a transcriptional activator protein required for the expression of a number of genes associated with virulence, including *tcp* genes. Thus, disruption of *toxR* should result in decreased expression of the *tcp-phoA* fusions isolated in Experiment 12. In general, following purification of the putative insertion mutants, the structure of the chromosomal insertion can be verified by Southern analysis or by rapid colony polymerase chain reaction (PCR) (used in this experiment).

The second experiment (Experiment 14B) utilizes a plasmid vector derived from pGP704, designated pKAS32 (Skorupski and Taylor 1995), that carries the *rpsL* gene (see Appendix G, Plasmids pKAS32 and pKAS46). Expression of *rpsL*, the gene encoding the

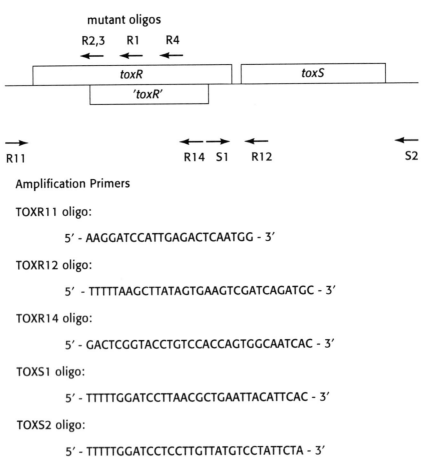

Amplification Primers

TOXR11 oligo:

    5′ - AAGGATCCATTGAGACTCAATGG - 3′

TOXR12 oligo:

    5′ - TTTTTAAGCTTATAGTGAAGTCGATCAGATGC - 3′

TOXR14 oligo:

    5′ - GACTCGGTACCTGTCCACCAGTGGCAATCAC - 3′

TOXS1 oligo:

    5′ - TTTTTGGATCCTTAACGCTGAATTACATTCAC - 3′

TOXS2 oligo:

    5′ - TTTTTGGATCCTCCTTGTTATGTCCTATTCTA - 3′

Figure 14.1 Locations of oligonucleotides within the *toxRS* operon.

ribosomal protein S12, in a streptomycin-resistant host renders that host streptomycin-sensitive because the wild-type allele is dominant over the mutant. Thus, this vector allows for selection of a merodiploid strain as with pGP704 (Ap^r) and then subsequent selection for streptomycin resistance, which allows for the isolation of colonies resulting from resolution of the merodiploid state through homologous recombination between flanking sequences within the cloned region. When the vector contains a cloned fragment harboring a mutation, a certain percentage of the cells that resolve the diploid state retain the mutation on the chromosome (see Section 4, Broad Host Range Allelic Exchange Systems). The percentage of colonies that retain the mutation approaches 50% if the lesion resides near the center of the cloned insert. This method is utilized in this experiment to analyze the site-directed *toxR* mutations constructed in Experiment 15 by placing them in unit copy on the chromosome. Analysis of such mutations in unit copy is especially important because gene dosage can greatly effect the phenotypes conferred by regulatory gene products. Locations of the oligonucleotides used to generate mutations and amplify portions of the *toxRS* operon are diagrammed in Figure 14.1. Below are listed the bacterial strains used in this experiment.

BACTERIAL STRAINS

TSM302	ϕ(*tcpA-phoA*)2-1(Hyb) (*tcpA2-1*::Tn*phoA*)*rpsL* Sm^r Km^r
TSM303	ϕ(*tcpC-phoA*)5-2(Hyb) (*tcpC5-2*::Tn*phoA*)*rpsL* Sm^r Km^r
TSM311	ϕ(*tcpA-phoA*)2-1(Hyb) (*tcpA2-1*::Tn*phoA*)*rpsL* Sm^r Km^r Δ*toxR43*
TSM343	RP4-2 Km^r *tet*::Mu λ*pir* (pVM55 [*oriR6K mobRP4 'toxR' *Ap^r])
TSM361	RP4-2 *tet*::Mu-*kan*::Tn*7* λ*pir* (pKAS37 [*oriR6K mobRP4 rpsL*$^+$ Sm^s *toxR*$^+$ Ap^r])
TSM362	RP4-2 *tet*::Mu-*kan*::Tn*7* λ*pir* (pKAS38 [*oriR6K mobRP4 rpsL*$^+$ Sm^s *toxR.R96K* Ap^r])
EXP12.1	Putative *tcp*::Tn*phoA* fusion strain #1 (see Experiment 12)
EXP12.2	Putative *tcp*::Tn*phoA* fusion strain #2 (see Experiment 12)

Experiment 14A: Interruption of *toxR* by Suicide Vector Insertion Mutation

MATERIALS

For culture media and supplements, see Appendix B.

LB-agar plates
LB-agar-Ap plates
LB-agar-Km plates
LB-agar-Sm-Ap-Glu-XP plates
LB-agar-Glu-XP plates
Vibrio broth
Vibrio broth-Ap
10x PCR amplification buffer (see Experiment 7)
dNTP mix (stock contains 1.25 mM of each)
Primers (each 50 μM stock): TOXR11, TOXR12, TOXS1, and
     TOXS2
Taq polymerase
Agarose
APase assay reagents (see Appendix E)
Thermocycler

> *Note:* Some thermocyclers require overlaying the sample with mineral oil or wax to prevent the condensation of water vapor on the upper surfaces of the tube. This can be avoided by using a thermocycler that maintains a higher temperature in the upper portion of the chamber than in the lower part (e.g., the PTC-100™ Thermal Cycler with a Hot Bonnet™ sold by MJ Research).

Sterile inoculating sticks

PROCEDURE

1. Prepare overnight streak plates of two potential *tcp*::Tn*phoA* mutants from Experiment 12 (EXP12.1 and EXP12.2) and control ToxR-regulated *tcp* fusion strains TSM302 and TSM303 to be used as recipients on LB-agar-Km plates. Streak donor strain TSM343 to LB-agar-Ap plates. Incubate overnight at 37ºC.

2. Using the fresh streaks of donors and recipients, set up one mating of TSM343 with each recipient strain (four matings) using the cross-streak technique on LB-agar (diagrammed in Experiment 12). Incubate overnight at 37ºC.

The PCR process for amplifying nucleic acids is covered by patents owned by Hoffmann-LaRoche.

3. Streak for transconjugants on LB-agar-Sm-Ap-Glu-XP plates. Use one plate per mating. Streak heavily using a large amount of bacteria from the mating for this streak (i.e., a large swipe with an inoculating stick) because the frequency of obtaining transconjugants, which in this case requires both acquisition of the plasmid and its integration into the chromosome, occurs at a low frequency. Incubate overnight at 30°C for enhanced ToxR-dependent gene expression.

4. Purify several transconjugants onto LB-agar-Sm-Ap-Glu-XP plates, especially trying to purify those that appear to be paler blue than the parent.

5. Compare the intensity of the blue color of the *toxR⁻* and the *toxR⁺* colonies with the blue color of wild-type and mutant controls. To do this, streak the colonies from step 4 as well as the parent strain (TSM302, TSM303, EXP12.1, or EXP12.2) on one LB-agar-Glu-XP plate. Having the parental colonies on the plate will provide an initial comparison to help identify the putative *toxR* insertion mutants. TSM311 provides an example of a *toxR⁻* derivative of TSM302. Incubate overnight at 30°C.

6. Confirm the insertion in *toxR* by rapid colony PCR.

   a. Resuspend a medium-size colony into 50 µl of sterile deionized H_2O for each of the following strains: one colony from TSM302 (wild-type *toxR*), one colony from TSM311 (small *toxR* deletion), and one colony from each of two putative *toxR* insertion mutations in TSM302, TSM303, EXP12.1, and EXP12.2 (10 colonies total).

   b. Use 5 µl of the colony suspension as template to set up the following 10 PCRs.

Sterile deionized H_2O	60 µl
10× PCR amplification buffer	10 µl
dNTP mix (1.25 mM of each)	16 µl
Colony suspension	5 µl
TOXR11 primer (50 µM stock)	2 µl
TOXR12 primer (50 µM stock)	2 µl
TOXS1 primer (50 µM stock)	2 µl
TOXS2 primer (50 µM stock)	2 µl
Taq polymerase	1 µl

   c. PCR amplify: 1 minute at 95ºC; 2 minutes at 55ºC; 1 minute at 72ºC for 30 cycles.

   d. Visualize 5 μl of each product on a 1% agarose gel using a φX174 *Hae*III standard. Insertion should result in a shift of the 1400-bp TOXR11–12 product to a much larger product (or potentially no product due to inefficient extension over long distances). The 670-bp TOXS1–2 product serves as a reaction control and should remain invariant.

7. Select a confirmed *toxR* insertion mutant in each strain background and compare the expression of the resident *phoA* fusion with expression in the *toxR*+ parental strain by alkaline phosphatase assay (see Appendix E, Alkaline Phosphatase Assays in Permeabilized Cells). In this case, grow *toxR*+ strains in Vibrio broth and *toxR*– strains in Vibrio broth-Ap at 30ºC to early stationary phase. ToxR-regulated fusions should show decreased expression in the mutant background.

Experiment 14B: Recombination of *toxR* Point Mutations into the Chromosome Using a Suicide Vector Expressing Ribosomal Protein S12

MATERIALS

For culture media and supplements, see Appendix B.

LB-agar plates
LB-agar-Ap plates
LB-agar-Km (45 μg/ml)-Ap (100 μg/ml) plates
LB-agar-Sm (1 mg/ml)-Glu-XP plates
Vibrio broth
10x PCR amplification buffer (see Experiment 7)
dNTP mix (stock contains 1.25 mM of each)
Primers (each 50 μM stock): TOXR11 and TOXR14
*Hha*I
10x *Hha*I buffer (see Appendix F)
10x Bovine serum albumin (BSA) (e.g., New England Bio-Labs)
Water bath (37ºC)

NuSieve 3:1 agarose gel
APase assay reagents (see Appendix E)
Thermocycler

> *Note:* Some thermocyclers require overlaying the sample with mineral oil or wax to prevent the condensation of water vapor on the upper surfaces of the tube. This can be avoided by using a thermocycler that maintains a higher temperature in the upper portion of the chamber than in the lower part (e.g., the PTC-100™ Thermal Cycler with a Hot Bonnet™ sold by MJ Research).

Sterile inoculating sticks

PROCEDURE

1. Prepare overnight streak plates of *tcpA* fusion recipient strain TSM302 on LB-agar and donor strains TSM361 and TSM362 on LB-agar-Ap. Incubate overnight at 37ºC.

2. Using fresh plates of donors and recipients, mate the *toxR* wild-type (TSM361) and *toxR* mutant (TSM362) donor strains with recipient TSM302 using the cross-streak technique on LB-agar (diagrammed in Experiment 12). Incubate overnight at 37ºC.

3. Streak for transconjugants onto LB-agar-Km-Ap plates. Use one plate per mating. Streak heavily using a large amount of bacteria from the mating for this streak (i.e., a large swipe with an inoculating stick) because the frequency of obtaining transconjugants, which in this case requires both acquisition of the plasmid and its integration into the chromosome, occurs at a low frequency. When streaking for isolated colonies, use the same stick for the entire streaking procedure. Incubate overnight at 37ºC.

4. Purify four transconjugants from each mating onto an LB-agar-Km-Ap plate. Incubate overnight at 37ºC.

5. Make a swipe through an area of confluent growth with an inoculating stick. Streak heavily onto an individual LB-agar-Sm-Glu-XP plate for each mating. For this mating, streak out for isolated colonies using the same stick for the entire streaking procedure. Incubate overnight at 30ºC.

6. Compare the intensity of blue color of the recombinant colonies. The colonies resulting from the mating with TSM361 should yield only wild-type colonies. For the TSM362 mating, both wild-type colonies and fainter blue colonies should be seen if the mutations alter activation of the *tcpA-phoA* fusion in *V. cholerae*. Purify several colonies from the TSM361 mating and several lighter and darker colonies from the TSM362 mating onto LB-agar-Sm-Glu-XP plates. Incubate overnight at 30°C.

7. Streak a colony from each purification onto an LB-agar-AP plate. Include Ap^r and Ap^s controls. Also streak parental *toxR*$^+$ recipient strain TSM302 onto LB-agar to provide a fresh control for colony PCR. Incubate all plates overnight at 37°C.

8. Perform rapid colony PCR and RFLP analysis (see below) on potential Ap^s mutants to confirm the presence of the point mutation.

   a. Prepare PCR cocktail for four reactions as follows. Aliquot 47.5 μl into each of four tubes.

Sterile deionized H$_2$O	128 μl
10x PCR amplification buffer	20 μl
dNTP mix (stock with 1.25 mM of each)	32 μl
TOXR11 primer (50 μM stock)	4 μl
TOXR14 primer (50 μM stock)	4 μl
Taq polymerase	2 μl

   b. Use a blue pipette tip to suspend three medium-size colonies as follows into 50 μl of sterile deionized H$_2$O each: a colony of TSM302, a light-blue colony from the TSM362 mating, and a darker-blue (wild-type) colony from the TSM362 mating.

   c. Inoculate 2.5 μl of each colony suspension into tubes 1–3. Add 2.5 μl of sterile deionized H$_2$O to tube 4 to serve as the water control.

   d. PCR amplify according to the following reaction conditions: 1 minute at 94°C, 1 minute at 59°C, 1 minute at 72°C for 30 cycles.

9. Perform RFLP analysis of PCR products. For each sample, perform the following (for the sterile deionized H_2O control, simply run as undigested).

   a. Drop microdialyze (see Appendix F) the 50-μl reaction for 30 minutes to 1 hour.

   b. Aliquot 11 μl for an undigested gel sample.

   c. Set up the following *Hha*I digest.

Product	11 μl
10× *Hha*I buffer	1.5 μl
10× BSA (bovine serum albumin)	1.5 μl
*Hha*I	1.0 μl

   d. Digest for 2 hours at 37°C.

   e. Resolve products on a 2% NuSieve 3:1 agarose gel using a φX174 *Hae*III standard. The undigested product should migrate at approximately 900 base pairs. For the digested samples, the mutants should yield an RFLP pattern that differs from the pattern derived from the wild-type PCR product. Arrange the lanes so that all the *undigested* products are adjacent to one another and all the *digested* products are adjacent to one another.

10. Select a confirmed *toxR* point mutant strain and compare the expression of the *tcpA-phoA* fusion with expression in the *toxR*+ parental strain TSM302 and *toxR*− strain TSM311 by alkaline phosphatase assay (Appendix E, Alkaline Phosphatase Assays in Permeabilized Cells). In this case, grow strains in Vibrio broth at 30°C to early stationary phase.

Oligonucleotide-directed Site-specific Mutagenesis

Experiment 15 introduces directed mutations into regions of *toxR* that may be important for its ability to activate expression of genes within the ToxR regulon (Ottemann et al. 1992). Initially, the effect of these *toxR* alleles on activation of *ctx* gene expression is monitored in *Escherichia coli* using a *ctx-lacZ* fusion present in a λ lysogen. A *tcpA-phoA* fusion then serves as a means to monitor mutant ToxR activation of additional promoters, required for activation of *tcp* gene expression in *Vibrio cholerae* (see Section 1, *Vibrio cholerae*). For this second assay, the site-directed mutation carrying plasmids are mobilized into *V. cholerae* and recombined into the chromosome, replacing the wild-type *toxR* by allelic exchange using a suicide plasmid vector as detailed in Experiment 14.

Although several generally used strategies are available to obtain a high yield of desired mutations (see Appendix F, Oligonucleotide-directed Site-specific Mutagenesis), we use the methodology utilized by the commercially available Altered Sites® system marketed by Promega as outlined in Figure F.1 in Appendix F. The Altered Sites system is based on the use of a second mutagenic oligonucleotide, in addition to the target gene mutagenic oligonucleotide, to confer antibiotic resistance to the mutated DNA strand. The pALTER-1 plasmid vector that carries the target gene to be mutagenized also carries two antibiotic resistance genes, an intact gene for Tc[r] and a second gene for Ap[r], which has been inactivated by a base substitution, yielding a Tc[r] Ap[s] phenotype. Annealing an

TOXR1 oligo: (R$_{96}$K)

Oligo 3′ C G G T A A A G C T G A G A C <u>T T</u> <u>T T</u> T T T A C G A G 5′
Wild Type 5′ G C C A T T T C G A C T C T G C G C A A A A T G C T C 3′
        A l a I l e S e r T h r L e u A r g L y s M e t L e u
    toxR921   A A A
    ToxR921   L y s
    (destroys *Hha*I site)

TOXR2 oligo: (D$_{85}$E)

Oligo 3′ C A A A A C T T C A G C T A C T <u>C</u> A G G T C G A A T T G 5′
Wild Type 5′ G T T T T G A A G T C G A T G A T T C C A G C T T A A C 3′
        l y P h e G l u V a l A s p A s p S e r S e r L e u T h
    toxR922   G A G
    ToxR922   G l u
    (destroys *Tfi*I site)

TOXR3 oligo: (D$_{85}$N)

Oligo 3′ C A A A A C T T C A G C T A <u>T T</u> <u>G</u> A G G T C G A A T T G 5′
Wild Type 5′ G T T T T G A A G T C G A T G A T T C C A G C T T A A C 3′
        l y P h e G l u V a l A s p A s p S e r S e r L e u T h
    toxR923   A A C
    ToxR923   A s n
    (destroys *Tfi*I site)

TOXR4 oligo: (Y$_{117}$F)

Oligo 3′ G C T T C G C G C C A A A <u>G</u> G T T A A C T A G C G G 5′
Wild Type 5′ C G A A G C G C G G T T A C C A A T T G A T C G C C 3′
        r o L y s A r g G l y T y r G l n L e u I l e A l a
    toxR924   T T C
    ToxR924   P h e
    (destroys *Bst*EII site)

Sequences of mutagenic oligonucleotides.

oligonucleotide with the Ap^r base correction during the in vitro extension reaction provides a selection for *E. coli* transformants which are progeny that have arisen from replication of the mutagenized DNA strand carrying the repaired *bla* gene and thus greatly enriches for coinheritance of the desired mutation. This recovery is enhanced by using a mismatch repair-defective (*mutS*) strain as the recipient. During the mutagenesis procedure, a third mutagenic oligonucleotide is incorporated that alters the Tc^r gene to yield a Tc^s phenotype. This procedure is useful for undergoing a second round of mutagenesis by selecting for repair of the Tc^s lesion in the same way as the original Ap^s repair enhanced recovery of mutated plasmids. This experiment utilizes the second round of mutagenesis to revert the original mutation back to the wild-type sequence to confirm that any mutant phenotype is due to the designed mutation and not the result of any additional aberrant event that might have occurred during the procedure. This avoids the need to sequence the entire cloned region carried on the pALTER-1 plasmid to ensure that any mutant phenotype is actually the consequence of the designed mutation. Although either single- or double-stranded DNA can be used for these mutagenic procedures, we use single-stranded templates, which leads to somewhat greater efficiencies. The structure of pRT651 and the positions of the mutagenic oligonucleotides to be used are shown in Figures 14.1 and 15.1. Below are listed the bacterial strains and bacteriophage used in this experiment.

BACTERIAL STRAINS

TSM344	(*recA1 endA1 gyrA96 thi hsdR17 supE44 relA1* Δ[*lac-proAB*] F′ *proAB$^+$ traD36 lacI^Q* Δ*lacZM15*)
TSM345	(TSM344 [pRT651 (*oriColE1 orif1 lacZα toxR$^+$* Tc^r Ap^s)])
TSM350	(Δ[*lac-proAB*] *rpoB gyrA recA56* λNFVM1 [*cI857*(ts)Sam100 *int$^+$ att$^+$* φ(*ctx-lacZ*)])
TSM351	(*mutS201*::Tn5)

BACTERIOPHAGE

R408	(Phagemid helper phage)

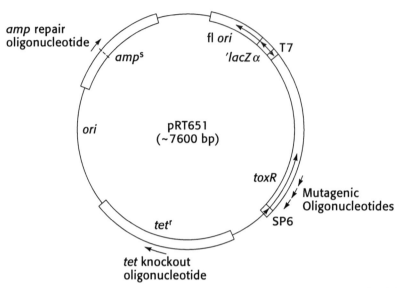

Figure 15.1 Structure of pRT651 and positions of mutagenic oligonucleotides.

Isolation of Single-stranded DNA Template

The protocols in this and the following sections are derived from the Promega Altered Sites® Technical Manual, where additional details can be found. For details on preparation of materials for this experiment, see Appendix F, Oligonucleotide-directed Site-specific Mutagenesis.

MATERIALS

TYP broth-Tc (see Appendix B)
25% PEG 8000/3 M NaCl precipitation solution
TES buffer (Appendix F)
Buffered phenol
3 M Sodium acetate (pH 4.8–6.0)
10 mM Tris-HCl (pH 7.5)
Ethanol (70% and 95%)
R408 single-stranded DNA
Agarose
10x Loading buffer
Kimwipes

PROCEDURE

1. Grow an overnight culture of TSM345 in TYP broth-Tc.

2. In the morning, inoculate 2 ml of TYP broth-Tc in a large tube (for good aeration) with 40 μl of overnight culture. Grow for 30 minutes at 37ºC.

3. Infect the culture with helper phage R408 at a multiplicity of infection (moi) of 10 (use ~20 μl of stock lysate). Continue incubation for 6–10 hours.

4. Pellet cells in a microcentrifuge at high speed for 1 minute at room temperature. Save supernatant.

5. Precipitate phage from 0.8 ml of supernatant by adding 175 μl of 25% PEG 8000/3 M NaCl solution. Mix by shaking several times. Let stand for 10 minutes at room temperature.

6. Pellet precipitate in a microcentrifuge at high speed for 5 minutes at room temperature.

7. Discard supernatant. It is important to remove all PEG. Place the tube upside down for several minutes and wipe lip with a clean Kimwipe.

8. Add 100 μl of TES buffer to pellet and resuspend by vortexing.

9. Add 100 μl phenol or phenol:chloroform and extract by vortexing for 20 seconds and spinning in a microcentrifuge for 2 minutes.

   Caution: Phenol can cause severe burns. Always wear gloves, protective clothing, and safety glasses and work in a chemical fume hood. Rinse any areas of skin that come in contact with phenol with a large volume of water and wash with soap and water; do not use ethanol!

10. Transfer upper phase to fresh tube.

11. Precipitate DNA by adding 5 μl of 3 M sodium acetate, mixing, and adding 250 μl of chilled (–20ºC) ethanol. Mix, let stand 10 minutes, and then spin in a microcentrifuge for 10 minutes.

12. Decant ethanol, add 0.3 ml of chilled 70% ethanol to rinse pellet, spin in a microcentrifuge for 2 minutes, decant, and dry pellet under vacuum. Pellet is often nearly invisible.

13. Resuspend pellet in 20 μl of 10 mM Tris-HCl (pH 7.5).

14. Run a 2-μl sample (2-μl sample + 7 μl of TE + 1 μl of 10x loading buffer) on a 0.7% agarose gel to ensure isolation of intact single-stranded DNA. The gel should reveal a prominent band that migrates near the faint R408 DNA band (can be included as a control lane). Since the DNA is single-stranded, its migration will not correlate with typical double-stranded size standards.

Phosphorylation of Mutagenic Oligonucleotide

For details on the preparation of materials for this experiment, see Appendix F, Oligonucleotide-directed Site-specific Mutagenesis.

MATERIALS

Mutagenic oligonucleotide (TOXR1, 2, 3, or 4; 100 pmoles)
10x Kinase buffer (see Appendix F)
T4 polynucleotide kinase

PROCEDURE

1. Combine the following components in a microcentrifuge tube:

Oligonucleotide (see Note)	5 μl (100 pmoles)
10x Kinase buffer (including 10 mM ATP)	2.5 μl
T4 polynucleotide kinase	1 μl (5 units)
Sterile deionized H_2O	16.5 μl
Total volume	25 μl

Note: Concentration of the stock oligonucleotide preparation can be determined by the formula: $OD_{260} \times 40$ = ng ssDNA/μl. A convenient way to determine the number of nanograms of oligonucleotide equivalent to 100 pmoles is with the formula: ng of oligonucleotide = $32.5 \times b$, where b is the number of bases in the oligonucleotide. For example, 100 pmoles of a 20 mer = 32.5×20 or 750 ng of DNA.

2. Incubate the reaction for 30 minutes at 37°C.

3. Incubate the reaction for 10 minutes at 70°C to inactivate the kinase.

4. The reaction products can be stored at −20°C until use.

Mutagenesis Reactions

For details on preparation of materials for this experiment, see Appendix F, Oligonucleotide-directed Site-specific Mutagenesis.

MATERIALS

Ampicillin repair oligonucleotide
Tetracycline knockout oligonucleotide
Kinased mutagenic oligonucleotides
10x Annealing buffer (see Appendix F)
10x Synthesis buffer (see Appendix F)
Water bath (70°C)
T4 DNA polymerase
T4 DNA ligase
Electroporation-competent strain TSM351
Plasmid broth (see Appendix B)
Ap stock
LB-agar-Ap plate (see Appendix B)

PROCEDURE

1. Set up mutagenesis annealing reaction:

pRT651 ssDNA	2 µl (0.05 pmole)
Ampicillin repair oligonucleotide*	1 µl (0.25 pmole)
Tetracycline knockout oligonucleotide*	1 µl (0.25 pmole)
Mutagenic oligonucleotide (phosphorylated)	2 µl (1.25 pmoles)
10x Annealing buffer	2 µl
Sterile H_2O	to final volume 20 µl

*Repair and knockout oligonucleotides are phosphorylated by the supplier.

2. Heat-anneal the reaction to 70°C for 5 minutes. Allow to cool slowly to room temperature (15–20 minutes) by removing the tube with some water from the water bath in a 100-ml beaker.

3. Extension and ligation: Place the reaction on ice and add the following:

10x Synthesis buffer	3 µl
T4 DNA polymerase (10 units/µl)	1 µl
T4 DNA ligase (2 units/µl)	1 µl
Sterile H_2O	to final volume 30 µl

   Incubate reaction for 90 minutes at 37°C.

4. Microdialyze ligated DNA against deionized H_2O (see Appendix F, Drop Dialysis) using a 0.025-µm filter for 45 minutes to decrease salt concentration prior to electroporation.

5. Electroporate 100 µl of TSM351 competent cells with one half of the reaction mix. Refrigerate the other half of the reaction (see Appendix D, Preparation of Electrocompetent Cells and Electroporation). A mock TSM351 electroporation control can be run in parallel and should result in no colonies on the plate in step 7.

6. Immediately add 4 ml of Plasmid broth and incubate for 1 hour at 37°C.

7. Place 1 ml of culture into a microcentrifuge tube and centrifuge at high speed for 1 minute at room temperature. Resuspend in 100 µl and plate 90 µl and 10 µl onto two halves of an LB-agar-Ap plate. This is a control to ensure that Ap[r] transformants are present. Do not use these colonies further. Add ampicillin to the remaining 3 ml of culture to a final concentration of 125 µg/ml (use 1.87 µl of 200 mg/ml stock). Grow overnight at 37°C.

Preparation of Plasmid from the Pooled *mutS* Transformants and Transformation into a *ctx-lacZ* Indicator Strain

MATERIALS

Culture of transformed TSM351 from step 7 of Mutagenesis
   Reactions (p. 440)
Plasmid DNA miniprep reagents (see Appendix F)
LB broth (see Appendix B)
MacConkey-Ap (50 µg/ml)-Lac plates (see Appendix B)
Transformation-competent TSM350
Microcentrifuge tubes

PROCEDURE

1. Place 1.5 ml of overnight culture from step 7 above into a microcentrifuge tube. Centrifuge at high speed for 2 minutes at room temperature, remove supernatant, and save pelleted cells.

2. Prepare plasmid DNA from the cell pellets following the protocol in Appendix F (Plasmid DNA Minipreps).

3. Use this DNA to transform *temperature-sensitive* strain TSM350 with selection for Ap^r as follows (see Appendix D, Preparation and Transformation of Competent Cells).

   a. Thaw a tube of 100 µl of competent TSM350 on ice. Add 2 µl of miniprep DNA. Incubate on ice for 30 minutes. Add 1 ml of LB broth and grow for 1 hour at 30ºC.

   b. Spin down cells and resuspend in approximately 0.1 ml of LB broth. Plate 1/10 of this mix on one MacConkey-Ap (50 µg/ml)-Lac plate and 9/10 on another MacConkey-Ap (50 µg/ml)-Lac plate to help ensure the formation of isolated colonies. Incubate overnight at 30ºC.

4. Examine transformants to identify colonies that carry pRT651 with the putative mutation that is altered in its activation of the *ctx-lacZ* fusion. Red colonies are *toxR*$^+$ while white to light-pink colonies harbor pRT651 with an altered *toxR* gene.

Reversion of the Site-directed Mutation

For details on preparation of materials for this experiment, see Appendix F, Oligonucleotide-directed Site-specific Mutagenesis.

MATERIALS

For culture media and supplements, see Appendix B.

LB-agar-Tc plates
LB-agar-Ap plates
Plasmid broth-Ap
TYP broth-Ap plates
Tc stock
MacConkey-Ap-Lac plates
MacConkey-agar-Lac-Tc (7.5 µg/ml) plates
TSM350 transformants carrying potential *toxR* mutations
Plasmid DNA miniprep reagents (see Appendix F)
R408 phage
pRT651.R96K ssDNA
Tetracycline repair oligonucleotide
TOXR5, 6, or 7 oligonucleotide (phosphorylated)
10x Annealing buffer (see Appendix F)

PROCEDURE

1. Purify 20 putative mutant plasmid-containing colonies (identified as white or pink colonies in step 4 [p. 441] above) on MacConkey-Ap-Lac plates. Incubate overnight at 30°C.

2. Streak colonies to LB-agar-Tc to identify clones in which the mutated pRT651 plasmid also carries a Tc^s gene. Incubate overnight at 30°C.

3. Grow 2-ml Plasmid broth-Ap overnight cultures of two Tc^s clones at 30°C.

4. Miniprep plasmid from each culture as described in Appendix F (Plasmid DNA Minipreps).

5. Use 1–2 μl of each miniprep to transform strain TSM344 as described in Appendix D (Preparation and Transformation of Competent Cells). Select transformants on LB-agar-Ap at 37°C.

6. Purify transformants on LB-agar-Ap. Incubate overnight at 37°C.

7. Grow overnight cultures of TSM344 containing each clone in TYP broth-Ap at 37°C.

8. In the morning, inoculate 2 ml of TYP broth-Ap in a large tube (for good aeration) with 40 μl of overnight culture. Grow for 30 minutes at 37°C.

9. Infect culture with helper phage R408 and prepare single-stranded DNA beginning with step 3 on p. 437.

10. Set up a reversion mutagenesis reaction for each mutant plasmid as in step 1 on p. 439 with the following changes. Substitute the Tc^r repair oligonucleotide for the Ap^r oligonucleotide. Substitute the appropriate phosphorylated repair oligonucleotide for the original mutagenic oligonucleotide according to the following table. Do not include the Tc^s oligonucleotide.

Mutagenic oligo	Repair oligo
TOXR1	TOXR5
TOXR2	TOXR6
TOXR3	TOXR6
TOXR4	TOXR7

Sample reversion mutagenesis reaction:

pRT651.R96K ssDNA	2 μl (0.05 pmole)
Tetracycline repair oligonucleotide	1 μl (0.25 pmole)
ToxR oligonucleotide 5 (phosphorylated)	2 μl (1.25 pmoles)
10× Annealing buffer	2 μl
Sterile H$_2$O	to final volume 20 μl

11. Complete the mutagenesis as described in Mutagenesis Reactions section (p. 439).

12. Electroporate strain TSM351 as described above except substitute Tc (12.5 µg/ml) wherever Ap is called for in the procedure.

13. Prepare DNA from TSM351 containing the potential reversion pool.

14. Transform strain TSM350 with selection on MacConkey-agar-Lac-Tc (7.5 µg/ml) at 30°C. Determine if the wild-type (Lac⁺, red) phenotype can be restored to the mutant plasmids. This indicates that the original mutant phenotype is due solely to the intended site-directed alteration.

Appendices

"I recall one experiment that [Jean] Weigle and I tried. We thought that one ought to be able to transfer genes from one bacterium to another via a temperate phage. [Max] Delbrück said that the idea was crazy and bet us (50 milkshakes to one) that our experiment would not work. And he won his bet. But the experiment failed not because the idea was crazy, but because it failed to incorporate what Delbrück calls 'the principle of limited sloppiness.' To keep the background of spontaneous mutants low, we used as donor and recipient a pair of bacterial strains that differed in *two* genes, and scored, as Lederberg and Tatum had done in the discovery of bacterial conjugation, only cases in which *both* genes were transferred. Not long afterward, Zinder and Lederberg did discover phage-mediated transduction, but found, of course, that only genes that are very closely linked (the ones we used had not been) are transferred jointly."

Seymour Benzer
Reprinted from Adventures in the rII Region.
In *Phage and the Origins of Molecular Biology* Expanded Edition
(ed. J. Cairns et al.), pp. 157–165
Cold Spring Harbor Laboratory Press, Cold Spring Harbor, NY (1992)

"In fact, any single selectable marker from almost any Salmonella strain could be transduced. . .We had clearly been using precisely the wrong procedure, a random collection of double markers, to find this phenomenon. Only two serendipitous occurrences allowed us to detect transduction at all. First, the original markers in LT-22 with which we found transduction were both in the pathway of aromatic amino acid biosynthesis and were probably linked so that they cotransduced. Second, the LT-22 strain carried a lysogenic transducing phage that could propagate on LT-2."

Norton D. Zinder
Reprinted, with permission, from Forty Years Ago: The Discovery of Bacterial Transduction.
Genetics 132: 291–294 (1992)

Bacterial Strains, Phage, and Plasmids

The strains, plasmids, and bacteriophages used in this manual are specifically designed to illustrate the experiments and most (such as the many *srl* mutants) will not find general use. However, many of these strains have been deposited in the appropriate stock centers, from which they are available for distribution to qualified investigators.

The strains listed here were given serial TSM number designations to facilitate their use in the Advanced Bacterial Genetics Course. However, publications reporting experiments using these strains should list the original strain designation, so that the reader is clear about the identity and source of each strain. For example, strain TSM101 should be designated as TT1704 in publications.

Salmonella Genetic Stock Center
Dr. Kenneth E. Sanderson
Department of Biological Sciences
University of Calgary
Calgary, Alberta T2N 1N4
Canada

Fax: 403-289-9311
Internet: kesander@acs.ucalgary.ca
World Wide Web:
    http://www.ucalgary.ca/~kesander

E. coli Genetic Stock Center
Dr. Mary Berlyn
Department of Biology, 255 OML
Yale University
P.O. Box 6666
New Haven, CT 06511-7444

Telephone: 203-432-9997
Internet: mary@cgsc.biology.yale.edu
World Wide Web:
    hhp://cgsc./biology.yale.edu/

American Type Culture Collection
12301 Parklawn Drive
Rockville, MD 20852

Telephone: 1-800-638-6597
Internet: request@atcc.org
World Wide Web:
    http://www.atcc.org/atcc.htm1

447

Strain List

Strain	AKA[a]	Genotype	Source
Salmonella typhimurium LT2			
TSM1	LT2	prototroph	1
TSM101	TT1704	Δ(*hisOE*)9533	2
TSM102	TT10289	*hisD9953*::MudJ *hisA9949*::Mud1	3
TSM103	VJSS095	pNK2881 (Ap^r, P$_{tac}$-*tnpA ats-1 ats-2*)	4
TSM104	TT10423	*proAB47*/F′ *pro$^+$ lac$^+$ zzf-1831*::Tn10d(Tc)	5
TSM105	VJSS068	*srl-251*::MudJ (Lac$^+$ Reg$^+$)	V. Stewart
TSM106	VJSS103	*zgc-7203*::Tn10d(Tc) (near *srl*)	V. Stewart
TSM107	VJSS098	*srl-251*::MudJ *zgc-1623*::Tn10d(Cm)	V. Stewart
TSM108	VJSS111	*srl-251*::MudJ/pMS421 (Sm^r Sp^r, *lacI*^{Q})/pNK2884 (Ap^r Cm^r, Tn10d[Cm] P$_{tac}$-*tnpA ats-1 ats-2*)	V. Stewart
TSM109	VJSS126	*srl-251*::MudJ *srl-252*::Tn10d(Cm) (Reg^c)	V. Stewart
TSM110	VJSS127	*srl-251*::MudJ *srlU253*::Tn10d(Cm) (Reg^c)	V. Stewart
TSM111	VJSS178	*zgx-7204*::Tn10d(Tc) (near *srlU*)	V. Stewart
TSM112	VJSS302	*zgc-7206*::Tn10d(Cm) (near *srl*)	V. Stewart
TSM113	VJSS185	DUP[[*nadB*]*MudA*[*cysH*]]	V. Stewart
TSM114	LB5010	*hsdL6* (r$^-$m$^+$) *hsdSA29* (r$^-$m$^+$) *hsdSB121* (r$^-$m$^+$) *galE856 xyl-404 leu-3121 metA22 metE551 trpC2 ilv-452 rpsL120*	6
TSM115	VJSS120	As TSM114 but *zxx-7201*::Mu cts hP1	V. Stewart
TSM116	VJSS367	As TSM114 but Δ(*srl-zgc*)267::MudA (Lac$^+$ Reg^c)	V. Stewart
TSM117	VJSS121	As TSM115 but pEG5005 (Ap^r Km^r, Mud5005)	V. Stewart
TSM118	TT10426	*proAB47*/F′ *pro$^+$ lac$^+$ zzf-1834*::Tn10d(Km)	5
TSM119	TT10604	*proAB47*/F′ *pro$^+$ lac$^+$ zzf-18361*::Tn10d(Cm)	5
TSM120	VJSS036	pNK972 (Ap^r, P$_{tac}$-*tnpA*)	4

TSM121	TT8790	*nadB224*::MudA (Lac⁻; orientation B)	J. Roth
TSM122	TT9643	*cys[HDC]1574*::MudA (Lac⁺; orientation B)	J. Roth
TSM123	TB293	Δ(*trpEA*)167	R. Bauerle
TSM124	VJSS323	*zgc-7205*::MudA	V. Stewart
TSM125	VJSS190	*mtl*::MudJ P22ʳ	V. Stewart
TSM126	VJSS069	*srl-254*::MudJ (Lac⁺ Reg⁺)	V. Stewart
TSM127	VJSS070	*srl-255*::MudJ (Lac⁺ Reg⁺)	V. Stewart
TSM131	VJSS337	Δ(*srl-zgc*)262::MudA (Lac⁺ Regᶜ)	V. Stewart
TSM134	VJSS342	Δ(*srl-zgc*)263::MudA (Lac⁺ Regᶜ)	V. Stewart
TSM135	VJSS343	Δ(*srl-zgc*)264::MudA (Lac⁺ Regᶜ)	V. Stewart
TSM136	VJSS344	Δ(*srl-zgc*)265::MudA (Lac⁺ Regᶜ)	V. Stewart
TSM139	VJSS347	Δ(*srl-zgc*)266::MudA (Lac⁺ Regᶜ)	V. Stewart
TSM140	VJSS348	Δ(*srl-zgc*)267::MudA (Lac⁺ Regᶜ)	V. Stewart
TSM141	VJSS349	Δ(*srl-zgc*)268::MudA (Lac⁺ Regᶜ)	V. Stewart
TSM144	VJSS352	Δ(*srl-zgc*)269::MudA (Lac⁺ Regᶜ)	V. Stewart
TSM145	VJSS371	*supD501*(Am) (*serU*) *cysA1348*(Am) *hisC527*(Am)(Cys⁺ His⁺) *zed-609*::Tn10 Δ(*srl-zgc*)267::MudJ	V. Stewart
TSM146	VJSS378	*supD501*(Am) (*serU*) *cysA1348*(Am) *hisC527*(Am)(Cys⁺ His⁺) *zed-609*::Tn10 *srl⁺ zgc-7206*::Tn10d(Cm)	V. Stewart
TSM147	VJSS380	*srl-270* (Am) (DES) *zgc-7206*::Tn10d(Cm)	V. Stewart
TSM148	VJSS381	*srl-271* (Am) (HA) *zgc-7206*::Tn10d(Cm)	V. Stewart
TSM149	VJSS382	*supD501*(Am) (*serU*) *cysA1348*(Am) *hisC527*(Am)(Cys⁺ His⁺) *zed-609*::Tn10 *srl-270* (Am) *zgc-7206*::Tn10d(Cm)	V. Stewart
TSM150	VJSS383	*supD501*(Am) (*serU*) *cysA1348*(Am) *hisC527*(Am)(Cys⁺ His⁺) *zed-609*::Tn10 *srl-271* (Am) *zgc-7206*::Tn10d(Cm)	V. Stewart
TSM201	MS1868	*leuA414*(Am) *hsdL* (r⁻ m⁺) Fels2⁻	M. Susskind

(Continued on following page.)

Strain List (continued)

Strain	AKA[a]	Genotype	Source
TSM202	MS1883	leuA414(Am) hsdL (r⁻m⁺) Fels2⁻ supE40	M. Susskind
TSM203	MST1762	As TSM201 but pMS421 (Smʳ Spʳ, lacIᵠ)	15
TSM204	MST2786	As TSM201 but pGW1700 (Tcʳ, mucA⁺B⁺)	13
TSM205	MS1582	As TSM202 but attP::(P22 sieA44 16[Am]H1455 Tpfr49)	7
TSM206	TH564	As TSM202 but attP::(P22 sieA44 Δ[mnt-a1 Tn1] Apˢ 9⁻ Tpfr184)	7
TSM207	MST2778	leuA414(Am) Fels2⁻ supE40/pPC36 (Spʳ Smʳ, P$_{tac}$-nac lacIᵠ)	S. Maloy
TSM208	MST2779	leuA414(Am) Fels2⁻ supE40 attP::(P22 sieA44 Δ[mnt-a1 Tn1] Apˢ Tpfr184)/pPC36 (Spʳ Smʳ, P$_{tac}$-nac lacIᵠ)	S. Maloy
TSM209	MST2780	As TSM202 but attP::(P22 sieA44 Δ[mnt-a1 Tn1] Apˢ Tpfr184)/ pPC36 (Spʳ Smʳ, P$_{tac}$-nac lacIᵠ)/pGW1700 (Tcʳ, mucA⁺B⁺)	S. Maloy
TSM210	MST2783	As TSM202 but pPC38 (Tcʳ; K. aerogenes put control region)	S. Maloy
TSM211	MST2784	As TSM202 but pPC39 (Tcʳ; K. aerogenes put control region)	S. Maloy
TSM213	MS1882	leuA414(Am) hsdL (r⁻m⁺) Fels2⁻ endA	M. Susskind
TSM214	MS1363	leuA414(Am) Fels2⁻ supE40	M. Susskind
TSM215	–	leuA414(Am) Fels2⁻ supE40 mutD200::Tn10dTc	S. Maloy
TSM216	–	leuA414(Am) Fels2⁻ supE40 mutS121::Tn10	S. Maloy
TSM1002	–	srl-256::MudJ (Lac⁺ Reg⁺)	ABG '91[b]
TSM1003	–	srl-257::MudJ (Lac⁺ Reg⁺)	ABG '91
TSM1004	–	srl-258::MudJ (Lac⁺ Reg⁺)	ABG '91
TSM1005	–	srl-259::MudJ (Lac⁺ Reg⁺)	ABG '91
TSM1008	–	srl-260::MudJ (Lac⁺ Reg⁺)	ABG '91
TSM1078	–	srl-261::MudJ (Lac⁺ Reg⁺)	ABG '92
TSM1141	–	srl-272::MudJ (Lac⁺ Reg⁺)	ABG '93

TSM1142	—	srl-273::MudJ (Lac⁺ Reg⁺)	ABG '93
TSM1143	—	Δ(srl-zgc)274::MudA (Lac⁺ Regᶜ)	ABG '93
TSM1144	—	Δ(srl-zgc)275::MudA (Lac⁺ Regᶜ)	ABG '93

Vibrio cholerae

TSM301	O395-N1	Classical, Ogawa; *ctxA* Smʳ	17
TSM302	CS2-1	O395 Φ(*tcpA-phoA*)2-1(Hyb) Smʳ Kmʳ	25
TSM303	KP5-2	O395 Φ(*tcpC-phoA*)5-2(Hyb) Smʳ Kmʳ	19
TSM304	CS7-1	O395 Φ(*tcpH-phoA*)7-1(Hyb) Smʳ Kmʳ	R. Taylor
TSM305	CS10-1	O395 Φ(*tcpC-phoA*)10-1(Hyb) Smʳ Kmʳ	R. Taylor
TSM306	KP8-85	O395 Φ(*tcpD-phoA*)8-85(Hyb) Smʳ Kmʳ	19
TSM307	KP2-21	O395 Φ(*tcpF-phoA*)2-21(Hyb) Smʳ Kmʳ	19
TSM308	KP4-2	O395 Φ(*tcpE-phoA*)4-2(Hyb) Smʳ Kmʳ	19
TSM309	KP3-41	O395 Φ(*tcpB-phoA*)3-41(Hyb) Smʳ Kmʳ	19
TSM310	RT104.11	O395 Φ(*tcpA-phoA*)104.11(Hyb) Smʳ Kmʳ	18
TSM311	—	O395 Δ*toxR43* Φ(*tcpA-phoA*)2-1(Hyb) Smʳ Kmʳ	R. Taylor
TSM312	—	As TSM311 but pVM7 (Apʳ, *toxR⁺*)	21

Escherichia coli K-12ᶜ

TSM191	JM105	*hsdR4* (r⁻m⁺) *thi rpsL endA sbcB15* Δ(*lac-proA*)X111 F' *traD36 proAB lacI*Q Δ(*lacZ*)M15	11
TSM192	VJS3834	As TSM191 but *srl*::Tn*10*	V. Stewart
TSM193	VJS3835	As TSM191 but pSU18 (Cmʳ)	16
TSM194	VJS3836	As TSM191 but pSU19 (Cmʳ)	16
TSM212	EM425	As TSM214 but pPY190 (Apʳ Tcˢ)	S. Maloy

(Continued on following page.)

Strain List (continued)

Strain	AKA[a]	Genotype	Source
TSM217	DH1	*thi-1 supE44 endA1 relA1 recA1 gyr-96 hsdR17* (r_k^- m_k^+)	8
TSM338	SM10	*thi-1 thr leu tonA lacY recA supE* (RP4-2 Km^r tet::Mu)	
TSM339	SM10/291	As TSM338 but pRT291 (Km^r Tc^r)	10
TSM340	SM10 λpir	As TSM338 but λpir	10
TSM343	SM10/55	As TSM340 but pVM55 (Ap^r)	14
TSM344	JM109	*recA1 endA1 gyrA96* (Nal^r) *thi hsdR17* (r_k^- m_k^+) *supE44 relA1* e14$^-$ (McrA$^-$) Δ(*lac-proAB*)/F' *proA$^+$B$^+$ traD36 lacI^Q ΔlacZM15*	
TSM345	–	As TSM344 but pRT651	R. Taylor
TSM347	CC118	*araD139* Δ(*ara-leu*)7697 ΔlacX74 ΔphoA20 galE galK thi rpsE (Sp^r) rpoB (Rif^r) argE(Am) recA1*	20
TSM348	MM294/RK	*endA hsdR pro supF*/pRK2013::Tn9 (Km^r Cm^r)	10
TSM349	MM294/PH	*endA hsdR pro supF*/pPH1JI (Gm^r Sp^r)	10
TSM350	VM2	*araD* Δ(*lac-proAB*) *argE*(Am) *rpoB* (Rif^r) *gyrA* (Nal^r) *recA56*/λ NFVM1 (cI857Sam100 *int$^+$ att$^+$* Φ[*ctx-lacZ*])	21
TSM351	ES1301	*lacZ53 mutS201*::Tn5 (Km^r) *thyA36 rha-5 metB1 deoCIN*(*rrnD–rrnE*)	22
TSM360	S17-1λpir	*recA thi pro hsdR-M$^+$*(RP4-2 Tc::Mu-Km::Tn7), λpir (Tp^r, Sm^r)	R. Taylor
TSM361	–	TSM360 (pKAS37)	R. Taylor
TSM362	–	TSM360 (pKAS38)	R. Taylor

Plasmid List

Plasmid	Resistance	Properties	Source
pCS9E11	Ap^r	pHC79 *tcp$^+$*	18
pEG5005	Ap^r Km^r	Mud5005	12
pGP704	Ap^r	pJM703.1 with polylinker	14

pGW1700	Tc^r	mucA^+B^+	13
pJM703.1	Ap^r	oriR6K mobRP4	14
pKAS32	Ap^r	pGP704, rpsL^+	26
pKAS37	Ap^r	pKAS32 toxR^+	26
pKAS38	Ap^r	pKAS32 toxR R96K	26
pMS421	Sm^r Sp^r	lacI^Q	15
pNK972	Ap^r	P_{tac}-tnpA	4
pNK2881	Ap^r	P_{tac}-tnpA ats-1 ats-2	4
pPC36	Sm^r Sp^r	P_{tac}-nac lacI^Q	S. Maloy
pPC38	Tc^r	pPY190 SmaI::SmaI-DraI K. aerogenes put control region	S. Maloy
pPC39	Tc^r	pPY190 SmaI::SmaI-HincII K. aerogenes put control region	S. Maloy
pPH1JI	Sp^r Gm^r	IncP1 tra^+	23
pPY190	Ap^r Tc^r	pBR322 EcoRI::P22 mnt-P_{ant}-SmaI/XmaI-ant	7
pRK2013::Tn9	Cm^r Km^r	IncP1 ColE1 tra^+	24
pRT1116	Ap^r Km^r	pSL1180 HindIII::HindIII-BamHI TnphoA asymmetric probe fragment::BamHI	R. Taylor
pRT291	Km^r, Tc^r	IncP1 TnphoA (white)	10
pRT651	Ap^r	pALTER-1 toxR^+	R. Taylor
pSU18	Cm^r	ori P15A, pUC18 polylinker	16
pSU19	Cm^r	ori P15A, pUC19 polylinker	16
pVM7	Ap^r	pBR322 toxR^+	21
pVM55	Ap^r	pJM703.1 'toxR'	14

[a] "Also known as"; original strain designation.

[b] Strain isolated by participants in the Advanced Bacterial Genetics course in the indicated year.

[c] All E. coli strains are F^- and λ^- unless otherwise indicated. (DES) indicates a mutation obtained by diethylsulfate mutagenesis and (HA) indicates a mutation obtained by hydroxylamine mutagenesis. Genetic nomenclature used is described in Sanderson et al. (1995).

References: (1) Zinder and Lederberg 1952; (2) Schmid and Roth 1980; (3) Hughes and Roth 1988; (4) Kleckner et al. 1991; (5) Elliott and Roth 1988; (6) Bullas and Ryu 1983; (7) Benson et al. 1986; (8) Hanahan 1983; (9) Simon et al. 1983; (10) Taylor et al. 1989; (11) Yanisch-Perron et al. 1985; (12) Grois- man and Casadaban 1986; (13) Perry and Walker 1982; (14) Miller and Mekalanos 1988; (15) Graña et al. 1988; (16) Bartolomé et al. 1991; (17) Mekalanos et al. 1983; (18) Shaw and Taylor 1990; (19) Peterson and Mekalanos 1988; (20) Manoil and Beckwith 1985; (21) Miller and Mekalanos 1985; (22) Siegel et al. 1982; (23) Ruvkun and Ausubel 1981; (24) Ditta et al. 1980; (25) Shaw et al. 1990; (26) Skorupski and Taylor 1995.

APPENDIX B
Culture Media and Supplements

Liquid Culture Media

10x E Salts

50x E salts	200 ml
Deionized H_2O	800 ml

Aliquot 100 ml to milk dilution (160 ml) bottles. Autoclave.

50x E Salts (Vogel-Bonner Minimal Salts)

Deionized H_2O	670 ml
$MgSO_4 \cdot 7H_2O$	10 g
Citric acid $\cdot H_2O$	100 g
K_2HPO_4	500 g
$NaHNH_4PO_4 \cdot 2H_2O$	175 g

Dissolve in order on a stirring hotplate (45°C). Adjust to 1000 ml with deionized H_2O. Store at room temperature.

LB Broth (Luria-Bertani)

Tryptone	10 g
Yeast extract	5 g
NaCl	5 g
Deionized H_2O	1000 ml

Aliquot 100 ml to milk dilution (160 ml) bottles. Autoclave.

LBEDO Broth (LB Broth with E Salts and Dextrose)

LB broth	100 ml
50x E salts	2 ml
1 M D-Glucose	1 ml

10x NCE Salts

50x NCE salts	200 ml
Deionized H_2O	800 ml

Aliquot 100 ml to milk dilution (160 ml) bottles. Autoclave.

50x NCE Salts (Citrate-free Minimal Salts)

Deionized H_2O	670 ml
KH_2PO_4	197 g
K_2HPO_4	250 g
$NaHNH_4PO_4{\cdot}2H_2O$	175 g

Dissolve in order on a stirring hotplate (45ºC). Adjust to 1000 ml with deionized H_2O. Store at room temperature.

P22 Broth

LBEDO broth	100 ml
P22 (~10^{11} pfu/ml)	1 ml

Store at 4ºC.

Plasmid Broth

Tryptone	12 g
Yeast extract	24 g
Glycerol	5 ml
Deionized H_2O	900 ml

Aliquot 90 ml to milk dilution (160 ml) bottles. Autoclave and then add 10 ml of sterile 1 M potassium phosphate buffer (pH 7.6) (see p. 464) to each bottle.

TY Broth (Tryptone–Yeast Extract)

Tryptone	8 g
Yeast extract	5 g
NaCl	5 g
Deionized H_2O	1000 ml

Aliquot 100 ml to milk dilution (160 ml) bottles. Autoclave.

TYP Broth (Tryptone–Yeast Extract–Phosphate)

Tryptone	16 g
Yeast extract	16 g
NaCl	5 g
K_2HPO_4	2.5 g
Deionized H_2O	1000 ml

Aliquot 100 ml to milk dilution (160 ml) bottles. Autoclave.

Vibrio Broth (For Testing Agglutination)

Prepare LB broth (see p. 455) as usual, but adjust pH to 6.5 with HCl before autoclaving.

Solid Culture Media

Bochner-Maloy Agar (For selecting Tc[s])

Flask A:

Tryptone	5 g
Yeast extract	5 g
Chlortetracycline	50 mg
Agar	15 g
Deionized H_2O	900 ml

Flask B:

NaCl	10 g
$NaH_2PO_4 \cdot H_2O$	10 g
Deionized H_2O	100 ml

Autoclave for *20 minutes*, mix, and then add

Fusaric acid (dissolved in dimethylformamide)	12 mg
20 mM $ZnCl_2$ (filter-sterilized)	5 ml

These plates are best when used fresh (<12 hours old), but they may be stored at 4°C and used effectively after 2 weeks.

Caution: N,N-dimethylformamide (DMF) is irritating to the eyes, skin, and mucous membranes. It can exert its toxic effects through inhalation, absorption through the skin, or ingestion. Chronic inhalation can cause liver and kidney damage. Wear gloves and safety glasses when handling DMF.

E Agar (Minimal Salts Agar)

Glucose (or other carbon source)	2 g
Agar	15 g
Deionized H_2O	980 ml
Acid-hydrolyzed casein (ACH) if needed	1 g

Autoclave and then add

5x E Salts (see p. 455)	20 ml

LB Agar

Tryptone	10 g
Yeast extract	5 g
NaCl	5 g
Agar	15 g
Deionized H_2O	1000 ml

MacConkey–Srl Agar

MacConkey agar *base*	40 g
Sorbitol (or 10 g of other test carbohydrate)	5 g
Deionized H_2O	1000 ml

Note: MacConkey agar usually contains 1% sugar. Srl is slightly inhibitory at this concentration for *Salmonella*, so it should be used at 0.5%.

NA-TTC Srl Agar (Nutrient Agar–Triphenyl Tetrazolium Chloride–Sorbitol)

Nutrient agar	23 g
NaCl	5 g
Sorbitol (or other test carbohydrate)	10 g
Deionized H_2O	1000 ml

Autoclave and then add

Triphenyl tetrazolium chloride (0.25%)	10 ml

NCE Agar (Citrate-free Minimal Salts Agar)

Carbon source	2 g
Agar	15 g
Deionized H_2O	900 ml

Autoclave and then add

10x NCE Salts (see p. 456)	100 ml
1 M $MgSO_4$	1 ml

Nutrient Agar

Nutrient agar	23 g
NaCl	5 g
Deionized H_2O	1000 ml

TBSA (Tryptone broth–soft agar)

Tryptone	10 g
NaCl	5 g
Agar	7 g
Deionized H_2O	1000 ml

Dissolve thoroughly in a microwave or steamer until melted. Aliquot into milk dilution bottles. Autoclave.

XP Indicator Agar for *V. cholerae*

Prepare LB agar (see p. 459). Autoclave. Add the following depending on the plate (see p. 462):

Glucose (20% stock)	15 ml
XP (20 mg/ml stock)	2 ml
Ap (100 mg/ml stock)	1 ml
Gm (30 mg/ml stock)	1 ml
Km (45 mg/ml stock)	1 ml
Sm (100 mg/ml stock)	1 ml
Tc (15 mg/ml stock)	1 ml

Culture Media Supplements

0.5 M CaCl$_2$

CaCl$_2$·2H$_2$O	74 g
Deionized H$_2$O	800 ml

Aliquot 100 ml to milk dilution (160 ml) bottles. Autoclave.

1 M EGTA (Ethylene Glycol-Bis[β-Aminoether]*N,N,N',N'*-Tetra-Acetic Acid)

EGTA	250 g
NaOH	53 g
Deionized H$_2$O	400 ml

Adjust pH to 7.0 with HCl. Adjust to 660 ml with H$_2$O. Filter. Aliquot 100 ml to milk dilution (160 ml) bottles. Autoclave. Use 1:100 in liquid and solid media (final 10 mM).

1 M MgSO$_4$

MgSO$_4$·7H$_2$O	246 g
Deionized H$_2$O	1000 ml

Aliquot 100 ml to milk dilution (160 ml) bottles. Autoclave.

XGal (5-Bromo-4-Chloro-3-Indolyl-β-D-Galactopyranoside)

40 mg/liter of medium. Prepare a 20 mg/ml stock solution in *N,N*-dimethylformamide (see *Caution* p. 458) in a polypropylene or glass tube. Store at 4°C in the dark.

XP (5-Bromo-4-Chloro-3-Indolyl-Phosphate; *p*-Toluidine Salt)

40 mg/liter of medium. Prepare a 20 mg/ml stock solution in *N,N*-dimethylformamide (see *Caution* p. 458) in a polypropylene or glass tube. Store at 4°C in the dark.

Antibiotics

Ap (Ampicillin, Sodium Salt)

100–200 mg/ml in deionized H_2O. Filter-sterilize and store at –20°C.

Note: Ap is unstable. Use fresh plates and store plates at 4°C. With older plates, Ap[r] colonies are surrounded by "satellites."

Cm (Chloramphenicol)

20–25 mg/ml in ethanol. Store at –20°C.

Caution: Cm may be a carcinogen. Wear gloves and work carefully.

Gm (Gentamicin Sulfate)

30 mg/ml in deionized H_2O. Filter-sterilize and store at 4°C.

Km (Kanamycin Sulfate)

45–50 mg/ml in deionized H_2O. Filter-sterilize and store at 4°C.

Sm (Streptomycin Sulfate)

100 mg/ml in deionized H_2O. Filter-sterilize and store at 4°C.

Sp (Spectinomycin Dihydrochloride)

50 mg/ml in deionized H_2O. Filter-sterilize and store at 4°C.

Tc (Tetracycline Hydrochloride)

15–20 mg/ml in ethanol. Store at –20°C in the dark.

Note: Tc is light-sensitive. Store plates protected from light.

Final concentrations, in μg/ml:

Antibiotic	Salmonella typhimurium		Vibrio cholerae
	defined medium	rich medium	rich medium
Ap[*]	100	200	100
Cm	5	20	–
Gm	–	–	30
Km	125	50	45
Sm	50	100	100
Sp	50	100	–
Tc	10	20	15

[*]This concentration of Ap is appropriate for selection for photocopy plasmids. Use 5 μg/ml in defined medium or 30 μg/ml in rich medium when selecting for single-copy Ap^r vectors.

Buffers

10x BS (Buffered Saline)

$Na_2HPO_4 \cdot 7H_2O$	110 g
KH_2PO_4	30 g
NaCl	8.5 g

Dissolve in 100 ml of deionized H_2O. Adjust to 200 ml with H_2O. Autoclave. Before use, dilute 10 ml into 100 ml of sterile H_2O for 1x BS.

Phage Buffer (T2 Buffer)

K_2SO_4	5 g
NaCl	4 g
KH_2PO_4	1.5 g
Na_2HPO_4	3 g
Deionized H_2O	1000 ml

Aliquot 100 ml to milk dilution (160 ml) bottles. Autoclave. Cool to room temperature and add

0.1% Gelatin	0.33 ml
1 M $MgSO_4$	33 µl
0.5 M $CaCl_2$	7 µl

1 M Potassium Phosphate Buffer (pH 7.6) (For Plasmid Broth)

KH_2PO_4	69 g
Deionized H_2O	300 ml

Adjust pH to 7.6 with KOH. Adjust to 500 ml with deionized H_2O. Aliquot 100 ml to milk dilution (160 ml) bottles. Autoclave. Add 10 ml/liter medium.

P22 Indicator Media

EBU Agar (Evans Blue–Uranine)

Tryptone	10 g
Yeast extract	5 g
NaCl	5 g
Glucose	2.5 g
Agar	15 g
Deionized H_2O	960 ml

Autoclave and then add

K_2HPO_4 (12.5%)	40 ml
Evans blue (1%)	1.25 ml
Uranine (1%) (sodium fluorescein)	2.5 ml

Prepare Evans blue and uranine by autoclaving. Store stock solutions at room temperature in the dark.

Green Agar

Flask A:

Tryptone	8 g
Yeast extract	1 g
NaCl	5 g
Alizarin yellow GG	630 mg
Methyl blue (aniline blue)	66 mg
Deionized H_2O	600 ml

Flask B:

Glucose	6.7 g
Deionized H_2O	400 ml

Autoclave and mix.

MOPS Culture Medium

Neidhardt et al. (1974) described this defined culture medium for physiological studies with enterobacteria. The medium is buffered with MOPS, a Good buffer with a pK_a of 7.2; enterobacteria will not use MOPS as a carbon source, and *Escherichia coli* strain B (but not *Salmonella typhimurium*) will use MOPS only poorly as a sulfur source. Tricine is present to keep Fe in solution. Bicarbonate provides a source of CO_2 during early stages of growth. For a modified version for growth of anaerobic cultures, see Stewart and Parales (1988).

Prepare all components as concentrated stock solutions and filter-sterilize to help provide reproducibility in medium composition (autoclaved solutions are often of uncertain concentration). The recipe given here is slightly modified from the original to allow manipulation of the nitrogen source (for additional details of preparation and use, see Neidhardt et al. 1974).

Autoclave ultrapure H_2O as approximately 110-ml aliquots in graduated milk dilution bottles. Aseptically remove water so that each bottle contains exactly 100 ml. For assembling 1x culture medium, remove water equivalent to the amount of stock solutions added, so that each bottle of medium contains exactly 100 ml. Low-glucose medium is used to prepare inoculum (for details of culture techniques, see Neidhardt et al. 1974). Substitute different nitrogen, phosphorus, or carbon sources and add additional nutritent or supplements, as desired (see Wanner et al. 1977; Stewart and Parales 1988).

1x MOPS Culture Medium (Normal Glucose)

10x MOPS	10 ml
50x Bicarbonate	2 ml
100x Ammonium	1 ml
100x Phosphate	1 ml
50x Glucose	2 ml
Ultrapure H_2O	84 ml

1x MOPS Culture Medium (Low Glucose)

10x MOPS	10 ml
50x Bicarbonate	2 ml
100x Ammonium	1 ml
100x Phosphate	1 ml
50x Glucose	0.1 ml
Ultrapure H_2O	86 ml

10x MOPS (500-ml Stock Solution)

MOPS	400 mM	41.9 g
Tricine	40 mM	3.6 g
NaCl	500 mM	14.6 g
$FeSO_4 \cdot 7H_2O$	0.1 mM	14 mg
$MgSO_4 \cdot 7H_2O$	5.5 mM	68 mg
1000x Micronutrients		0.5 ml

Dissolve MOPS and tricine in 380 ml of ultrapure H_2O. Adjust pH to 7.4 with solid NaOH. Dissolve remaining components. Adjust to 500 ml with ultrapure H_2O. Filter-sterilize and store cold and dark.

100x Ammonium (150-ml Stock Solution)

NH_4Cl	1.43 M	11.5 g

Dissolve and adjust to 150 ml with ultrapure H_2O. Filter-sterilize and store at room temperature.

50x Bicarbonate (150-ml Stock Solution)

$NaHCO_3$	1 M	12.6 g

Dissolve and adjust to 150 ml with ultrapure H_2O. Filter-sterilize and store at room temperature.

50x Glucose (150-ml Stock Solution)

| D-Glucose | 1 M | 54 g |

Dissolve and adjust to 150 ml with ultrapure H_2O. Filter-sterilize and store at room temperature.

1000x Micronutrients (100-ml Stock Solution)

H_3BO_3	4 mM	24.8 mg
$CaCl_2 \cdot 2H_2O$	5 mM	73.6 mg
$CoCl_2 \cdot 6H_2O$	0.3 mM	7.2 mg
$CuSO_4 \cdot 5H_2O$	0.1 mM	2.4 mg
$MnCl_2 \cdot 4H_2O$	0.8 mM	15.8 mg
$ZnSO_4 \cdot H_2O$	0.1 mM	2.8 mg

Dissolve and adjust to 100 ml with ultrapure H_2O. Filter-sterilize and store at room temperature.

100x Phosphate (150-ml Stock Solution)

| K_2HPO_4 | 88 mM | 2.3 g |

Dissolve and adjust to 150 ml with ultrapure H_2O. Filter-sterilize and store at room temperature.

APPENDIX C
Strain Collections

Storing Strains

For short-term storage of cultures, use agar stabs; stabs also provide a convenient method for transporting or shipping strains. For long-term storage, cryopreservation is preferred, particularly for strains with unstable characteristics (e.g., plasmids, transposon insertions, and duplications).

For long-term storage, grow a fresh broth culture of the strain, mix equal portions of culture and glycerol (a cryoprotectant) in a small screw-cap tube, and place the tube in an ultra-low freezer (–70ºC or below). To recover a strain, simply remove the tube from the freezer, scrape a bit of frozen culture from the top, spot the culture on an appropriate plate, and immediately return the tube to the freezer. If care is taken to ensure that the frozen culture does not thaw, strains stored in this manner can be subcultured and kept viable for decades. If the freezer fails, simply refreeze the stocks—most strains will withstand a few complete thaw-freeze cycles, albeit with reduced viability. Some laboratories use 7% DMSO as a cryoprotectant, but most prefer glycerol.

5 M Glycerol

Mix 230 g of glycerol and deionized H_2O to a final volume of 500 ml. Aliquot 100 ml to milk dilution (160 ml) bottles. Autoclave. Mix 0.5 ml of 5 M glycerol and 0.5 ml of fresh culture in a sterile 2.0-ml vial.

Stab Agar

Agar	0.6 gm
Nutrient Broth	1 gm
NaCl	0.8 gm
Deionized H_2O	100 ml

Autoclave in a milk dilution (160 ml) bottle. Dispense 1 ml into sterile 2-ml screw cap vials.

Shipping Strains

Sending strains to other investigators is an essential activity of bacterial geneticists. Researchers who publish on a particular strain, phage, or plasmid are obligated to provide samples to qualified investigators upon request. Shipping pathogenic strains requires special packaging materials, labels, and procedures, which are comprehensively described by Sanderson and Zeigler (1991). However, nonpathogenic strains are conveniently distributed on filter disks, wrapped in sterile plastic wrap. This procedure was invented by phage geneticists for sending phage strains, but it also works well with bacterial cultures, and is employed by the *E. coli* Genetic Stock Center.

1. Alternate approximately 20 layers of plain paper and plastic wrap (microwave-grade Saran Wrap works well).

2. Cut the layered paper/wrap into squares of approximately 4 inches (10 cm) and stack in a 15-cm petri dish.

3. Place filter disks in a separate glass petri dish. We use 0.5-inch-diameter penicillin assay disks, from Schleicher and Schuell, but cut-up Whatman No. 1 paper also works.

4. Autoclave the petri dishes and dry in an incubator. The paper-interleaved plastic wrap tends to shrink and curl when autoclaved. Place the interleaves in the lid of the dish and cover with the bottom of the dish inverted so that the interleaves are held flat between the two surfaces. Place a weight on top of the entire assembly to maintain integrity during autoclaving.

5. Grow a fresh verified culture of the strain to be shipped.

6. Use sterile forceps to remove a plastic-paper heterodimer from the stack. The plastic wrap tends to stick to the paper, facilitating this procedure.

7. Place a disk in the center of the wrap. Spot 0.1 ml of culture (or phage lysate) on the disk. Wrap the disk with the plastic wrap and affix to an index card with tape.

8. To recover the strain, simply cut open the wrap, extract the disk with sterile forceps, place on an appropriate agar plate, add a drop of diluent, and streak for colonies. For phage, place the disk on a seeded overlay plate and streak for plaques.

Strain Records

Maintaining accurate strain collection records is a major challenge for any microbial genetics laboratory and particularly for those in which many different people are depositing strains. For the Advanced Bacterial Genetics Course, we maintain strain records on the Culture Collection Record Sheet (Fig. C.1), which records the strain number (TSM designation), species, genotype, method of construction (e.g., P22 transduction), and parental strains. Considerations for maintaining culture collections and records are elaborated by Sanderson and Zeigler (1991).

Culture Collection Record Sheet

Page _____

Strain No.	Sp.	Genotype/Phenotype	Method	Donor	Recipient

Figure C.1 Culture Collection Record Sheet.

APPENDIX D
Genetic Exchange and Mapping

Generalized Transduction with Phage P22

For all culture media and supplements used here, see Appendix B.

Preparation of P22 HT Lysates

1. Inoculate each donor strain into 0.5 ml of LB broth in a large culture tube.

2. Aerate at 30ºC or 37ºC until the culture is saturated (at least 6 hours; overnight cultures can be used).

3. Add 2.0 ml of P22 broth (LBEDO broth containing ~10^9 pfu/ml of P22 HT 105/1 *int-201*).

4. Continue aeration for at least 5–8 hours at 30ºC or 37ºC (overnight cultures can be used). The culture will be quite dense, but cell debris and "strings" are often visible when lysis has proceeded well.

5. Decant 1.5 ml of the lysed culture into a microcentrifuge tube.

6. Sediment the cells and debris in a microcentrifuge at 14,000 rpm for 2 minutes.

7. Label a sterile 2-ml screw-cap tube for each lysate. Add two drops of chloroform to each tube.

   Caution: Chloroform is a carcinogen and may damage the liver and kidneys. Do not mouth pipette and avoid contact with skin.

8. Remove most of the supernatant with a sterile pasteur pipette and transfer to the screw-cap tube.

9. Invert the tube several times to thoroughly disperse the chloroform throughout the lysate.

10. Store the lysate on ice until use. Allow enough time for the chloroform and debris to settle.

Transduction Protocol (Delayed Expression)

1. Grow saturated cultures of the desired recipient strains (including strain TSM123 and other control strains) each in 2.5 ml of LB broth aerated at 30ºC or 37ºC. Store the cultures on ice until use.

2. For each transduction, mix 0.1 ml of recipient culture and the appropriate amount (usually 10 µl) of phage lysate in a microcentrifuge tube.

3. Incubate for 10–15 minutes at 30ºC or 37ºC.

4. Add 0.9 ml of LB broth-EGTA (10 mM) and gently mix.

5. Incubate for 15–20 minutes at 30ºC or 37ºC.

6. Plate 0.1 ml of each transduction mixture on the appropriate selective medium.

7. Plate 0.1 ml of each recipient culture alone on the same medium (cells-only control).

8. Spot 10 µl of each *undiluted* phage lysate on an LB plate (lysate-sterility control).

9. For each lysate used, mix 0.05 ml of strain TSM123 with 2.5 ml of TBSA (tryptone broth–soft agar) and pour onto an LB plate. Spot on 6/.02, 7/.02, 8/.02, and 9/.02 dilutions of the lysate (see Section 2, Microbiological Procedures) and allow the spots to dry.

10. For each lysate used, spread 0.1 ml of strain TSM123 culture on an E-Glc-ACH (acid-hydrolyzed casein) plate. Spot on 0/.02, 1/.02, 2/.02, and 3/.02 dilutions of the lysate and allow the spots to dry.

11. Incubate the experimental plates overnight at 30–37ºC. Incubate the TSM123 transduction control plates overnight at 37ºC.

Transduction Protocol (Plate Transduction)

1. Grow saturated cultures of the desired recipient strains (including strain TSM123 and other control strains) each in 2.5 ml of LB broth aerated at 30ºC or 37ºC. Store the cultures on ice until use.

2. For each transduction, pipette 0.1 ml of recipient culture and the appropriate amount (usually 10 µl) of phage lysate directly onto the appropriate selective medium.

3. Spread the mixture on the surface of the agar until dry.

4. Plate 0.1 ml of each recipient culture alone on the same medium (cells-only control).

5. Spot 10 µl of each *undiluted* phage lysate on an LB plate (lysate-sterility control).

6. For each lysate used, mix 0.05 ml of strain TSM123 with 2.5 ml of TBSA (tryptone broth–soft agar) and pour onto an LB plate. Spot on 6/.02, 7/.02, 8/.02, and 9/.02 dilutions of the lysate (see Section 2, Microbiological Procedures) and allow the spots to dry.

7. For each lysate used, spread 0.1 ml of strain TSM123 culture on an E-Glc-ACH plate. Spot on 0/.02, 1/.02, 2/.02, and 3/.02 dilutions of the lysate and allow the spots to dry.

8. Incubate the experimental plates at 30–37ºC. Incubate the TSM123 transduction control plates at 37ºC.

Rapid Mapping in *S. typhimurium* with Mud-P22 Prophages

For all culture media and supplements used here, see Appendix B.

Preparation of Lysates

1. Grow cultures of Mud-P22 strains overnight in 2.5 ml of LB broth.

2. Inoculate 0.1 ml each into 5 ml of LBEDO broth. (Add supplements for *cys*, *gua*, and *pur* auxotrophs.)

3. Aerate for 90 minutes at 37°C.

4. Add 5 µl of mitomycin C (2 mg/ml stock; Sigma, 2-g quantities). Resuspend in sterile H_2O and store at –20°C. Rehydrated mitomycin C is stable for at least 1 year.

   Caution: Mitomycin C is a mutagen and possible carcinogen. Wear gloves and work carefully.

5. Continue aeration for approximately 3 hours or until the cultures clear.

6. Sediment debris in a clinical centrifuge for 5–10 minutes.

7. Decant supernatant into a sterile screw-cap tube.

8. Add 30 µl of tail preparation (see below) and mix by vortexing.

9. Incubate overnight at room temperature.

10. Add chloroform (~0.3 ml) and mix by vortexing.

    Caution: Chloroform is a carcinogen and may damage the liver and kidneys. Do not mouth pipette and avoid contact with skin.

11. Determine approximate transducing titers by diluting 1:50 and mapping known mutations. Lysates still contain small amounts of mitomycin C. Be careful.

Preparation of Tails

1. Culture strain PY13579 overnight in 3 liters of LBEDO broth + Ap.

2. Sediment the cells at 10,000g for 10 minutes at 4°C.

3. Resuspend approximately 6 g cell pellet in 10 ml of 0.1 M Tris-HCl (pH 7.5) + 10 mM EDTA.

4. Transfer suspension to an Oak Ridge (35 ml) centrifuge tube.

5. Add 10 mg of lysozyme.

6. Incubate for approximately 4 hours at 65°C.

7. Sediment debris at 20,000–25,000g for 70 minutes at 4°C.

8. Store supernatant at 4°C. This supernatant (~10 ml) should contain about 10^{14} tails per milliliter. Some precipitate may form during storage.

Mapping Protocol

1. Grow recipient strain overnight in 2.5 ml of LB broth.

2. Plate 0.1 ml on selective medium; allow the plate to dry completely.

3. Aliquot the Mud-P22 lysates as 1:50 dilutions in a microtiter dish.

4. Spot approximately 5 μl of each lysate on the plate in a grid pattern. Use a "frog" (48-pin replica plate; Sigma R 2383) designed for microtiter dishes. Lysates still contain small amounts of mitomycin C. Be careful.

5. Incubate plates at the appropriate temperature.

Basic Set of Strains for Mud-P22 Mapping[a]

No.	Centisome[b]	Strain	Genotype[c]	Orientation[d]
1	0.0	TT15223	thr-469::MudP	A
2	0.0	TT15224	thr-465::MudQ	B
3	7.8	TT15229	proA692::MudQ	A
4	7.8	TT15231	proA692::MudP	B
5	12.6	TT15232	purE2154::MudQ	A
6	12.6	TT15235	purE2154::MudP	B
7	17.2	TT15238	nadA219::MudP	A
8	17.2	TT15629	nadA219::MudQ	B
9	25.6	TT15239	putA1019::MudQ	A
10	25.6	TT15240	putA1019::MudP	B
11	30.5	TT15244	aroD561::MudP	A
12	30.5	TT15243	aroD561::MudQ	B
13	37.6	TT15245	pyrF2690::MudQ	A
14	37.6	TT15246	pyrF2690::MudP	B
15	40.2	TT15630	treA152::MudP	A
16	40.2	TT15625	treA152::MudQ	B
17	44.8	TT17163	hisH9962::MudP	A
18	44.8	TT17164	hisH9962::MudQ	B
19	53	TT15254	cysA1586::MudP	A
20	53	TT15258	cysA1586::MudQ	B
21	61.4	TT15261	proW1884::MudP	A
22	61.4	TT15628	proW1884::MudQ	B
23	67	TT16706	zgh-3715::MudP	A
24	67	TT16707	zgh-3715::MudQ	B
25	72	TT17190	zhc-3717::MudP	A
26	72	TT17191	zhc-3717::MudQ	B
27	75.5	TT15264	cysG1573::MudP	A
28	75.5	TT15631	cysG1573::MudQ	B
29	81.4	TT15267	pyrE2419::MudP	A
30	81.4	TT15266	pyrE2419::MudQ	B
31	86.1	TT15270	metE2131::MudQ	A
32	86.1	TT15271	metE2131::MudP	B
33	88.3	TT15273	pnuE41::MudQ	A
34	88.3	TT15272	pnuE41::MudP	B
35	93.8	TT15276	mel-396::MudP	A
36	93.8	TT15275	mel-396::MudQ	B

Additional Set of Strains for Mud-P22 Mapping[a]

No.	Centisome[b]	Strain	Genotype[c]	Orientation[d]
37	3.6	TT15226	*nadC218*::MudP	A
38	3.6	TT15227	*nadC218*::MudQ	B
39	8.9	TT15230	*proC963*::MudP	A
40	15.6	TT15237	*cobD498*::MudP	A
41	15.6	TT15236	*cobD498*::MudQ	B
42	44	TT15249	*zee-3666*::MudQ	A
43	44	TT15250	*zee-3666*::MudP	B
44	54.4	TT15632	*guaA5641*::MudQ	A
45	54.4	TT15255	*guaA5641*::MudP	B
46	56.3	TT15256	*purG2149*::MudP	A
47	56.3	TT15257	*purG2149*::MudQ	B
48	64.2	TT15263	*cys[HIJ]1547*::MudP	B
49	68	TT16708	*zgi-3716*::MudP	A
50	68	TT16709	*zgi-3716*::MudQ	B
51	76	TT17165	*envZ1005*::MudQ	A
52	76	TT15265	*envZ1005*::MudP	B
53	85.3	TT15269	*ilvA2648*::MudQ	A
54	85.3	TT15268	*ilvA2648*::MudP	B
55	95.1	TT15277	*purA1881*::MudP	A
56	97	TT15638	*zjh-3725*::MudP	A
57	97	TT15633	*zjh-3725*::MudQ	B

[a]The basic set should be sufficient for mapping most unknown mutations. The additional strains can be used for more precise mapping if desired.

[b]Genetic map positions are calibrated according to Sanderson et al. (1995). Revised locations of *zxx* insertions are estimated by extrapolation from the previously reported map positions (Benson and Goldman 1992).

[c]Additional markers: *leuA414*(Am) *hsdL* (r⁻ m⁺) Fels2⁻.

[d]A and B, clockwise and counterclockwise packaging, respectively.

Auxiliary Strains for Mud-P22 Mapping

Strain	Genotype
PY13518	*leuA414*(Am) *hsdL* (r⁻ m⁺) Fels2⁻/F′ $_{114}$(Ts) *lac*⁺ *zzf-20*::Tn*10* *zzf-3551*::MudP
PY13579	MC1061 (*E. coli*)/pPB13 (Ap^r, P22 gene *9* constitutively expressed)
PY13757	*leuA414*(Am) *hsdL* (r⁻ m⁺) Fels2⁻/F′ $_{114}$(Ts) *lac*⁺ *zzf-20*::Tn*10* *zzf-3553*::MudQ
TT16716	*his-644 pro-621*/F′ $_{128}$ *pro*⁺ *lacZ477*::Tn*10*d(Tc)

Additional Considerations

- Mud-P22 strains are described by Benson and Goldman (1992) and may be obtained from the *Salmonella* Genetic Stock Center.

- Selection for Tc^s transductants of a Tn*10*-carrying strain provides a convenient and general method for mapping. Mud-*lac* insertion strains can be transduced with *lacZ*::Tn*10*d(Tc) (strain TT16716) to place Tc^r at the site of the Mud insertion. Homologous recombination between the *lacZ* sequences in the Mud-*lac* and the donor *lac*::Tn*10* transducing fragment occurs at low but detectable frequency in P22-mediated crosses, due to the limited amount of homology. Verify that the newly constructed Mud-*lac*::Tn*10*d(Tc) strain is Lac$^-$ and then map by selection for Tc^s.

- Spontaneous rearrangements involving standard Tn*10* occur at relatively high frequency, so it is best to culture such strains in broth supplemented with Tc before plating on Bochner-Maloy medium, to reduce the background. This also ensures that the *tetA* gene is fully induced, leading to maximal sensitivity. Plate 1:10 dilutions of such cultures to further reduce background. Strains with Tn*10*d(Tc) elements have less spontaneous background, but it helps to use Tc broth for these strains as well. For best results, use freshly prepared Bochner-Maloy plates.

- A new derivative of MudJ (Mud*Sac*I) carrying the counterselectable *sacB* gene is now available (Lawes and Maloy 1995). This provides an alternative means for mapping Mud insertions, by transduction and selection for sucrose resistance (i.e., loss of the Mud*Sac*I).

- Alternatively, Mud insertions can be mapped by replacing the resident Mud with MudP or MudQ by transductional exchange using strains PY13518 or PY13757 as donors. The newly constructed Mud-P22 strain can then be used as the donor in a series of crosses with known auxotrophic or Tn*10* insertion recipients.

- Mud-P22 lysates provide a source of DNA that is highly enriched for a relatively small region of the chromosome. This is convenient to use for cloning a known gene.

Phage P22 Lysates

The following procedures are for preparing P22 lysates for challenge phage experiments. For preparation of P22 HT *int* transducing phage lysates, see Generalized Transduction with P22 above, and for all culture media and supplements, see Appendix B.

Preparation of P22 Mini-Lysates

1. Plug a well-isolated plaque using a sterile pasteur pipette.

2. Add the agar plug to 2 ml of LBEDO broth. Vortex.

3. Add 0.2 ml of an overnight culture of a sensitive host (such as TSM201).

4. Incubate with thorough aeration for approximately 3 hours at 37°C or until the culture lyses.

5. Centrifuge in a microcentrifuge for 1 minute to pellet the cells.

6. Pour the supernatant into a clean tube. Add several drops of chloroform and vortex. Store the lysates at 4°C.

   Caution: Chloroform is a carcinogen and may damage the liver and kidneys. Do not mouth pipette and avoid contact with skin.

Preparation of High-titer P22 Lysates

1. For each phage lysate, add 0.1 ml of a sensitive *Salmonella* strain (such as TSM202) to a sterile test tube with 5 ml of LBEDO broth. Incubate with thorough aeration for approximately 1–2 hours at 37°C or until the cells reach early exponential phase.

2. Use a sterile pasteur pipette to plug an independent, well-isolated plaque and add 1 plaque to each tube of TSM202.

3. Incubate with thorough aeration for approximately 3 hours at 37°C or until lysed.

4. Add several drops of chloroform, vortex, and continue aeration for approximately 5 minutes.

5. Allow the chloroform to settle and then transfer the supernatant (avoiding the chloroform) to microcentrifuge tubes. Centrifuge in a microcentrifuge for 2 minutes to pellet the cell debris.

6. Transfer the supernatant from each phage lysate into clean microcentrifuge tubes (for 5 ml of lysate, use four microcentrifuge tubes with 1.25 ml of lysate per tube). Centrifuge in a microcentrifuge for 30 minutes at 4°C to pellet the phage.

7. Carefully pour off the supernatant. Add 0.1 ml of BS (Buffered Saline) to each phage pellet and vortex to resuspend the phage. Combine the supernatants from a single phage into one tube. Add one drop of chloroform and vortex. Store the concentrated lysate at 4°C.

Determining Phage Titers

The following procedure uses microtiter dishes to dilute the phage lysates and a micropipettor to spot small volumes of the diluted lysates on a lawn of cells. It is also possible to dilute the phage in test tubes and to use a single plate for each phage dilution, but using microtiter dishes requires substantially fewer supplies and is much faster.

1. Melt TBSA (tryptone broth–soft agar) in a microwave. Add 3 ml of the melted top agar to a test tube and place in a 50°C heating block.

2. After the top agar cools to about 50°C, add 0.2 ml of an overnight culture of TSM202, swirl, and quickly pour onto an LB-agar plate. Allow the top agar to solidify for 10–15 minutes. Dry the plates at 37°C with the lids ajar until there is no visible moisture on the surface (usually between 15 and 60 minutes, depending on the wetness of the plates).

3. Add 45 μl of sterile 0.85% NaCl to the first ten columns of a microtiter dish using a multichannel pipettor (see figure below). Add 5 μl of each phage to be titered to separate wells in the first column of the microtiter dish. Mix by gently pipetting the solution up and down with the pipettor. Dilute the phage by sequentially transferring 5 μl into each subsequent column of the microtiter dish. Use new pipette tips for each dilution.

	#1	#2	#3	#4	#5	#6	#7	#8	#9	#10	#11	#12
ϕ1	NaCl	NaCl	NaCl	NaCl	NaCl	NaCl	NaCl	NaCl	NaCl			
ϕ2	NaCl	NaCl	NaCl	NaCl	NaCl	NaCl	NaCl	NaCl	NaCl			
ϕ3	NaCl	NaCl	NaCl	NaCl	NaCl	NaCl	NaCl	NaCl	NaCl			
ϕ4	NaCl	NaCl	NaCl	NaCl	NaCl	NaCl	NaCl	NaCl	NaCl			
ϕ5	NaCl	NaCl	NaCl	NaCl	NaCl	NaCl	NaCl	NaCl	NaCl			
ϕ6	NaCl	NaCl	NaCl	NaCl	NaCl	NaCl	NaCl	NaCl	NaCl			
ϕ7	NaCl	NaCl	NaCl	NaCl	NaCl	NaCl	NaCl	NaCl	NaCl			
ϕ8	NaCl	NaCl	NaCl	NaCl	NaCl	NaCl	NaCl	NaCl	NaCl			

10^{-1} 10^{-2} 10^{-3} 10^{-4} 10^{-5} 10^{-6} 10^{-7} 10^{-8} 10^{-9}

4. Use a micropipettor to spot 5 μl of the 10^{-6}, 10^{-7}, 10^{-8}, and 10^{-9} dilutions onto the TSM202 lawn (a multichannel pipettor greatly simplifies this task).

5. Allow the drops of phage to dry (~30 minutes) and then incubate the plates upside down overnight at 37ºC.

6. Count the number of plaques in each spot. Each viable phage (pfu) will produce one plaque. Calculate the phage titer as follows:

$$\frac{\text{number of plaques}}{5 \text{ μl} \times \text{dilution factor}} \quad \times \quad \frac{1000 \text{ μl}}{\text{ml}} \quad = \quad \text{pfu/μl}$$

Purification of Phage DNA from High-titer Lysates

1. Add 0.2 ml of each concentrated phage lysate (avoiding the chloroform) to a microcentrifuge tube. Thoroughly mix a

DEAE-cellulose slurry (see p. 485) and then add 0.2 ml of the slurry to each tube. Invert 30–40 times to mix. Centrifuge in a microcentrifuge for 5 minutes.

2. Carefully remove the supernatant and place it in a clean microcentrifuge tube. Centrifuge in a microcentrifuge for 5 minutes.

3. Remove 0.3 ml of the supernatant and place it in a clean microcentrifuge tube. Add 30 µl of 5 M NaCl and 200 µl of isopropanol. Invert to mix. Place on ice for at least 10 minutes and centrifuge in a microcentrifuge for 5 minutes.

4. Pour off the supernatant. Overlay the DNA pellet with 200 µl of cold 70% ethanol and centrifuge in a microcentrifuge for 5 minutes.

5. Pour off the supernatant and invert the tube on a lint-free tissue (Kimwipe) to remove any excess liquid. Resuspend the pellet in 50 µl of TE.

6. Add 50 µl of phenol:chloroform (1:1) saturated with TE. Vortex thoroughly. Centrifuge in a microcentrifuge for 3 minutes.

   Caution: Phenol can cause severe burns. Wear gloves and safety glasses.

   Chloroform is a carcinogen and may damage the liver and kidneys. Do not mouth pipette and avoid contact with skin.

7. Remove the upper aqueous layer and place in a clean microcentrifuge tube. Repeat the phenol extraction until no more white material (denatured protein) is visible at the interphase.

8. Add 50 µl of chloroform to the aqueous layer. Vortex. Centrifuge in a microcentrifuge for 3 minutes.

9. Remove the upper aqueous layer and place in a clean microcentrifuge tube. Check the volume of the aqueous layer and adjust to 50 µl with TE. Add 5 µl of 5 M NaCl and 150 µl of cold 95% ethanol. Invert to mix. Place on ice for at least 10

minutes and then centrifuge in a microcentrifuge for 15 minutes.

10. Pour off the supernatant. Overlay the DNA pellet with 0.2 ml of cold 70% ethanol and centrifuge in a microcentrifuge for 5 minutes.

11. Resuspend the DNA pellet in 20 μl of TE.

DEAE-CELLULOSE SLURRY (SILHAVY ET AL. 1984)

1. Suspend DE52 (Whatman) in several volumes of 0.05 N HCl to give a final pH of less than 4.5.

2. Place the slurry on a stirrer and slowly add concentrated NaOH until the pH increases to 6.8.

3. Allow the slurry to settle and then decant the supernatant.

4. Add several volumes of LB broth and mix.

5. Decant the supernatant and replace with LB broth. Repeat several times until the pH of the supernatant is identical to that of the LB broth.

6. Resuspend the final slurry to give approximately 75% resin and 25% LB broth. Add 0.1% sodium azide to prevent contamination.

Hydroxylamine Mutagenesis of Plasmid DNA

In addition to in vitro mutagenesis of phage as described in Experiment 9, hydroxylamine can be used to mutate purified plasmid DNA. The procedure for mutagenesis of plasmids is essentially identical to the procedure for phage, except that after mutagenesis, the hydroxylamine is removed by dialysis prior to transforming cells (Klig et al. 1988). The transformants are selected on antibiotic plates and the colonies are then replica plated onto appropriate media to screen for the desired mutations. It is a good idea to assay mutagenesis of another plasmid gene as a control. Some plasmids seem to be harder to mutagenize than others. This may be due to secondary structure in the DNA, which strongly inhibits mutagenesis by hydroxylamine (Drake 1970). If this is a problem, it may be necessary to partially denature the DNA by using a higher temperature for the mutagenesis (Humphreys et al. 1976).

1. Use highly purified plasmid DNA (e.g., DNA purified from CsCl, miniprep DNA treated with LiCl or RNase and phenol extracted, or miniprep DNA purified from an affinity system). Add the following to a sterile microcentrifuge tube:

Plasmid DNA	5 µl
Sterile deionized H_2O	30 µl
Phosphate-EDTA buffer	20 µl
7 M Hydroxylamine (prepared fresh)	45 µl

2. As a control, add the following to another sterile microcentrifuge tube:

Plasmid DNA	5 µl
Sterile deionized H_2O	75 µl
Phosphate-EDTA buffer	20 µl

3. Incubate overnight at 37°C.

4. Drop dialyze the DNA to remove the hydroxylamine and salts and transform or electroporate into an appropriate recipient.

BUFFERS AND SOLUTIONS

Hydroxylamine/NaOH (Prepare Fresh)

Hydroxylamine (NH_2OH)	350 mg
4 M NaOH	0.56 ml

Adjust to 5 ml with sterile deionized H_2O.

Phosphate-EDTA Buffer (0.5 M KPO_4 [pH 6], 5 mM EDTA)

When exposed to oxygen, hydroxylamine solutions form by-products that are toxic to cells (probably peroxides and free radicals). This nonspecific toxicity is decreased by EDTA.

KH_2PO_4	6.81 g
Deionized H_2O	70 ml

Dissolve on a stirrer. Adjust to pH 6 with 1 N KOH. Adjust to 99 ml with deionized H_2O. Add 1 ml of 0.5 M EDTA. Autoclave.

Preparation and Transformation of Competent Cells

A variety of protocols have been developed for preparation of competent *E. coli* cells (Hanahan 1983, 1987). This rapid transformation procedure (Chung et al. 1989) is convenient to use on a routine basis, whereas the more elaborate methods may be used only when maximum transformation efficiency is desired (e.g., when constructing a gene library).

This protocol is written for a total of five separate transformations. Adjust culture volumes as necessary for more or fewer transformations. For culture media and supplements, see Appendix B.

1. Inoculate a single colony of each desired strain into two tubes each of 3 ml of LB broth.

2. Aerate at 30–37ºC until the cultures reach early exponential phase.

3. Aliquot 1 ml of culture into each of five microcentrifuge tubes on ice.

4. Sediment in a microcentrifuge at 14,000 rpm for 30 seconds at 4ºC.

5. Decant the supernatant and gently resuspend the pellet in 0.05 ml of LB broth.

6. Incubate on ice for 5 minutes.

7. Add 0.05 ml of 2x TSS (see p. 489) and gently mix.

8. Incubate on ice for 5 minutes.

9. Add 5 µl of DNA to each of four tubes, as shown below.

Tube	Cells (ml)	DNA	Comments
A	0.1	plasmid 1	
B	0.1	plasmid 2	
C	0.1	plasmid 3	
D	0.1	vector	transformation control
E	0.1	none	cells-only control

10. Incubate on ice for 5–60 minutes.

11. Incubate for 90 seconds at 42ºC (heat shock). Some laboratories find this step unnecessary.

12. Incubate on ice for 1–5 minutes.

13. Add 1 ml of LB broth and incubate for 45–60 minutes at 30–37ºC. This allows time for expression of antibiotic resistance.

14. Sediment in a microcentrifuge at 14,000 rpm for 30 seconds at 4ºC.

15. Decant the supernatant and suspend the pellet in 0.1 ml of LB broth.

16. Plate on selective medium.

17. Incubate at 30–37ºC.

BUFFERS AND SOLUTIONS

2x TSS (Transformation and Storage Solution)

Tryptone	0.8 g
Yeast extract	0.5 g
NaCl	0.5 g
Polyethylene Glycol 8000	20 g
Dimethylsulfoxide	10 ml
1 M MgSO$_4$·7H$_2$O	10 ml

Dissolve in 80 ml of deionized H$_2$O. Adjust pH to 6.5 (6.2 to 6.8). Adjust to 100 ml with deionized H$_2$O. Autoclave. Store at 4ºC.

Preparation of Electrocompetent Cells and Electroporation

Preparation of Large Volumes of Frozen Electrocompetent Cells

This procedure yields sufficient electrocompetent cells for 15 electroporations. The cells may be stored at -70°C for many months. Aliquots may be thawed as needed.

1. Subculture 2 ml of a fresh overnight culture into 200 ml of LB broth (see Appendix B) in a 500-ml flask.

2. Grow at 37°C with vigorous shaking to mid exponential phase (~70–100 Klett units).

3. Pour the cells into sterile centrifuge bottles. Centrifuge the cells in a large rotor at 6000 rpm for 10 minutes at 4°C.

4. Resuspend the cells in 200 ml of ice-cold sterile deionized H_2O and then centrifuge the cells at 6000 rpm for 10 minutes at 4°C.

5. Resuspend the cells in 100 ml of ice-cold sterile deionized H_2O and then centrifuge the cells at 6000 rpm for 10 minutes.

6. Resuspend the cells in 5 ml of ice-cold sterile 10% glycerol and transfer the cell suspension to a sterile centrifuge tube.

7. Centrifuge the cells in an SS-34 rotor at 10,000 rpm for 5 minutes at 4°C.

8. Resuspend the cells in 0.75 ml of ice-cold sterile 10% glycerol. The final cell concentration should be approximately 10^{10}/ml.

9. Place 50-μl aliquots of cells in small sterile microcentrifuge tubes and freeze at -70°C.

10. Before use, remove the cells from the freezer, thaw briefly at room temperature, and then place on ice before electroporation.

Preparation of Small Volumes of Electrocompetent Cells

This procedure yields sufficient electrocompetent cells for two electroporations. The procedure may be proportionally scaled up for intermediate volumes of electrocompetent cells.

1. Subculture 0.2 ml of a fresh overnight culture into 5 ml of LB broth (see Appendix B) and grow with aeration at 37°C. For many healthy strains of E. coli and Salmonella, a fresh overnight culture works well, but for certain compromised strains (e.g., recA mutants), mid exponential phase cells are required for efficient electrotransformation.

2. Pellet the cells by centrifuging in a microcentrifuge for 30 seconds.

3. Pour off the supernatant. Resuspend the pellet in 1 ml of sterile cold deionized H_2O. Vortex.

4. Centrifuge the cells in a microcentrifuge for 30 seconds.

5. Pour off the supernatant. Resuspend the pellet in 1 ml of sterile cold deionized H_2O. Vortex.

6. Centrifuge the cells in a microcentrifuge for 30 seconds.

7. Pour off the supernatant. Resuspend the pellet in 1 ml of sterile cold 10% glycerol. Vortex.

8. Centrifuge the cells in a microcentrifuge for 30 seconds.

9. Pour off the supernatant. Resuspend the cells in 100 µl of sterile cold 10% glycerol. Store the recipient cells on ice until use.

Electroporation

These conditions work well for Salmonella, E. coli, and most related enteric bacteria. Optimal electroporation conditions may differ for other bacteria. Technical bulletins from companies that sell electroporators provide a good source of suggested conditions for different organisms.

1. Mix 2 µl of plasmid DNA with 40 µl of electrocompetent cells in a sterile microcentrifuge tube.

2. Transfer the cells + DNA into a precooled 0.1 mM electroporation cuvette.

3. Shock the cells in an electroporator set at 25 µF, 200 ohms, and 1.5 kV.

4. Add 1 ml of Plasmid broth (see Appendix B) to the cuvette. Transfer the cell suspension to a test tube and incubate with aeration for 1–2 hours at 37°C.

5. As controls, add 1 ml of Plasmid broth to 40 µl of the recipient cells without plasmid (cell control). Add 1 ml of Plasmid broth to 2 µl of plasmid DNA without cells (DNA control). Incubate with aeration for 1–2 hours at 37°C.

6. Centrifuge the electroporation and control cultures in a microcentrifuge for 30 seconds to pellet the cells.

7. Pour off the supernatant. Resuspend the cell pellet in 50 µl of LB broth (see Appendix B). Spot the entire sample on a selective plate, allow to dry, and streak for isolation (this will yield isolated colonies even if the electroporation was very efficient). Spread the rest of the cell suspension on another selective plate (this should yield isolated colonies if the electroporation was less efficient). Incubate the plates overnight at 37°C.

APPENDIX E
Enzyme Assays

β-Galactosidase Assay in Permeabilized Cells

This convenient assay for β-galactosidase activity was developed by Lederberg (1950), refined by Monod and co-workers (Pardee et al. 1959), and formalized by Miller (1972). Cells are permeabilized by exposure to chloroform/SDS, allowing entry of the chromogenic substrate, ONPG. The reaction is started by adding ONPG, and stopped by adding sodium carbonate, which raises the pH of the reaction to a point where the enzyme is no longer active. This raise in pH also intensifies the yellow color of ONP. The OD_{420} (ONP plus light scattering) and OD_{550} (light scattering only) values are measured, and the OD_{550} value is used to correct the OD_{420} value. This eliminates the need to remove the cells from the suspension prior to measurement. The OD_{660} of the cell suspension is also measured to provide a basis for calculating specific activity. The resulting units ("Miller units"), although arbitrary, are commonly used and quite familiar.

1. Grow cultures to the mid exponential phase.

2. Sediment and resuspend in Working Buffer (see p. 495).

3. Store cell suspensions on ice.

4. Mix cell suspension (10 μl to 1 ml, depending on activity) with Working Buffer for a final volume of 1 ml in a 100 X 13-mm enzyme assay tube. Set up two or three tubes for each culture, each with different amounts of cell suspension (e.g., 10, 20, and 30 μl; or 0.2, 0.4, and 0.6 ml).

5. Add 2 drops of chloroform and 1 drop of 0.1% SDS (see p. 495).

   Caution: Chloroform is a carcinogen and may damage the liver and kidneys. Do not mouth pipette and avoid contact with skin.

6. Vortex each tube for 5 seconds. Be consistent, so that each sample is permeabilized to the same extent.

7. Incubate the tubes for 5–10 minutes at room temperature.

8. To start the assay, add 0.2 ml of ONPG (see p. 495) and mix well. Record the time.

9. Incubate at 28ºC until yellow color is clearly visible. Try to keep the assays between 0.1 and 0.4 OD_{420} units.

10. To stop the assay, add 0.5 ml of sodium carbonate (see p. 495) and mix well. Record the time.

11. Measure the OD_{420} and OD_{550} of each reaction.

12. Measure the OD_{660} of each cell suspension. Dilute the suspension as necessary so that the measured OD_{660} is between 0.1 and 0.4.

13. Calculate the activity (amount of ONPG hydrolyzed per minute as a function of cell density) using the formula:

$$\text{Units} = 1000 \times \frac{OD_{420} - (1.75 \times OD_{550})}{t \times V \times OD_{660}}$$

where t = time of the reaction (minutes) and V = volume of cell suspension in the assay (ml).

BUFFERS AND SOLUTIONS

1 M DTT (Dithiothreitol; Cleland's Reagent)

Dissolve 3.1 g of DTT in 20 ml of 10 mM sodium acetate (pH 5.2). Filter-sterilize. Aliquot 1-ml portions into screw-cap tubes. Store at –20ºC.

β-Galactosidase Assay Buffer (pH 7.0)

61 mM Na_2HPO_4	8.7 g
39 mM NaH_2PO_4	5.4 g
10 mM KCl	750 mg
10 mM $MgSO_4 \cdot 7H_2O$	246 mg

Adjust to 1000 ml with deionized H_2O. Do not autoclave. Store at 4ºC.

ONPG (o-Nitrophenyl-β-D-Galactopyranosideside)

Dissolve 100 mg of ONPG in 25 ml of enzyme assay buffer. Prepare fresh at least once a week. Protect from light. Store at –20ºC.

0.1% SDS (Sodium Dodecyl Sulfate)

Add 0.1 ml of 10% SDS to 9.9 ml of deionized H_2O. Mix well. Prepare fresh daily.

1 M Sodium Carbonate

1 M Na_2CO_3	21.2 g

Dissolve and adjust to 200 ml with deionized H_2O. Autoclave. Store at room temperature.

Working Buffer

Add 40 µl of 1 M DTT to 100 ml of enzyme assay buffer. Prepare fresh at least once a week. Store at 4ºC.

Alkaline Phosphatase Assay in Permeabilized Cells

A simple and reproducible assay has been developed by Beckwith and colleagues for the quantitation of alkaline phosphatase activity (Brickman and Beckwith 1975; Michaelis et al. 1983). The procedure for this assay is very similar to that of the β-galactosidase assay described above. Cells are permeabilized by exposure to chloroform/SDS, allowing entry of the chromogenic substrate, PNPP. The reaction is started by adding PNPP, and stopped by adding a vast excess of KH_2PO_4, which prevents further hydrolysis of PNPP by effectively competing for the available enzyme. The OD_{420} (PNP plus light scattering) and OD_{550} (light scattering only) values are measured, and the OD_{550} value is used to correct the OD_{420} value. This eliminates the need to remove the cells from the suspension prior to measurement. The OD_{600} of the cell suspension is also measured to provide a basis for calculating specific activity. The resulting units of specific activity are calculated using the same equation as the β-galactosidase assay.

A modified assay has also been developed that includes addition of the sulfhydryl alkylating agent, iodoacetamide, to buffer used for resuspending the cells after growth and for performing the assay (Derman and Beckwith 1995). The modification prevents alkaline phosphatase retained in the cytoplasm from slowly achieving an active disulfide-bonded conformation in cells that are no longer growing. This may be critical when using assay measurements to infer the subcellular localization of alkaline phosphatase. The original assay can be performed according to the method below.

1. Grow culture to desired density.

2. Centrifuge an aliquot (e.g., 0.5 ml) of culture in a microcentrifuge for 30 seconds at room temperature to collect bacterial pellet.

3. Resuspend the pellet in original volume (e.g., 0.5 ml) of 1 M Tris-HCl (pH 8.0; see p. 498).

4. Store cell suspensions on ice.

5. Add an aliquot of resuspended cells (e.g., 0.05 ml, depending on activity) to 1 M Tris (pH 8.0) such that the final volume is 1.0 ml. Set up two or three tubes for each culture, each with different amounts of cell suspension (e.g., 10, 20, and 30 μl; or 0.2, 0.4, and 0.6 ml).

6. Add 0.05 ml (1 drop) of 0.1% SDS (see p. 498) and 0.05 ml (2 drops) of chloroform.

   Caution: Chloroform is a carcinogen and may damage the liver and kidneys. Do not mouth pipette and avoid contact with skin.

7. Vortex each tube for 5 seconds. Be consistent, so that each sample is permeabilized to the same extent.

8. Equilibrate tubes for 5 minutes in a 37°C water bath.

9. Add 0.1 ml of 0.4% PNPP (see p. 498) solubilized in 1 M Tris (pH 8.0), vortex briefly, and incubate at 37°C while timing the reaction.

10. After development of yellow color (try to keep the reaction between 0.1 and 0.4 OD_{420} units), stop the reaction with the addition of 0.1 ml of 1 M KH_2PO_4 (see p. 498). Record the time (in minutes). Longer reaction times (e.g., >10 minutes) allow for greater accuracy and reproducibility.

11. Measure the OD_{420} and OD_{550} of the reaction.

12. Measure the OD_{600} of each cell suspension. Dilute the suspension as necessary so that the measured OD_{600} is between 0.1 and 0.4.

13. Calculate the activity (amount of PNPP hydrolyzed per minute as a function of cell density) using the formula:

$$\text{Units} = 1000 \times \frac{OD_{420} - (1.75 \times OD_{550})}{t \times V \times OD_{600}}$$

where t = time of the reaction (minutes) and V = volume of cell suspension in the assay (ml).

BUFFERS AND SOLUTIONS

1 M KH_2PO_4

Dissolve 13.61 g of KH_2PO_4 in deionized H_2O and adjust to a total volume of 100 ml with deionized H_2O. Store at room temperature.

0.4% PNPP (*p*-Nitrophenyl Phosphate)

Dissolve 0.04 g of PNPP in 10 ml of 1 M Tris-HCl (pH 8.0). Store protected from light at 4ºC. Solution should be made fresh weekly.

0.1 % SDS (Sodium Dodecyl Sulfate)

Add 0.1 ml of 10% SDS to 9.9 ml of deionized H_2O. Mix well. Store at room temperature.

1 M Tris-HCl (pH 8.0)

Dissolve 121.1 g of Trizma Base into 900 ml of deionized H_2O. Adjust pH to 8.0 with concentrated HCl. Adjust volume up to 1 liter with deionized H_2O. Store at room temperature.

APPENDIX F
Recombinant DNA Methods

Plasmid DNA Minipreps

Cloning experiments involve numerous small-scale preparations of plasmid DNA, and a variety of procedures have been developed. This time-tested protocol is relatively fast and consistently provides clean DNA suitable for restriction analysis and transformation (Birnboim and Doly 1979).

1. Inoculate desired strains into 2 ml of Plasmid broth (see Appendix B) plus antibiotic.

2. Aerate at 30–37°C until the cultures are fully saturated (8 hours to overnight).

3. Pour approximately 1.4 ml of culture into a microcentrifuge tube.

4. Sediment the cells in a microcentrifuge for 30 seconds.

5. Remove the spent culture medium with a water aspirator equipped with a micropipette tip or with a pasteur pipette. Remove as much liquid as possible.

6. Resuspend the pellet in 0.1 ml of PEB (see p. 502) with gentle vortexing. Make sure that the pellet is well-dispersed. Clumped cells yield decreased DNA.

7. Add 0.2 ml of ALM (see p. 501) and mix gently by inverting the tube. The solution will clear and become viscous.

8. Add 0.15 ml of 3 M sodium acetate (pH 4.8; see p. 502). Mix well by inverting and shaking the tube. A flocculent white precipate will form.

9. Sediment the precipitate at 14,000 rpm for 5 minutes. The pellet contains chromosomal DNA and cell debris; the plasmid is in the supernatant.

10. Recover 400 μl of supernatant into a clean tube containing 0.8 ml of cold 95% ethanol. The insoluble "pellet" tends to float; avoid taking big chunks.

11. Chill for 30 minutes to overnight at –20ºC. A light-brown DNA precipitate will form. This is a good stopping point; the ethanol precipitate can be left indefinitely at –20ºC.

12. Sediment the precipitate in a microcentrifuge at 14,000 rpm for 5 minutes.

13. Discard the supernatant and rinse the pellet with 1 ml of cold 95% ethanol.

14. Sediment the pellet in a microcentrifuge at 14,000 rpm for 2 minutes and remove the residual ethanol with a micropipette.

15. Dry the pellet in a rotary evaporator for 5 minutes or in air for 30–60 minutes.

16. Dissolve the pellet in 0.05 ml of TE (see p. 502) and store at 4ºC. Allow 60 minutes or more for the DNA to go completely into solution.

   Note: Steps 17–26 are optional, but recommended. Treatment with LiCl removes much of the rRNA and residual material, and thus the DNA prep is much cleaner and probably more stable for long-term storage (Pelham 1985). However, for preliminary screening, or for preparing DNA for transformation, these steps are not essential.

17. Add 0.05 ml of 5 M LiCl (see p. 502) and mix well.

18. Incubate on ice for 5–10 minutes.

19. Sediment the precipitate in a microcentrifuge at 14,000 rpm for 5 minutes. The pellet contains RNA and residual cell debris; the plasmid is in the supernatant.

20. Recover 100 µl of supernatant into a clean tube containing 0.2 ml of cold 95% ethanol.

21. Chill for 60 minutes to overnight at –20°C. A brown DNA precipitate will form. This is a good stopping point; the ethanol precipitate can be left indefinitely at –20°C.

22. Sediment the precipitate in a microcentrifuge at 14,000 rpm for 5 minutes.

23. Discard the supernatant and rinse the pellet with 1 ml of cold 95% ethanol.

24. Sediment the pellet in a microcentrifuge at 14,000 rpm for 2 minutes and remove the residual ethanol with a micropipette.

25. Dry the pellet in a rotary evaporator for 5 minutes or in air for 30–60 minutes.

26. Dissolve the pellet in 0.05 ml of TE. The plasmid DNA is now ready for restriction analysis; 5 µl per digest is usually sufficient. The DNA is fairly stable at 4°C, but it can be frozen at –20°C for long-term storage.

BUFFERS AND SOLUTIONS

ALM (Alkaline Lysis Mix)

1 N NaOH (0.2 N)	1.0 ml
10% SDS (1.0%)	0.5 ml
Deionized H$_2$O	3.5 ml

Prepare fresh every week. Store at room temperature.

5 M Lithium Chloride

LiCl	21.2 g
Deionized H_2O	100 ml

Autoclave. Store at room temperature.

PEB (Plasmid Extraction Buffer)

Trizma base (25 mM)	0.6 g
Glucose (50 mM)	1.8 g
Na_2EDTA (10 mM)	0.74 g
Deionized H_2O	180 ml

Adjust pH to 8.0 with HCl. Adjust to 200 ml with H_2O. Aliquot 50 ml to 60-ml bottles. Autoclave. Store at 4ºC.

3 M Sodium Acetate

$NaCO_2CH_3 \cdot H_2O$	81.6 g
Deionized H_2O	30 ml

Dissolve completely. Adjust pH to 4.8 with glacial acetic acid (~60 ml). Adjust to 200 ml with H_2O. Distribute 50 ml to 60-ml bottles. Autoclave. Store at room temperature.

TE (50x Stock Solution)

Trizma base (500 mM)	12.1 g
Na_2EDTA (50 mM)	3.8 g
Deionized H_2O	180 ml

Adjust pH to 8.2 with HCl. Adjust to 200 ml with H_2O. Autoclave. Dilute 1:50 in sterile H_2O for working solution. Store at room temperature.

Restriction Endonuclease Buffers

Most suppliers of restriction endonucleases provide 10x buffers with each order; each buffer is usually optimized for that particular enzyme. It is sometimes convenient to use more "universal" buffers, for example, when doing multiple digests. The following recipes are good general-purpose buffers that are suitable for most restriction endonucleases.

Intermediate Salt

	Stock	Amount of stock
NaCl (50 mM)	5 M	1.0 ml
Tris-HCl (pH 7.4; 6 mM)	1 M	0.6 ml
MgCl$_2$ (6 mM)	1 M	0.6 ml
β-Mercaptoethanol (6 mM)	14.3 M	42 μl
Deionized H$_2$O		7.8 ml

High Salt

	Stock	Amount of stock
NaCl (150 mM)	5 M	3.0 ml
Tris-HCl (pH 7.4; 6 mM)	1 M	0.6 ml
MgCl$_2$ (6 mM)	1 M	0.6 ml
β-Mercaptoethanol (6 mM)	14.3 M	42 μl
Deionized H$_2$O		5.8 ml

Caution: β-Mercaptoethanol may be fatal if inhaled or absorbed through the skin and is harmful if swallowed. High concentrations are extremely destructive to the mucous membranes, the upper respiratory tract, the skin, and the eyes. Wear gloves and safety glasses and work in a chemical fume hood.

Ethanol Precipitation

Ammonium acetate is used as a counterion in the following procedure, but a variety of alternative counterions can be substituted as described in Section 7.

1. Add 0.5 volume of 7.5 M ammonium acetate. For low concentrations of DNA, add 1 μl of glycogen (35 mg/ml; molecular biology grade available from Calbiochem-Novabiochem or Boehringer Mannheim).

2. Add 2.5 times the total volume (volume solution + volume ammonium acetate) of 95% ethanol.

3. Place on ice for 15–30 minutes. Let stand overnight at 4°C to increase the yield somewhat, especially if the DNA concentration is low.

4. Spin in a microcentrifuge for 15–30 minutes at room temperature.

5. Pour off the supernatant. Add 1 ml of cold 70% ethanol. Spin in a microcentrifuge for 10 minutes.

6. Pour off the supernatant. Invert the microcentrifuge tube on a clean Kimwipe to blot off any excess ethanol.

7. Dry off any residual ethanol in a SpeedVac or a vacuum desiccator.

8. Resuspend the pellet in a small volume of 1x TE buffer (see p. 502).

Phenol Extraction

Caution: Phenol can cause severe burns. Always wear gloves, protective clothing, and safety glasses and work in a chemical fume hood. Rinse any areas of the skin that come in contact with phenol with a large volume of water and wash with soap and water—do not rinse with ethanol!

1. Add an equal volume of the TE-saturated phenol (see p. 506) to the DNA sample. Mix thoroughly to form an emulsion. Centrifuge in a microcentrifuge for 3–5 minutes to separate the phases.

2. Carefully remove the upper aqueous phase, avoiding the interphase or phenol phase. Discard the used phenol in a glass bottle. A good rule of thumb is that if any white material can be seen at the interphase, do another extraction.

3. Mix an equal volume of TE-saturated phenol with chloroform just before use. Add an equal volume of the phenol:chloroform to the DNA sample. Mix thoroughly to form an emulsion. Centrifuge in a microcentrifuge for 3–5 minutes to separate the phases.

4. Carefully remove the upper aqueous phase, avoiding the interphase or phenol phase. Discard the used phenol:chloroform in a glass bottle. Repeat the extraction if any white material can be seen at the interphase.

5. Any residual phenol in the aqueous phase can severely inhibit enzymes used in later steps. Ether extraction of the aqueous phase can be done to remove any traces of phenol (Maniatis et al. 1982); however, ether is potentially hazardous and this step is usually not necessary.

Phenol (TE-saturated)

Saturated phenol purchased from commercial suppliers is slightly more expensive than less pure grades but significantly decreases the need to handle this potentially toxic reagent. The following procedure is for preparation of TE-saturated phenol from ultrapure phenol stored as a solid at −20°C.

Caution: Phenol can cause severe burns. Always wear gloves, protective clothing, and safety glasses and work in a chemical fume hood. Rinse any areas of the skin that come in contact with phenol with a large volume of water and wash with soap and water—do not rinse with ethanol!

1. Remove redistilled phenol from the freezer and allow to warm to room temperature. Warm the phenol in a beaker placed in a 43–60°C water bath to melt the crystals. The phenol should be colorless. Do not use if the phenol is colored.

2. Add 8-hydroxyquinoline to 0.1% final concentration (yellow colored).

3. Add an equal volume of 1 M Tris-HCl (pH 8) to saturate the phenol, mix, and remove the aqueous phase. Check pH of aqueous phase. If necessary, continue extracting the phenol until the pH of the aqueous phase is greater than 7.

4. Aliquot 1 ml of the phenol solution into microcentrifuge tubes. Add 0.3 ml of 1x TE buffer (see p. 502) to each tube. Mix by inverting. Store the tubes in a brown bottle at −20°C until use.

Spin Columns

Sephadex G-50 Spin Columns

Sephadex is a gel filtration matrix that separates molecules on the basis of size and shape. Large molecules elute from the column quickly, whereas small molecules become entrapped in the Sephadex beads and elute very slowly. Sephadex "spin columns" prepared in microcentrifuge tubes are inexpensive, easy to use, and have a variety of applications (e.g., removing NaOH from denatured DNA for sequencing reactions, removing excess labeled nucleotides from probes, and removing low-molecular-weight contaminants from DNA samples).

1. Prepare the Sephadex as follows:

    a. Hydrate the resin by adding 10 g of Sephadex G-50 (medium or coarse) to 100 ml of 1x TE buffer (see p. 502) and autoclaving for 15 minutes.

    b. Allow to cool at room temperature. Swirl gently, allow to settle a few minutes, and then pour off excess TE buffer and any fine Sephadex particles that do not sediment.

    c. Add fresh TE buffer and store at 4°C until use.

2. To prepare Sephadex G-50 columns:

    a. Remove the cap from a 500-µl microcentrifuge tube, punch a small hole in the bottom with a 23-gauge needle, and add approximately 50 µl of glass beads (212–300 µm, sterilized by autoclaving).

    b. Swirl the flask of Sephadex to form a uniform suspension. Remove 600 µl of the suspension and add it to the microcentrifuge tube.

    c. Make sure the column flows freely.

3. To run the sample:

    a. Place the column inside a 1.5-ml microcentrifuge tube and centrifuge at 3000 rpm for 2 minutes. (The Sephadex should look as dry as the glass beads.)

b. Transfer the column to a new 1.5-ml microcentrifuge tube.

c. Add 25–30 µl of sample to the top of the column and centrifuge at 3000 rpm for 2 minutes. The volume recovered should be approximately the same as the volume loaded.

Drop Dialysis

1. Pour approximately 10 ml of 1x TE buffer (see p. 502) into a sterile petri dish.

2. Use forceps to remove a 25-mm type-VS Millipore filter. Float the filter on the surface of the TE buffer with the shiny side up. Allow the filter to become completely wet (~2 minutes).

3. Use a micropipettor to carefully place 20–100 μl of DNA in the center of the filter.

4. Cover the petri dish and put it where it will not be disturbed.

5. Let stand for 1–2 hours at room temperature.

6. *Carefully* remove the DNA with a micropipettor.

Agarose Gels

1. Place the gel apparatus on a level surface. Adjust the comb so that the teeth are a few millimeters above the bed of the gel tray.

2. Pipette or pour cool (50–55°C) agarose (see p. 511) into the gel apparatus. If pipetting, use a bulb. Use approximately 15 ml for a 5 × 7.5-cm apparatus and approximately 40 ml for a 13.5 × 7.5-cm apparatus.

3. Insert the comb near the anode end. Avoid bubbles.

4. Allow the gel to solidify (~15–20 minutes).

5. Fill the buffer reservoirs with running buffer (see p. 512), such that the gel is only 1–2 mm submerged in buffer.

6. Carefully and gently remove the comb. Avoid tearing the bottoms of the wells.

7. Gently rinse each well with a microcapillary pipette to remove residual agarose.

8. Load the gel with a drawn-out microcapillary pipette or with a Pipetman. Place the tip of the pipette near the bottom of the well to avoid spilling the sample into adjacent wells.

9. Connect the electrodes to a power supply. Note that DNA is negatively charged and will migrate toward the cathode.

10. Run at 25 mA maximum (80–100 V) for a 5 × 7.5-cm apparatus and at 60 mA maximum (60–80 V) for a 13.5 × 7.5-cm apparatus.

11. When the tracking dyes have migrated the appropriate distance, adjust the power supply to minimum current, turn the power supply off, and disconnect the electrodes.

12. Wear gloves! Remove the gel for photography.

BUFFERS AND SOLUTIONS

Agarose solution

Agarose	"X" g (see below)
10x TBE (see p. 512)	20 ml
Deionized H_2O	180 ml

Add reagents to a 250-ml flask. Heat in microwave oven until agarose is completely dissolved. Cool to 50–55ºC. Add 12.5 µl of 10 mg/ml ethidium bromide. Store in a 50–55ºC water bath.

Agarose Concentrations

Concentration	Amount ("X")	Separation (kbp)
0.3%	0.6 g	60–5
0.6%	1.2 g	20–1
0.7%	1.4 g	10–0.8
0.9%	1.8 g	7–0.5
1.2%	2.4 g	6–0.4
1.5%	3.0 g	4–0.2
2.0%	4.0 g	3–0.1

10 mg/ml Ethidium Bromide

Ethidium bromide	1 g
Deionized H_2O	100 ml

Store in a brown bottle at room temperature.

Caution: Ethidium bromide is a powerful mutagen and moderately toxic. Always wear gloves when working with solutions that contain this dye. After use, decontaminate and dispose of these solutions in accordance with the safety practices established by your institution's Safety Office.

Loading Dye Solution (5x)

Bromophenol blue	125 mg
Xylene cyanol FF	125 mg
Glycerol	25 ml
Deionized H_2O	25 ml

Store at room temperature. Add DNase-free RNase at 10 µg/ml for a working volume of 1 ml of solution. Store the working solution in a microcentrifuge tube at room temperature (RNase is very stable).

Running Buffer

10x TBE	50 ml
10 mg/ml Ethidium bromide	25 µl
Deionized H_2O	450 ml

Store at room temperature. Running buffer is quite stable and may be reused for many gels.

Caution: Ethidium bromide is a powerful mutagen and moderately toxic. Always wear gloves when working with solutions that contain this dye. After use, decontaminate and dispose of these solutions in accordance with the safety practices established by your institution's Safety Office.

10x TBE (Tris-Borate-EDTA)

Tris base (Sigma 7-9) (890 mM)	108 g
H_3BO_3 (890 mM)	55 g
Na_2EDTA (20 mM)	7.4 g
Deionized H_2O	1000 ml

Final pH should be 8.3 without adjustment. Autoclave in a 2000 ml flask. Store at room temperature.

Nondenaturing Polyacrylamide Gel Electrophoresis of DNA

1. Assemble gel plates with side spacers. Seal the sides and bottom with tape and clamp.

2. Prepare the polyacrylamide solution by mixing the following solutions in a flask:

Solution	Final acrylamide/bis concentration		
	5%	8%	12%
38% Acrylamide/2% bis	12.5 ml	20 ml	30 ml
Deionized H_2O	77.5 ml	70 ml	60 ml
10x TBE	10.0 ml	10 ml	10 ml

Volumes are for a 16 cm X 20 cm X 0.8 mm gel.

3. Add 600 μl of 10% ammonium persulfate (see p. 515) and 30 μl of TEMED (*N,N,N,'N,'*-tetramethylenediamine).

4. Swirl gently and immediately pour into the gel plates. Save a small amount of acrylamide in a pasteur pipette to check the polymerization.

5. Insert the comb and place a clamp over it.

6. Allow to polymerize for approximately 1 hour.

7. Remove the comb. Remove the tape from the bottom of the gel. Place the gel in an electrophoresis set-up. Fill the chambers with 1x TBE (see p. 512). Thoroughly rinse the wells with 1x TBE.

8. Use a micropipettor to slowly load the sample (the sample should contain Loading dye solution as described on p. 512).

9. Electrophorese at 200 V (Caution: High Voltage) until the bromophenol blue reaches the bottom of the gel (approximately 3 hours).

10. Turn off the power supply and unplug the leads. Remove the gel and place it on a flat surface. Separate the plates by gently prying them apart with a spatula.

11. With the gel attached to one plate, stain it in a tray with 0.5 µg/ml ethidium bromide (see p. 511) for approximately 30–60 minutes.

12. Photograph the gel on a UV transilluminator.

BUFFERS AND SOLUTIONS

Acrylamide/Bis-acrylamide Stock (38:2)

Caution: Acrylamide is a potent neurotoxin and is absorbed through the skin. The effects of acrylamide are cumulative. Wear gloves and a mask when weighing powdered acrylamide and methylenebisacrylamide and when handling solutions containing these chemicals. Although polyacrylamide is considered to be nontoxic, handle it with care because it may contain small quantities of unpolymerized acrylamide.

Purchasing stock solutions from commercial suppliers eliminates the need to handle the potentially toxic reagents. The following protocol describes how to prepare the stock solution from the solid.

1. Dissolve 76 g of ultrapure acrylamide and 4 g of bisacrylamide in approximately 150 ml of deionized H_2O.

   Note: Cheaper grades of acrylamide and bisacrylamide are often contaminated with metal ions. Stock solutions of acrylamide can be easily purified by stirring overnight with about 0.2 volume of monobed resin (MB-1, Mallinckrodt), followed by filtration through Whatman No. 1 paper.

2. Adjust to 200 ml with deionized H_2O.

3. Filter through Whatman No. 1 filter paper.

4. Store in a brown bottle at 4°C. During storage, acrylamide and bisacrylamide are slowly converted to acrylic acid and bisacrylic

acid. This deamination reaction is catalyzed by light and alkali. Check that the pH of the acrylamide solution is 7.0 or less and store the solution in dark bottles at room temperature. Fresh solutions should be prepared every few months.

Ammonium Persulfate (100 mg/ml)

Caution: Ammonium persulfate is extremely destructive to tissue of the mucous membranes and upper respiratory tract, eyes, and skin. Inhalation may be fatal. Exposure can cause gastrointestinal disturbances and dematitis. Wear gloves and safety glasses and work in a chemical fume hood. Wash thoroughly after handling.

1. Dissolve 1 g of ammonium persulfate in 9 ml of deionized H_2O.

2. Aliquot and store at $-20^{\circ}C$.

Useful DNA Molecular Weight Standards

λ *Hind*III (0.2 μg/μl)

Use 1 μl of λ *Hind*III + 8 μl of TE (p. 502) + 2 μl of 5× Loading dye solution (p. 512) per lane.

Fragment	Size (kb)
1	23.13
2	9.42
3	6.56
4	4.36
5	2.32
6	2.03
7	0.56
8	0.13*

*This band may not be visible on agarose gels.

φX174 *Hae*III (1 μg/μl)

Use 1 μl of φX174 *Hae*III + 8 μl of TE (p. 502) + 2 μl of 5× Loading dye solution (p. 512) per lane.

Fragment	Size (kb)
1	1353
2	1078
3	872
4	603
5	310
6a	281*
6b	271*
7	234
8	194
9	118
10	72

*These two bands usually appear as a doublet.

Purification and Quantitation of PCR Products

Contaminants in polymerase chain reaction (PCR) products (e.g., primers, unincorporated nucleotides, polymerase, and mineral oil) can interfere with many subsequent reactions. Some contaminants can be simply removed by extraction with organic solvents and ethanol precipitation. However, this is often inadequate for applications that require very clean DNA. The following are two methods for removing contaminants from PCR products: (1) separate the DNA on a polyacrylamide gel and then purify the DNA using ultrafiltration or (2) selectively bind the DNA to affinity agents such as silica or diatomaceous earth, wash the DNA complex, and then elute the DNA off the affinity agent. Numerous commercial kits are available for this purpose. Several methods are described below.

Purification of PCR Products by Extraction with Organic Solvents

1. Centrifuge the DNA in a microcentrifuge for 1 minute.

2. Add 100 μl of chloroform. Vortex. Centrifuge in a microcentrifuge for 2 minutes.

   Caution: Chloroform is a carcinogen and may damage the liver and kidneys. Do not mouth pipette and avoid contact with skin.

3. Remove the upper aqueous layer and place in a new microcentrifuge tube. Adjust the volume to 100 μl with sterile deionized H_2O.

4. Add 100 μl of 4 M ammonium acetate. Mix.

5. Add 200 μl of isopropanol. Mix. Let stand for 5–10 minutes at room temperature.

6. Centrifuge in a microcentrifuge for 10 minutes.

7. Carefully remove the supernatant with an automatic pipettor.

8. Overlay the pellet with cold 70% ethanol and centrifuge in a microcentrifuge for 5 minutes.

9. Dry the pellet for approximately 5 minutes under vacuum or air-dry for approximately 30 minutes.

10. Resuspend the pellet in 10 µl of TE buffer (see p. 502).

Purification of PCR Products from Acrylamide Gels Using Millipore MC Filters

1. Assemble polyacrylamide gel plates with side spacers, seal with tape, and clamp the outside edges.

2. Prepare a 38%/2% polyacrylamide gel in 1x TBE (see p. 513, Nondenaturing Polyacrylamide Gel Electrophoresis of DNA). Save a small amount of acrylamide in a pasteur pipette to check the polymerization.

3. Insert the comb and place a clamp over it. Allow the gel to polymerize approximately 1 hour.

4. Remove the comb and the tape. Place the gel in an electrophoresis set-up. Fill the chambers with 1x TBE. Thoroughly rinse out the wells with 1x TBE.

5. Add 5 µl of Loading dye solution (see p. 513) to each 100-µl sample. Slowly load the samples into the wells with an automatic pipettor. Include a lane with ϕX174 DNA standards (see p. 516, Useful DNA Molecular Weight Standards).

6. Separate the DNA fragments by electrophoresis at 200 V (Caution: High Voltage) until the bromophenol blue reaches the bottom of the gel (~3 hours).

7. Turn off the power supply and unplug the leads. Remove the gel and place it on a flat surface. Separate the plates by gently prying them apart with a spatula.

8. With the gel attached to one plate, stain it in a tray with 0.5 µg/ml ethidium bromide (see p. 511) for 30–60 minutes.

   Caution: Ethidium bromide is a powerful mutagen. Always wear gloves when working with solutions that contain this dye. After use, decontaminate and dispose of these solutions in accordance with the safety practices established by your institution' Safety Office.

9. View the gel on a transilluminator. For each PCR, cut out the desired band with a razor blade and place on a piece of Parafilm. Thoroughly macerate the gel fragment.

10. Transfer the macerated gel fragment from the Parafilm to a Millipore Ultrafree MC filter unit (0.45 µm). Totally immerse the gel fragments in TE buffer (see p. 502) (~100–200 µl). Elute overnight at 4ºC.

11. Let the filter unit stand for 1 hour at room temperature.

12. Centrifuge in a microcentrifuge for 20 minutes.

13. Remove the upper filter assembly and store the DNA in a microcentrifuge tube at 4ºC.

Purification of PCR Products using Qiagen® Affinity Columns

Qiagen Affinity Column kits are available from Qiagen Inc.

1. Add an equal volume of QP Buffer to the DNA.

2. Add 0.6 ml of QP Buffer to the column. Centrifuge the column in a microcentrifuge at 8000g for 1 minute. Pour the buffer out of the lower tube and immediately replace the column (do not allow the column to touch anything).

3. Add the DNA sample to the column. Centrifuge for 1 minute and then discard the solution in the lower tube as described in step 2.

4. Add 0.6 ml of isopropanol to the column. Centrifuge for 1 minute and then discard the solution in the lower tube as described in step 2.

5. Repeat step 4.

6. Add 0.6 ml of QB Buffer to the column. Centrifuge for 1 minute and then discard the solution in the lower tube as described in step 2.

7. Repeat step 6 two more times.

8. Add 0.4 ml of QXE Buffer to the column. Centrifuge for 1 minute and as described in step 2, add an additional 0.4 ml of QXE Buffer to the column and recentrifuge. This step elutes the DNA from the column.

9. Add 10 µl of QIAEX resin. Mix. Let stand for 10 minutes at room temperature.

10. Centrifuge in a microcentrifuge for 30 seconds.

11. Wash the pellet with 0.5 ml of QX3 Buffer.

12. Repeat step 11.

13. Centrifuge for an additional 30 seconds. Immediately remove any traces of ethanol.

14. Dry the pellet for 5 minutes in a vacuum or air-dry for 10 minutes.

15. Elute the DNA from the QIAEX with 20 µl of TE buffer (see p. 502).

Purification of PCR Fragments using Wizard™ Preps

The Wizard™ PCR Preps DNA Purification System is available from Promega.

1. Transfer each completed PCR to a clean microcentrifuge tube.

2. Aliquot 100 μl of Direct Purification Buffer into a 1.5-ml microcentrifuge tube. Add the entire contents (100 μl) of the PCR. Vortex briefly to mix.

3. Add 1 ml of resin and vortex briefly three times over a 1-minute period.

4. Prepare one Wizard Minicolumn per PCR. Remove the plunger from a 3-ml disposable syringe and set aside. Attach the syringe barrel to the Luer-Lok extension of each Minicolumn.

5. Pipette the resin/DNA mix from step 2 into the syringe barrel. Insert the syringe plunger and slowly push the slurry into the Minicolumn by applying gentle pressure on the plunger.

6. Detach the syringe from the Minicolumn and remove the plunger from the syringe. Reattach the syringe barrel to the Minicolumn. Add 2 ml of 80% isopropanol into the syringe to wash the column. Insert the syringe plunger and gently push the isopropanol through the Minicolumn.

7. Remove the syringe and transfer the Minicolumn to a 1.5-ml microcentrifuge tube. Centrifuge the Minicolumn for 20 seconds to dry the resin.

8. Transfer the Minicolumn to a clean microcentrifuge tube. Add 50 μl of TE buffer (see p. 502) to the Minicolumn and wait 1 minute. Centrifuge the Minicolumn for 2 minutes to elute the bound DNA fragment.

9. Remove and discard the Minicolumn. Store the purified DNA at 4°C until use.

Determining the DNA Concentration of PCR Products

1. Spot 5 μl of each DNA sample on a piece of Parafilm.

2. Spot 5 μl containing 0, 1, 2, 4, 8, and 16 μg/ml of λ *Hind*III DNA standards on the Parafilm.

3. Add 5 μl of 2 μg/ml ethidium bromide. Mix by gently pipetting up and down several times with a micropipettor.

   Caution: Ethidium bromide is a powerful mutagen and moderately toxic. Always wear gloves when working with solutions that contain this dye. After use, decontaminate and dispose of these solutions in accordance with the safety practices established by your institution's Safety Office.

4. Photograph the spots on a UV transilluminator.

   Caution: UV is a mutagen and carcinogen, as well as an eye-damaging agent. Always wear safety glasses and work in a glass-enclosed hood if possible.

5. Estimate the concentration of DNA from the photograph by comparing the intensity of fluorescence in the sample with the intensity of the standards.

6. Use the following formula to calculate the molar concentration of double-stranded DNA:

$$\frac{\mu g\ DNA}{ml} \times \frac{1\ pmole\ DNA}{(6.6 \times 10^{-4}\ \mu g\ DNA) \times (no.\ bp)} = \frac{pmole\ DNA}{ml}$$

Purification of DNA Bands from Agarose by QIAquick™ Gel Extraction

Many molecular biology applications are enhanced by using a purified DNA fragment. A number of commercial kits and agarose types are available for such purification schemes. An example of one such method is the QIAquick kit (Qiagen) used with the following procedure. This kit is designed for purification of DNA from TAE, TBE, regular, and low-melt agarose gels.

1. Excise the DNA fragment from the agarose gel with a clean, sharp scalpel or razor blade.

2. Weigh the gel slice in a tube. Add 3 volumes of QX1 Buffer (supplied) to 1 volume of gel. Assume that each 1 mg of gel is equal to 1 μl (i.e., for a 100-mg slice, add 300 μl of QX1 Buffer).

   Note: For gels with more than 2% agarose, add 6 volumes of QX1 Buffer. The maximum loading volume for the column in step 5 is 800 μl.

3. Incubate for 10 minutes at 50ºC to dissolve the gel. To help dissolve the gel, mix by flicking the tube and inverting two or three times.

   Note: For greater than 2% agarose gels, incubate for 20 minutes.

4. Check the pH of the sample. It should be less than or equal to pH 7.5 for efficient binding to the resin in the spin column in the next step. If the pH is greater than 7.5, add 10 μl of 3 M sodium acetate (pH 5.0) to the sample and mix by inverting.

5. Place a QIAquick spin column into a 2-ml collection tube and load the sample.

   Note: For DNA fragments greater than 5 kb, add 1 gel volume of sterile deionized H_2O to the solubilized gel slice and mix by flicking and inverting the tube before applying the sample to the column.

6. Centrifuge in a microcentrifuge at high speed for 60 seconds at room temperature.

7. Drain flow-through fraction from the collection tube and return the column to the same collection tube.

8. Wash by adding 0.75 ml of PE buffer (supplied) to the column and centrifuging in a microcentrifuge at high speed for 30–60 seconds at room temperature. If the fragment is to be used for low-salt restriction, blunt-end ligation, or direct sequencing, allow the column to stand for 5 minutes prior to centrifuging.

9. Drain wash flow-through from collection tube.

10. Centrifuge column for an additional minute to remove residual PE buffer.

11. Place column in a clean 1.5-ml microcentrifuge tube with the cap cut off.

12. Elute DNA by adding 50 μl of 10 mM Tris-HCl (pH 8.5) to the middle (important) of the column membrane and centrifuge in a microcentrifuge at high speed for 1 minute at room temperature. For more concentrated DNA, add 30 μl of elution buffer to the column and let the column stand for 1 minute before centrifuging.

Southern Hybridization Solutions

20x SSC (pH 7.4)

0.3 M NaCitrate (m.w. 294.1)
88.2 g/liter   264.6 g/3 liters   705.6 g/8 liters

3.0 M NaCl (m.w. 58.5)
175.5 g/liter   526.5 g/3 liters   1404 g/8 liters

Adjust pH if needed. Use 20x SSC as a stock for 2x SSC and 5x SSC.

0.25 N HCl

HCl (standard 12 N)	20 ml
Deionized H$_2$O	940 ml

Denaturation Solution

0.5 M NaOH (m.w. 40)	20 g/liter
1.5 M NaCl (m.w. 58.5)	87.75 g/liter

Neutralization Buffer (pH 7.0)

3.0 M NaCl (m.w. 58.5)	175.5 g/liter
0.5 M Tris base (m.w. 121.1)	60.6 g/liter

Adjust pH to 7.0 using standard 12 N HCl.

50x Denhardt's

Ficoll	5 g
Polyvinylpyrrolidone	5 g
Bovine serum albumin (Fraction 5)	5 g
Deionized H$_2$O	to 500 ml

Nonspecific Competitor DNA

Mix DNA (calf thymus, salmon sperm, or herring sperm) at 10 mg/ml in deionized H_2O. Shear through a 20-gauge needle until the viscosity is reduced to the point where small amounts can be pipetted. Store at $-20°C$.

Hybridization Solutions

Solution	5 ml	10 ml	20 ml	50 ml
50% Formamide (very high stringency)				
Formamide	2.5	5	10	25
20X SSC	1.25	2.5	5	12.5
20% SDS	25 μl	50 μl	100 μl	0.25
0.5 M EDTA	10 μl	20 μl	40 μl	0.1
50X Denhardt	0.1	0.2	0.4	1.0
Deionized H_2O	1.115	2.23	4.46	11.15
37.5% Formamide (high stringency for most uses; very clean)				
Formamide	1.875	3.75	7.5	22.5
20X SSC	1.25	2.5	5	12.5
20% SDS	25 μl	50 μl	100 μl	0.25
0.5 M EDTA	10 μl	20 μl	40 μl	0.1
50X Denhardt	0.1	0.2	0.4	1.0
Deionized H_2O	1.74	3.48	6.96	13.65
25% Formamide (medium stringency for many uses)				
Formamide	1.25	2.5	5	12.5
20X SSC	1.25	2.5	5	12.5
20% SDS	25 μl	50 μl	100 μl	0.25
0.5 M EDTA	10 μl	20 μl	40 μl	0.1
50X Denhardt	0.1	0.2	0.4	1.0
Deionized H_2O	2.365	4.73	9.46	23.65
20% Formamide (low stringency)				
Formamide	1.0	2.0	4.0	10.0
20X SSC	1.25	2.5	5	12.5
20% SDS	25 μl	50 μl	100 μl	0.25
0.5 M EDTA	10 μl	20 μl	40 μl	0.1
50X Denhardt	0.1	0.2	0.4	1.0
Deionized H_2O	2.615	5.23	10.46	26.15

Southern Solutions: Genius™ System

Hybridization Solution

5x SSC	50 ml (20x)
0.1% Sarcosyl	2 ml (10%)
0.02% SDS	1 ml (1%)
2% Block (supplied)	40 ml (10%)
50% Formamide	100 ml

Adjust to 200 ml with deionized H_2O. Store at 4ºC for 2 weeks.

Genius Buffer 1

150 mM NaCl	50 ml (3 M NaCl)
100 mM Tris (pH 7.5)	100 ml (1 M Tris, pH 7.5)

Adjust to 1 liter with deionized H_2O. For long storage, filter through a 0.45-μm filter.

Genius Buffer 2

150 mM NaCl	5 ml (3 M NaCl)
100 mM Tris (pH 7.5)	10 ml (1 M Tris, pH 7.5)
2% Blocking Reagent	20 ml (10%)

Adjust to 100 ml with deionized H_2O. Store at 4ºC for 2 weeks.

Genius Buffer 3

100 mM NaCl	33.3 ml (3 M NaCl)
100 mM Tris (pH 9.5)	100 ml (1 M Tris, pH 9.5)

Adjust to 1 liter with deionized H_2O.

Alkaline Stripping Solution

0.2 N NaOH
0.1% SDS

Incubate for 30 minutes at 37ºC; rinse in 2x SSC thoroughly; proceed with prehybridization step.

Mapping Fusion Joints by Combining Southern and Western Analyses

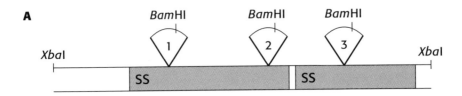

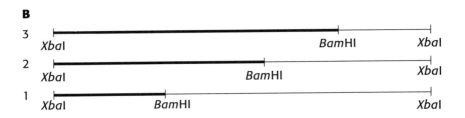

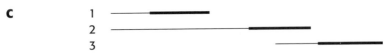

Schematic diagram of the strategy used to physically map genes carrying Tn*phoA* insertions. (*A*) In this example, the fragment carries two genes (*shaded boxes*) encoding proteins with signal sequences (SS), and the three independent Tn*phoA* inserts (numbered pie slices) are fused to these two genes in the same orientation as indicated by the asymmetric *Bam*HI site. A single *Xba*I chromosomal fragment of DNA (in which three different Tn*phoA* inserts have been localized) is depicted. (*B*) The three lines depict fragments detected in a Southern blot analysis of chromosomal DNA digested with *Xba*I plus *Bam*HI and hybridized to the TP-1 probe (a Tn*phoA* probe that recognizes DNA flanking both sides of the *Bam*HI site described in Experiment 13). The lines to the left of the *Bam*HI sites are thicker to show that the TP-1 probe contains more homology with these upstream junction fragments containing *phoA* sequences and thus produces a darker autoradiographic band. (*C*) The three lines represent expected PhoA fusion proteins detectable by immunoblot with anti-PhoA antibodies. The thicker portion of these lines represents PhoA polypeptide, whereas the total length of the lines corresponds to the expected size of the fusion protein. (Adapted from Peterson and Mekalanos 1988.)

Oligonucleotide-directed Site-Specific Mutagenesis

Mutational analysis of the gene being investigated or of a previously analyzed homolog will sometimes provide enough data to predict which bases or resulting protein residues influence regulation or function. At such times, the design of specific mutations or the generation of many random mutations within a specific region of a gene may be desirable. Site-directed mutagenesis is a widely used and valuable tool employed in such cases. The technique generally involves in vitro hybridization of single-stranded DNA to a synthetic oligonucleotide that is complementary to the single-stranded template except for the region of desired mismatch near the center. Following hybridization, the oligonucleotide is extended with a DNA polymerase to complete a double-stranded molecule. The ligated molecule is transformed either into *E. coli* for isolation prior to subsequent analysis or directly into the strain of interest with selection or screening for the desired mutation. The theoretical yield of mutations arising from this technique should be approximately 50% due to semiconservative DNA replication, but it is often lower, due to primer displacement during in vitro synthesis and host-directed mismatch repair of the unmethylated, newly synthesized DNA strand. Although there are several generally used strategies to avoid these problems (see Section 4, Mutagenesis), the one described here and utilized for Experiment 15 is the commercially available Altered Sites® system marketed by Promega.

The Altered Sites system is based on the use of a second mutagenic oligonucleotide, in addition to the target gene mutagenic oligonucleotide, to confer antibiotic resistance to the mutated DNA strand. The pALTER-1 plasmid vector that carries the target gene to be mutagenized also carries two antibiotic resistance genes, an intact gene for Tc^r, and a second for Ap^r which has been inactivated by a base substitution, yielding a $Tc^r Ap^s$ phenotype. Annealing an oligonucleotide with the Ap^r base correction during the in vitro extension reaction provides a selection of *E. coli* transformants replicating the mutated product. This recovery is enhanced by using a mismatch repair defective (*mutS*) strain as the recipient. During the mutagenesis procedure, a third mutagenic oligonucleotide can be incorporated that alters the Tc^r gene to yield a Tc^s phenotype. This procedure is useful for undergoing a second round of mutagenesis by selecting for repair of the Tc^s lesion in the same way as the original Ap^s repair enhanced recovery of mutated plasmids. This experiment is utilized to add additional mutations or to revert the original mutation back to the wild-type sequence to confirm that any mutant phenotype is due to the designed mutation and not the result of any additional aberrant event that might have occurred during the procedure. The latter application avoids the need to sequence the entire cloned region carried on the pALTER-1 plasmid. An additional feature of the pALTER-1 plasmid that is useful for a broad host range is that the plasmid is a pBR322 derivative which retains the cis-active region required for its mobilization into many bacterial species using the helper plasmid pRK2013.

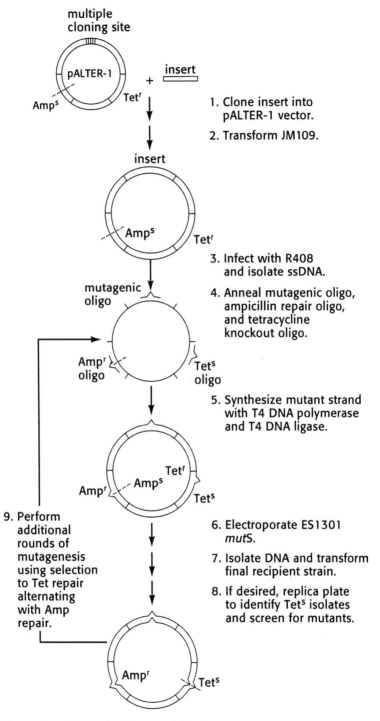

Figure F.1 Schematic diagram of the Altered Sites in vitro mutagenesis procedure using the pALTER-1 Vector as an example. (Modified from Promega to include the use of an ssDNA template and electroporation of ES1301.)

BUFFERS AND SOLUTIONS

For Single-stranded DNA Phage Stocks

PEG/NaCl Precipitation Buffer
    25% PEG 8000
    3 M NaCl

TES Buffer	
Tris-HCl (pH 7.5)	20 mM
NaCl	20 mM
EDTA	1 mM

Oligonucleotide Phosphorylation

10x Kinase Buffer	
Tris-HCl (pH 7.5)	500 mM
$MgCl_2$	100 mM
Dithiothreitol	50 mM
Spermidine	1.0 mM
ATP	10 mM

Mutagenesis Buffers

10x Annealing Buffer	
Tris-HCl (pH 7.5)	200 mM
$MgCl_2$	100 mM
NaCl	500 mM

10x Synthesis Buffer	
Tris-HCl (pH 7.5)	100 mM
dNTPs	5 mM
ATP	10 mM
Dithiothreitol	20 mM

APPENDIX G
Plasmid and Transposon Restriction Maps

This section contains restriction maps of plasmids and transposons used in the Advanced Bacterial Genetics Course 1991–1995. *Many* alternative plasmids and transposons are available for various uses. Indeed, the choice of a specific plasmid or transposon is often determined by its ready availability.

As with all biological reagents described in this manual, these plasmids and transposons were chosen to illustrate specific concepts in the experiments. Furthermore, many of these plasmids and transposons are restricted to the enterobacteria, because their replication origins or transposases do not function in other bacterial groups. Thus, when using this manual to establish experiments in different systems, it is critical to determine which types of plasmids and transposons will be suitable, both for the organism and for the specific application.

Plasmid pALTER-1

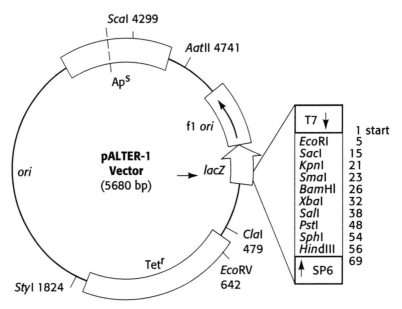

Plasmid pALTER-1 for site-directed mutagenesis. (Reprinted, with permission, from Promega.)

Plasmid pEG5005

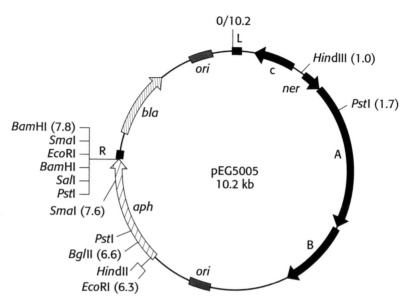

Plasmid pEG5005. Bacteriophage Mud5005 is inserted into plasmid pBC0, a pUC9 derivative. (Redrawn, with permission, from Groisman and Casadaban 1986.) Bacteriophage Mu sequences are black, pMB1 origins of replication (*ori*) are gray, Km^r and Ap^r (*aph* and *bla*) are hatched. The left (L) and right (R) ends of Mu are indicated. Coordinates for selected restriction endonuclease sites are given in kilobases.

Plasmids pSU18 and pSU19

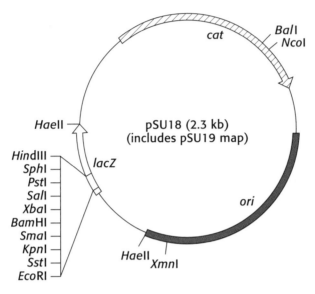

Polylinker Sequences

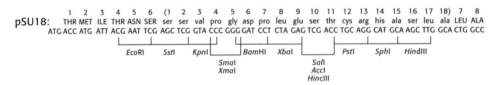

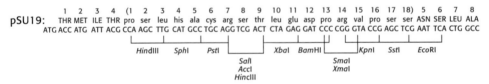

Plasmids pSU18 and pSU19. The plasmid backbone is from plasmid pACYC184. The *lacZ* α genes are from the pUC plasmid series. The pSU18 and pSU19 polylinker sequences are from plasmids pUC18 and pUC19, respectively. (Redrawn, with permission, from Bartolomé et al. 1991.)

Plasmid pGP704

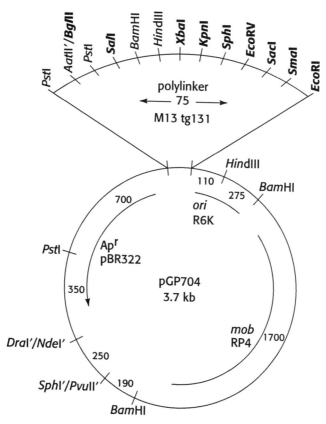

Restriction map of suicide vector plasmid pGP704 (Miller and Mekalanos 1988). Unique restriction enzyme sites (*bold letters*) that can be used for insertions are *Bgl*II, *Eco*RI, *Eco*RV, *Kpn*I, *Sac*I, *Sal*I, *Sma*I, *Sph*I, and *Xba*I. Enzymes that cut multiple times or are in essential genes are *Bam*HI, *Hin*dIII, and *Pst*I. (Adapted, with permission, from Pearson 1989.)

Plasmids pKAS32 and pKAS46

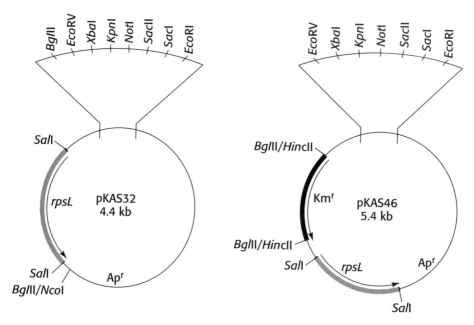

Restriction maps of suicide vectors pKAS32 and pKAS46 (Skorupski and Taylor 1995). These plasmids are derived from pGP704 and contain the *E. coli rpsL* gene on a 715-bp *Sal*I fragment along with a portion of the polylinker (from the *Sac*I site to the *Nco*I site) of plasmid pSL1180 (Pharmacia). Restriction enzyme sites known to be unique in pKAS32 are *Bgl*II, *Eco*RV, *Xba*I, *Kpn*I, *Not*I, *Sac*II, *Sac*I, and *Eco*RI. With the exception of *Bgl*II, the unique sites are the same in pKAS46.

Plasmid pPY190

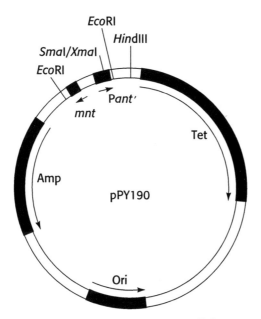

Plasmid pPY190 contains a small fragment from the *immI* region of P22 cloned into *Eco*RI–*Hin*dIII sites of pBR322. Total size is about 4.83 kb. *Eco*RI digestion yields two fragments of 341 bp and 4492 bp. The *Sma*I/*Xma*I site was constructed by site-directed mutagenesis. Cloning at *Sma*I places the insert at –3 relative to P_{ant} and cloning at *Xma*I places the insert at –1 relative to P_{ant} (Maloy and Youderian 1994).

Plasmid pPC36

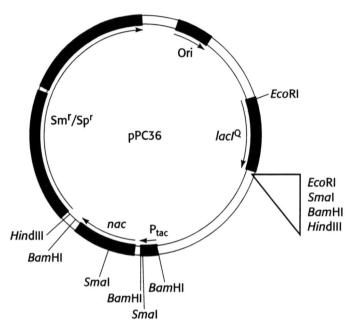

pEC205.1 was derived from pEC205 by transposition of Tn*5tac1* in front of the *nac* gene (Best and Bender 1990). pPC36 was constructed by cloning the *Hind*III fragment from pEC205.1 carrying P$_{tac}$-*nac* into the *Hind*III site of pMS421. Nac expression is regulated by *lacI*^{Q}. pPC36 is approximately 9 kb (L.-M. Chen and S.R. Maloy, in prep.).

Plasmid pMS421

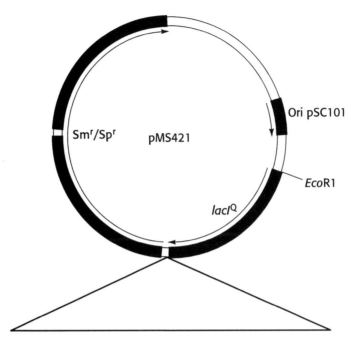

*Hind*III – *Pst*I – *Hinc*II – *Sal*I – *Bam*HI – *Sma*I – *Eco*RI

pMS421 was constructed by cloning a 1.1-kb fragment carrying *lacI*^Q from *E. coli* into the *Eco*RI site of pGB2. pGB2 is a derivative of pSC101 (Sm^r Sp^r) that has no homology with pBR322 and is compatible with pBR322 (Churchward et al. 1984). pMS421 is approximately 6.2 kb and carries resistance to 50 µg/ml streptomycin and 100 µg/ml spectinomycin (Graña et al. 1988).

Plasmid pGW1700

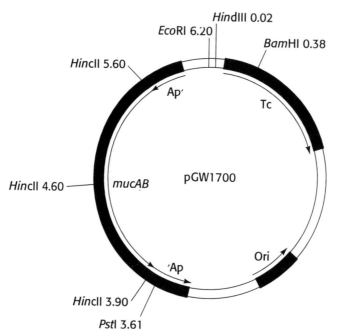

A 1.9-kb DNA fragment carrying the *mucAB* genes from pKM101 cloned into the Ap gene of a pBR322 derivative. Stimulates SOS mutagenesis in *Salmonella* (from Perry and Walker 1982).

Bacteriophage Mu Derivatives

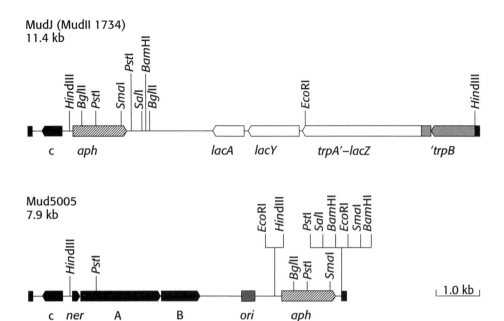

(*Top*) MudJ (MudI 1734) is used for insertional mutagenesis and *lacZ* operon fusion analysis. (Redrawn, with permission, from Castilho et al. 1984.) (*Small black rectangles*) Ends of bacteriophage Mu. The *aph* gene (Km[r]) is from transposon Tn5. The ′ *trpB trpA* ′ *-lacZ* element is from the W209 *trp-lac* fusion (Casadaban and Cohen 1980; Aksoy et al. 1984). The structure of the right end is described by Zieg and Kolter (1989). (*Bottom*) Mud5005 is used for in vivo cloning. (Redrawn, with permission, from Groisman and Casadaban 1986.) (*Small black rectangles*) Ends of bacteriophage Mu. The bacteriophage Mu *c*, *ner*, *A* and *B* gene structures are described by Priess et al. (1987). The plasmid origin of replication (*ori*) is from pMB1 (ColE1-type origin). The *aph* gene (Km[r]) is from transposon Tn5.

Transposon Tn*10* Derivatives

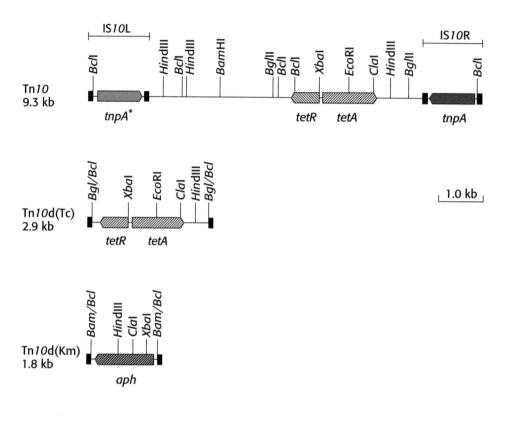

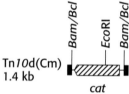

(*Small black rectangles*) Ends of transposon Tn*10*. The *tnpA* (transposase) gene in IS*10*L is nonfunctional. The *aph* (Km^r) gene is from transposon Tn*903*. The *cat* (Cm^r) gene is from transposon Tn*9*. (Redrawn, with permission, from Kleckner et al. 1991.)

Transposon Tn*phoA*

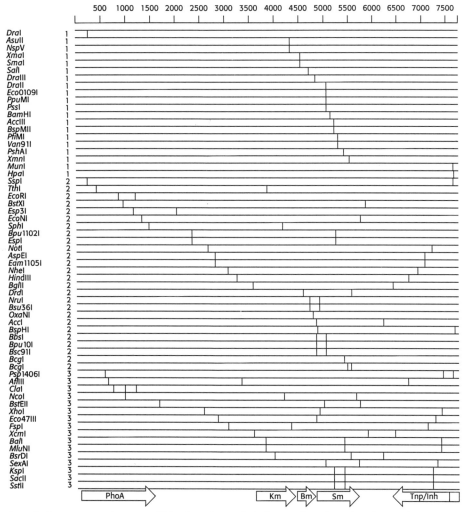

Restriction map and genes of Tn*phoA* (7737 bp) (Manoil and Beckwith 1985). Commercially available restriction enzymes that cut Tn*phoA* once, twice, or three times are indicated at the left side of the map. Enzymes are grouped according to number of sites and ordered with respect to position, beginning with the most leftward on the map. The locations of the genes encoding alkaline phosphatase (PhoA), transposase (Tnp), transposase inhibitor (Inh), and resistance to kanamycin (Km), bleomycin (Bm), and species-dependent streptomycin resistance (Sm) are indicated at the bottom. Restriction enzymes for which there are no sites in Tn*phoA* are *Aat*II, *Acc*65I, *Afl*II, *Apa*I, *Apa*LI, *Asc*I, *Ase*I, *Asp*718, *Avr*II, *Bsi*WI, *Bsp*120I, *Bsp*1407I, *Bst*1107I, *Ecl*136II, *Eco*72I, *Eco*ICRI, *Eco*RV, *Kpn*I, *Mlu*I, *Nde*I, *Nsi*I, *Pac*I, *Pma*CI, *Pme*I, *Ppu*10I, *Pvu*I, *Sac*I, *Sca*I, *Sfi*I, *Sgr*AI, *Sna*BI, *Spe*I, *Spl*I, *Srf*I, *Sse*8387I, *Sst*I, *Stu*I, *Sun*I, *Swa*I, *Vsp*I, and *Xba*I. (Redrawn, with permission, from Schifferli 1995.)

REFERENCES

"Also, the literature is now enormous and you can only follow it by living in a sort of cooperative, a reading cooperative. Or, as they now are, xeroxing cooperatives. Because hardly anybody reads any more—people only xerox things. I once asked a student who had a big xerox bill whether he'd tried neuroxing some papers. So he asked me what that was. I said 'It's a very easy and cheap process. You hold the page in front of your eyes and you let it go through there into the brain. It's much better than xeroxing.' You see, most people will xerox things on the grounds that if they have the paper they can read it any time. So they don't have to read it now. It's on the same grounds as someone in Rome never goes to see anything there—it's only tourists who see things. Perhaps the only people reading in a field are the ones who don't know anything about it and aren't working on it. The ones who are in it are just xeroxing the papers."

Sydney Brenner
Reprinted, with permission of Oxford University Press, as quoted by
Lewis Wolpert and Alison Richards.
In *A Passion for Science*, p. 104.
Oxford University Press, Oxford, U.K. (1988)

Abshire, K.Z. and F.C. Neidhardt. 1993. Growth rate paradox of *Salmonella typhimurium* within host macrophages. *J. Bacteriol.* **175:** 3744–3748.

Adams, J.N. 1965. Automotive pistons for use as bases in velveteen replication. *J. Bacteriol.* **89:** 1627.

Adams, M. 1959. *Bacteriophages.* Interscience Publishers, New York.

Adhin, M.R. and J. van Duin. 1990. Scanning model for translational reinitiation in eubacteria. *J. Mol. Biol.* **213:** 811–818.

Aksoy, S., C.L. Squires, and C. Squires. 1984. Evidence for antitermination in *Escherichia coli* rRNA transcription. *J. Bacteriol.* **159:** 260–264.

Alifano, P., F. Rivellini, D. Limauro, C.B. Bruni, and M.S. Carlomagno. 1991. A consensus motif common to all Rho-dependent prokaryotic transcription terminators. *Cell* **64:** 553–563.

Amman, E., J. Brosius, and M. Ptashne. 1983. Vectors bearing a hybrid *trp-lac* promoter useful for regulated expression of cloned genes in *Escherichia coli. Gene* **25:** 167–178.

Ausubel, F. 1994. *Current protocols in molecular biology.* Greene/Wiley, New York.

Bachmann, B.J. 1990. Linkage map of *Escherichia coli* K-12, edition 8. *Microbiol. Rev.* **54:** 130–197.

Bardwell, J.C. and J. Beckwith. 1993. The bonds that tie: Catalyzed disulfide bond formation. *Cell* **74:** 769–771.

Bardwell, J.C., A.K. McGovern, and J. Beckwith. 1991. Identification of a protein required for disulfide bond formation in vivo. *Cell* **67:** 581–589.

Barnes, W. 1994. PCR amplification of up to 35-kb DNA with high fidelity and high yield from λ bacteriophage templates. *Proc. Natl. Acad. Sci.* **91:** 2216–2220.

Barras, F. and M.G. Marinus. 1989. The great GATC: DNA methylation in *E. coli. Trends Genet.* **5:** 139–143.

Bartolomé, B., Y. Jubete, E. Martinez, and F. de la Cruz. 1991. Construction and properties of a family of pACYC184-derived cloning vectors compatible with pBR322 and its derivatives. *Gene* **102:** 75–78.

Beachey, E.H. and H.S. Courtney. 1987. Bacterial adherence: The attachment of group A streptococci to mucosal surfaces. *Rev. Infect. Dis.* **9:** S475–S481.

Beck, C.F., R. Mutzel, J. Barbé, and W. Müller. 1982. A multifunctional gene (*tetR*) controls Tn*10*-encoded tetracycline resistance. *J. Bacteriol.* **150:** 633–642.

Beckwith, J. 1978. *lac:* The genetic system. In *The operon* (ed. J.H. Miller and W.S. Reznikoff), pp. 11–30. Cold Spring Harbor Laboratory, Cold Spring Harbor, New York.

———. 1991. Stratagies for finding mutants. *Methods Enzymol.* **204:** 3–17.

Bender, J. and N. Kleckner. 1986. Genetic evidence that Tn*10* transposes by a nonreplicative mechanism. *Cell* **45:** 801–815.

———. 1992. IS*10* transposase mutations that specifically alter target site recognition. *EMBO J.* **11:** 741–750.

Benson, N.R. and B.S. Goldman. 1992. Rapid mapping in *Salmonella typhimurium* with Mud-P22 prophages. *J. Bacteriol.* **174:** 1673–1681.

Benson, N., P. Sugiono, S. Bass, L. Mendelman, and P. Youderian. 1986. General selection for specific DNA-binding activities. *Genetics* **114**: 1–14.

Benzer, S. 1955. Fine structure of a genetic region in bacteriophage. *Proc. Natl. Acad. Sci.* **41**: 344–354.

———. 1957. The elementary units of heredity. In *The chemical basis of heredity* (ed. W.D. McElroy and B. Glass), pp. 70–93. The Johns Hopkins University Press, Baltimore.

———. 1959. On the topology of the genetic fine structure. *Proc. Natl. Acad. Sci.* **45**: 1607–1620.

———. 1961. On topography of the genetic fine structure. *Proc. Natl. Acad. Sci.* **47**: 403–415.

———. 1966. Adventures in the rII region. In *Phage and the origins of molecular biology* (ed. J. Cairns et al.), pp. 157–165. Cold Spring Harbor Laboratory, Cold Spring Harbor, New York.

Berg, B.L. and V. Stewart. 1990. Structural genes for nitrate-inducible formate dehydrogenase in *Escherichia coli* K-12. *Genetics* **125**: 691–702.

Berg, D.E., A. Weiss, and L. Crossland. 1980. Polarity of Tn5 insertions mutants in *Escherichia coli. J. Bacteriol.* **142**: 439–446.

Berget, P. and A. Poteete. 1980. Structure and functions of the bacteriophage P22 tail protein. *J. Virol.* **34**: 234–243.

Berkowitz, D., J. Hushon, H. Whitfield, J.R. Roth, and B.N. Ames. 1968. Procedure for identifying nonsense mutations. *J. Bacteriol.* **96**: 215–220.

Berlyn, M.B. and S. Letovsky. 1992. COTRANS: A program for cotransduction analysis. *Genetics* **131**: 235–241.

Bertani, G. 1951. Studies on lysogenesis. I. The mode of phage liberation by lysogenic *Escherichia coli. J. Bacteriol.* **62**: 293–300.

Best, E.A. and R.A. Bender. 1990. Cloning of the *Klebsiella aerogenes nac* gene, which encodes a factor required for nitrogen regulation of the histidine utilization (*hut*) operons in *Salmonella typhimurium. J. Bacteriol.* **172**: 7043–7048.

Biek, D. and J.R. Roth. 1980. Regulation of Tn5 transposition in *Salmonella typhimurium. Proc. Natl. Acad. Sci.* **77**: 6047–6051.

Birnboim, H.C. and J. Doly. 1979. A rapid alkaline extraction procedure for screening recombinant plasmid DNA. *Nucleic Acids Res.* **7**: 1513–1523.

Birren, B. and E. Lai. 1994. Rapid pulsed field separation of DNA molecules to to 250 kb. *Nucleic Acids Res.* **22**: 5366–5370.

Blakesley, R. 1983. Diagnostic dideoxy DNA sequencing. *BRL Focus* **5**: 1–14.

Bliska, J., J. Balan, and S. Falkow. 1993. Signal transduction in the mammalian cell during bacterial attachment and entry. *Cell* **73**: 903–920.

Blomfield, I.C., V. Vaughn, R.F. Rest, and B.I. Eisenstein. 1991. Allelic exchange in *Escherichia coli* using the *Bacillus subtilis sacB* gene and a temperature sensitive pSC101 replicon. *Mol. Microbiol.* **5**: 1447–1457.

Bochner, B.R. 1984. Curing bacterial cells of lysogenic viruses by using UCB indicator plates. *BioTechniques* **2**: 234–240.

Bochner, B.R. and M.A. Savageau. 1977. Generalized indicator plate for genetic, metabolic, and taxonomic studies with microorganisms. *Appl. Environ. Microbiol.* **33**: 434–444.

Bochner, B.R., H.-C. Huang, G.L. Schieven, and B.N. Ames. 1980. Positive selection for loss of tetracycline resistance. *J. Bacteriol.* **143:** 926–933.

Bolivar, F. 1978. Construction and characterization of new cloning vehicles. III. Derivatives of plasmid pBR322 carrying unique EcoRI sites for selection of EcoRI generated recombinant DNA molecules. *Gene* **4:** 121–136.

Bolivar, F., R.L. Rodriquez, P.J. Greene, M.C. Betlach, H.L. Heynecker, H.W. Boyer, and S. Falkow. 1977. Construction and characterization of new cloning vehicles. II. A multipurpose cloning system. *Gene* **2:** 95–113.

Bordo, D. and P. Argos. 1991. Suggestions for "safe" residue substitutions in site-directed mutagenesis. *J. Mol. Biol.* **217:** 721–729.

Borodovsky, M.E., V. Koonin, and K.E. Rudd. 1994. New genes in old sequence: A strategy for finding genes in the bacterial genome. *Trends Biochem. Sci.* **19:** 309–313.

Bossi, L. 1985. Informational suppression. In *Genetics of bacteria* (ed. J. Scaife et al.), pp. 49–64. Academic Press, New York.

Botstein, D. 1992. Discovery of the bacterial transposon Tn*10*. In *The dynamic genome* (ed. N. Federoff and D. Botstein), pp. 225–232. Cold Spring Harbor Laboratory Press, Cold Spring Harbor, New York.

———. 1993 A phage geneticist turns to yeast. In *The early days of yeast genetics* (ed. M.N. Hall and P. Linder), pp. 361–373. Cold Spring Harbor Laboratory Press, Cold Spring Harbor, New York.

Botstein, D. and R. Maurer. 1982. Genetic approaches to the analysis of microbial development. *Annu. Rev. Genet.* **16:** 61–83.

Bradley, D. 1967. Ultrastructure of bacteriophages and bacteriocins. *Bacteriol. Rev.* **31:** 230–314.

Bradley, D.E., D.E. Taylor, and D.R. Cohen. 1980. Specificiation of surface mating systems among conjugative drug resistance plasmids in *Escherichia coli* K12. *J. Bacteriol.* **143:** 1466–1470.

Bramucci, M.G. and R.K. Holmes. 1978. Radial passive immune hemolysis assay for detection of heat-labile enterotoxin produced by individual colonies of *Escherichia coli* or *Vibrio cholerae. J. Clin. Microbiol.* **8:** 252–255.

Braus, G., M. Argast, and C.F. Beck. 1984. Identification of additional genes on transposon Tn*10: tetC* and *tetD. J. Bacteriol.* **160:** 504–509.

Brenner, D.J. 1984. *Enterobacteriaceae.* In *Bergey's manual of systematic bacteriology* (ed. N.R. Krieg and J.R. Holt), vol. 1, pp. 408–420. Williams and Wilkins, Baltimore.

Brickman, E. and J. Beckwith. 1975. Analysis of the regulation of *Escherichia coli* alkaline phosphatase synthesis using deletions and ɸ80 transducing phages. *J. Mol. Biol.* **96:** 307–316.

Brock, T.D. 1990. *The emergence of bacterial genetics,* pp. 189–191. Cold Spring Harbor Laboratory Press, Cold Spring Harbor, New York.

Brown, R.C. and R.K. Taylor. 1995. Organization of *tcp, acf,* and *toxT* genes within a ToxT-dependent operon. *Mol. Microbiol.* **16:** 425–439.

Buchmeier, N. and F. Heffron. 1989. Intracellular survival of wild-type *Salmonella typhimurium* and macrophage-sensitive mutants in diverse populations of macrophages. *Infect. Immun.* **57:** 1–7.

Bukhari, A.I., J.A. Shapiro, and S.L. Adhya. 1977. *DNA insertion elements, plasmids, and episomes.* Cold Spring Harbor Laboratory, Cold Spring Harbor,

New York.

Bullas, L.R. and J.-I. Ryu. 1983. *Salmonella typhimurium* LT2 strains which are r⁻ m⁺ for all three chromosomally located systems of DNA restriction and modification. *J. Bacteriol.* **156:** 471–474.

Burkhardt, H.-J., G. Riess, and A Puhler. 1979. Relationship of group P1 plasmids revealed by heteroduplex experiments. RP1, RP4, R68 and RK2 are identical. *J. Gen. Microbiol.* **114:** 341–348.

Butterton, J.R., J.A. Stoebner, S.M. Payne, and S.B. Calderwood. 1992. Cloning, sequencing, and transcriptional regulation of *viuA*, the gene encoding the ferric vibriobactin receptor of *Vibrio cholerae*. *J. Bacteriol.* **174:** 3729–3738.

Cadwell, R. and G. Joyce. 1992. Randomization of genes by PCR mutagenesis. *PCR Methods Appl.* **2:** 28–33.

Campos, J.M., J. Geisselsoder, and D.R. Zusman. 1978. Isolation of bacteriophage Mx4, a generalized transducing phage for *Myxococcus xanthus*. *J. Mol. Biol.* **119:** 167–178.

Cantor, C.R., A. Gaal, and C.L. Smith. 1988. High resolution separation and accurate size determination in pulsed field gel electrophoresis. III. Effect of electrical field shape. *Biochemistry* **27:** 9216–9221.

Cantor, C.R., C.L. Smith, and M.K. Mathew. 1987. Pulsed field gel electrophoresis of very large DNA molecules. *Annu. Rev. Biophys. Biophys. Chem.* **17:** 287–304.

Carlomagno, M.S., L. Chiarotti, P. Alifano, A.G. Nappo, and C.B. Bruni. 1988. Structure and expression of the histidine operons of *Escherichia coli* and *Salmonella typhimurium*. *J. Mol. Biol.* **203:** 585–606.

Carlson, K., E.A. Raleigh, and S. Hattman. 1994. Restriction and modification. In *Molecular biology of bacteriophage T4* (ed. J.D. Karam), pp. 1369–1381. American Society for Microbiology Press, Washington, D.C.

Casadaban, M. 1976a. Transposition and fusion of the *lac* genes to selected promoters in *Escherichia coli* using bacteriophage Lambda and Mu. *J. Mol. Biol.* **104:** 541–555.

———. 1976b. Regulation of the regulatory gene for the arabinose pathway, *araC*. *J. Mol. Biol.* **104:** 557–566.

Casadaban, M.J. and J. Chou. 1984. *In vivo* formation of gene fusions encoding hybrid β-galactosidase proteins in one step with a transposable Mu-*lac* transducing phage. *Proc. Natl. Acad. Sci.* **81:** 535–539.

Casadaban, M.J. and S.N. Cohen. 1979. Lactose genes fused to exogenous promoters in one step using a Mu-*lac* bacteriophage: *In vivo* probe for transcriptional control sequences. *Proc. Natl. Acad. Sci.* **76:** 4530–4533.

———. 1980. Analysis of gene control signals by DNA fusion and cloning in *Escherichia coli*. *J. Mol. Biol.* **138:** 179–207.

Casjens, S. and M. Hayden. 1988. Analysis *in vivo* of the bacteriophage P22 headful nuclease. *J. Mol. Biol.* **199:** 467–474.

Casjens, S. and C. Manoil. 1991. Thomas Hunt Morgan medal: Dale Kaiser. *Genetics* **128:** s12–s13.

Casjens, S., W.M. Huang, M. Hayden, and R. Parr. 1987. Initiation of bacteriophage P22 DNA packaging series. Analysis of a mutant that alters the DNA target specificity of the packaging apparatus. *J. Mol. Biol.* **194:** 411–

422.

Cassel, D. and T. Pfeuffer. 1978. Mechanism of cholera toxin action: Covalent modification of the guanyl nucleotide binding protein of the adenylate cyclase system. *Proc. Natl. Acad. Sci.* **75**: 2669–2673.

Castilho, B.A., P. Olfson, and M.J. Casadaban. 1984. Plasmid insertion mutagenesis and *lac* gene fusion with mini-Mu bacteriophage transposons. *J. Bacteriol.* **158**: 488–495.

Chan, R.K., D. Botstein, T. Watanabe, and Y. Ogata. 1972. Specialized transduction of tetracycline resistance by phage P22 in *Salmonella typhimurium*. *Virology* **50**: 883–898.

Chang, A.C.Y. and S.N. Cohen. 1978. Construction and characterization of amplifiable multicopy DNA cloning vehicles derived from the P15A cryptic miniplasmid. *J. Bacteriol.* **134**: 1141–1156.

Chassy, B., A. Mercenier, and J. Flickinger. 1988. Transformation of bacteria by electroporation. *Trends Biotechnol.* **6**: 303–309.

Chatfield, S.N., C.J. Dorman, C. Hayward, and G. Dougan. 1991. Role of *ompR*-dependent genes in *Salmonella typhimurium* virulence: Mutants deficient in both OmpC and OmpF are attenuated in vivo. *Infect. Immun.* **59**: 449–452.

Chauthaiwale, V.M., A. Therwath, and V.V. Deshpande. 1992. Bacteriophage lambda as a cloning vector. *Microbiol. Rev.* **56**: 577–591.

Chelala, C.A. and P. Margolin. 1974. Effects of deletions on cotransduction linkage in *Salmonella typhimurium*: Evidence that bacterial chromosome deletions affect the formation of transducing DNA fragments. *Mol. Gen. Genet.* **131**: 97–112.

———. 1976. Evidence that HT mutants of bacteriophage P22 retain an altered form of substrate specificity in the formation of transducing particles in *Salmonella typhimurium*. *Genet. Res.* **27**: 315–322.

Chen, L.-M. and S. Maloy. 1991. Regulation of proline utilization in *Klebsiella aerogenes*: Cloning of the *put* operon and characterization of the *put* control region. *J. Bacteriol.* **173**: 783–790.

Christopher, F., H. Franklin, and R. Spooner. 1989. Broad-host-range cloning vectors. In *Promiscuous plasmids of gram-negative bacteria* (ed. C.M. Thomas), pp. 247–267. Academic Press, Oxford.

Chung, C.T. and R.H. Miller. 1988. A rapid and convenient method for the preparation and storage of competent bacterial cells. *Nucleic Acids Res.* **16**: 3580.

Chung, C.T., S.L. Niemela, and R. Miller. 1989. One-step transformation of *Escherichia coli*: Transformation and storage of bacterial cells in the same solution. *Proc. Natl. Acad. Sci.* **86**: 2172–2175.

Churchward, G., D. Belin, and Y. Nagamine. 1984. A pSC101-derived plasmid which shows no sequence homology to other commonly used cloning vectors. *Gene* **31**: 165–171.

Ciampi, M.S. and J.R. Roth. 1988. Polarity effects in the *hisG* gene of *Salmonella* require a site within the coding sequence. *Genetics* **118**: 193–202.

Ciampi, M.S., M.B. Schmid, and J.R. Roth. 1982. Transposon Tn*10* provides a promoter for transcription of adjacent sequences. *Proc. Natl. Acad. Sci.* **79**: 5016–5020.

Ciampi, M.S., P. Alifano, A.G. Nappo, C.B. Bruni, and M.S. Carlomagno. 1989.

Features of the rho-dependent transcription termination polar element within the *hisG* cistron of *Salmonella typhimurium*. *J. Bacteriol.* **171:** 4472–4478.

Clark, V.L. and P.M. Bavoil. 1994. Bacterial pathogenesis: Identification and regulation of virulence factors. *Methods Enzymol.* **235:** 1–789.

Clewell, D.B. 1993. *Bacterial conjugation.* Plenum Press, New York.

Cohn, M. 1957. Contributions of studies on the β-galactosidase of *Escherichia coli* to our understanding of enzyme synthesis. *Bacteriol. Rev.* **21:** 140–168.

Cole, S.T. and I. Saint-Girons. 1994. Bacterial genomics. *FEMS Microbiol. Rev.* **14:** 139–160.

Collado-Vides, J., B. Magasanik, and J. Gralla. 1991. Control site location and transcriptional regulation in *Escherichia coli*. *Microbiol. Rev.* **55:** 371–394.

Collins, L.A., S.M. Egan, and V. Stewart. 1992. Mutational analysis reveals functional similarity between NARX, a nitrate sensor in *Escherichia coli* K-12, and the methyl-accepting chemotaxis proteins. *J. Bacteriol.* **174:** 3667–3675.

Cossart, P. and C. Kocks. 1994. The actin-based motility of the facultative intracellular pathogen Listeria monocytogenes. *Mol. Microbiol.* **13:** 403–416.

Crawford, I.P. 1989. Evolution of a biosynthetic pathway: The tryptophan paradigm. *Annu. Rev. Microbiol.* **43:** 567–600.

Crick, F.H.C. and L. Orgel. 1964. The theory of interallelic complementation. *J. Mol. Biol.* **8:** 161–165.

Crick, F., L. Barnett, S. Brenner, and R. Watts-Tobin. 1961. General nature of the genetic code for proteins. *Nature* **192:** 1227–1232.

Csonka, L.N. and A.J. Clark. 1979. Deletions generated by the transposon Tn*10* in the *srl recA* region of the *Escherichia coli* K12 chromosome. *Genetics* **93:** 321–343.

Csonka, L.N., M.M. Howe, J.L. Ingraham, L.S. Pierson III, and C.L. Turnbough, Jr. 1981. Infection of *Salmonella typhimurium* with coliphage Mu *d*1 (Ap^r *lac*): Construction of *pyr::lac* gene fusions. *J. Bacteriol.* **145:** 299–305.

Cutrecasas, P. 1973. Gangliosides and membrane receptors for cholera toxin. *Biochemistry* **12:** 3558–3566.

Dammel, C.S. and H.F. Noller. 1995. Supression of a cold-sensitive mutation in 16S rRNA by overexpression of a novel ribosome-binding factor, R6fA. *Genes Dev.* **9:** 626–637.

Daniels, D.L. 1990. The complete *Avr*II restriction map of the *Escherichia coli* genome and comparisons of several laboratory strains. *Nucleic Acids Res.* **18:** 2649–2651.

Darzins, A. and M.J. Casadaban. 1989. In vivo cloning of *Pseudomonas aeruginosa* genes with mini-D3112 transposable bacteriophage. *J. Bacteriol.* **171:** 3917–3925.

Das, A. and C. Yanofsky. 1989. Restoration of a translational stop-start overlap reinstates translational coupling in a mutant *trpB'-trpA* gene pair of the *Escherichia coli* tryptophan operon. *Nucleic Acids Res.* **17:** 9333–9340.

Davies. J. 1994. Inactivation of antibiotics and the dissemination of resistance genes. *Science* **264:** 375–382.

Davis, B.D. 1948. Isolation of biochemically deficient mutants of bacteria by penicillin. *J. Am. Chem. Soc.* **70:** 4267.

——. 1987. Mechanism of bactericidal action of aminoglycosides. *Microbiol. Rev.* **51:** 341–350.

Davis, R.H., D. Botstein, and J.R. Roth. 1980. *Advanced bacterial genetics.* Cold Spring Harbor Laboratory, Cold Spring Harbor, New York.

Dean, D. 1981. A plasmid cloning vector for the direct selection of strains carrying recombinant plasmids. *Gene* **15:** 99–102.

de Lorenzo, V., M. Herrero, U. Jakubzik, and K.N. Timmis. 1990. Mini-Tn5 transposon derivatives for insertion mutagenesis, promoter probing, and chromosomal insertion of cloned DNA in gram-negative eubacteria. *J. Bacteriol.* **172:** 6568–6572.

Demerec, M. and H. Ozeki. 1959. Tests for alleleism among auxotrophs of *Salmonella typhimurium. Genetics* **44:** 269–278.

Demerec, M., I. Blomstrand, and Z.E. Demerec. 1955. Evidence of complex loci in *Salmonella. Proc. Natl. Acad. Sci.* **41:** 359–364.

Derman, A.I. and J. Beckwith. 1995. *Escherichia coli* alkaline phosphatase localized to the cytoplasm slowly acquires enzymatic activity in cells whose growth has been suspended: A caution for gene fusion studies. *J. Bacteriol.* **177:** 3764–3770.

Derman, A.I., W.A. Prinz, D. Belin, and J. Beckwith. 1993. Mutations that allow disulfide bond formation in the cytoplasm of *Escherichia coli. Science* **262:** 1744–1748.

Dila, D. and S. Maloy. 1987. Proline transport in *Salmonella typhimurium: putP* permease mutants with altered substrate specificity. *J. Bacteriol.* **168:** 590–594.

DiRita, V.J. 1992. Co-ordinate expression of virulence genes by ToxR in *Vibrio cholerae. Mol. Microbiol.* **6:** 451–458.

DiRita, V.J. and J.J. Mekalanos. 1991. Periplasmic interaction between two membrane regulatory proteins, ToxR and ToxS, results in signal transduction and transcriptional activation. *Cell* **64:** 29–37.

DiRita, V.J., C. Parsot, G. Jander, and J.J. Mekalanos. 1991. Regulatory cascade controls vitulence in *Vibrio cholerae. Proc. Natl. Acad. Sci.* **88:** 5403–5407.

Ditta, G., S. Stanfield, D. Corbin, and D.R. Helinski. 1980. Broad host range DNA cloning system for gram-negative bacteria: Construction of a gene bank of *Rhizodium meliloti. Proc. Natl. Acad. Sci.* **77:** 7347–7351.

Donnenberg, M.S. and J.B. Kaper. 1991. Construction of an *eae* deletion mutant of enteropathogenic *Escherichia coli* by using a positive-selection suicide vector. *Infect. Immun.* **59:** 4310–4317.

Doolittle, R.F. 1985. Proteins. *Sci. Am.* **253:** 88–99.

——. 1986. *Of urfs and orfs. A primer on how to analyze derived amino acid sequences.* University Science Books, Mill Valley, California.

Dorman, C.J. 1991. DNA supercoiling and environmental regulation of gene expression in pathogenic bacteria. *Infect. Immun.* **59:** 745–749.

Dorman, C. and N.N. Bhriain. 1993. Coordination of gene expression in pathogenic *Salmonella typhimurium.* In *Biology of* Salmonella (ed. F. Cabello et al.), pp. 41–50. Plenum Press, New York.

Dougan, G., D. Maskell, D. Pickard, and C. Hormaeche. 1987. Isolation of stable *aroA* mutants of *Salmonella typhi* Ty2: Properties and preliminary characterisation in mice. *Mol. Gen. Genet.* **207:** 402–405.

Drake, J. 1970. *The molecular basis of mutation*, pp. 152–155. Holden-Day, San Francisco.

Drake, J. and L. Ripley. 1994. Mutagenesis. In *Bacteriophage T4* (ed. J.D. Karam), pp. 98–124. American Society for Microbiology, Washington, D.C.

Ebel-Tsipis, J., M.S. Fox, and D. Botstein. 1972. Generalized transduction by phage P22 in *Salmonella typhimurium*. II. Mechanism of integration of transducing DNA. *J. Mol. Biol.* **71:** 449–469.

Echols, H. and M. Goodman. 1991. Fidelity mechanisms in DNA replication. *Annu. Rev. Biochem.* **60:** 477–511.

Egan, S.M. and V. Stewart. 1990. Nitrate regulation of anaerobic respiratory gene expression in *narX* deletion mutants of *Escherichia coli* K-12. *J. Bacteriol.* **172:** 5020–5029.

———. 1991. Mutational analysis of nitrate regulatory gene *narL* in *Escherichia coli* K-12. *J. Bacteriol.* **173:** 4424–4432.

Eggertsson, G. and D. Söll. 1988. Transfer-ribonucleic acid-mediated suppression of termination codons in *Escherichia coli*. *Microbiol. Rev.* **52:** 354–374.

Elledge, S. and R. Davis. 1989. Position and density effects on repression by stationary and mobile DNA-binding proteins. *Genes Dev.* **3:** 185–197.

Elliott, T. 1992. A method for constructing single-copy *lac* fusions in *Salmonella typhimurium* and its application to the *hemA-prfA* operon. *J. Bacteriol.* **174:** 245–253.

Elliott, T. and J.R. Roth. 1988. Characterization of Tn*10*d-Cam: A transposition-defective Tn*10* specifying chloramphenicol resistance. *Mol. Gen. Genet.* **213:** 332–337.

Elsinghorst, E.A., L.S. Baron, and D.J. Kopecko. 1989. Penetration of human intestinal epithelial cells by *Salmonella*: Molecular cloning and expression of *Salmonella typhi* invasion determinants in *Escherichia coli*. *Proc. Natl. Acad. Sci.* **86:** 5173–5177.

Ely, B. and R.C. Johnson. 1977. Generalized transduction in *Caulobacter crescentus*. *Genetics* **87:** 391–399.

Ely, B. and L. Shapiro. 1989. The molecular genetics of differentiation. *Genetics* **123:** 427–429.

Enomoto, M. and B.A.D. Stocker. 1974. Transduction by phage P1*kc* in *Salmonella typhimurium*. *Virology* **60:** 503–514.

Ernst, R.K., D.M. Dombroski, and J.M. Merrick. 1990. Anaerobiosis, type 1 fimbriae, and growth phase are factors that affect invasion of HEp-2 cells by *Salmonella typhimurium*. *Infect. Immun.* **58:** 2014–2016.

Escalante-Semerena, J.C., M.G. Johnson, and J.R. Roth. 1992. The CobII and CobIII regions of the cobalamin (vitamin B_{12}) biosynthetic operon of *Salmonella typhimurium*. *J. Bacteriol.* **174:** 24–29.

Everiss, K.D., K.J. Hughes, M.E. Kovach, and K.M Peterson. 1994. The *Vibrio cholerae acfB* colonization determinant encodes an inner membrane protein that is related to a family of signal-transducing proteins. *Infect. Immun.* **62:** 3289–3298.

Falkow, S. 1988. Molecular Koch's postulates applied to microbial pathogenicity. *Rev. Infect. Dis.* (suppl.) **10**: 274–276.

―――. 1994. A look through the retrospectoscope. In *Molecular genetics of bacterial pathogenesis. A tribute to Stanley Falkow* (ed. V.L. Miller et al.), pp. xxiii–xxxix. ASM Press, Washington, D.C.

Falkow, S., R.R. Isberg, and D.A. Portnoy. 1992. The interaction of bacteria with mammalian cells. *Annu. Rev. Cell Biol.* **8**: 333–363.

Fasano, A., B. Baudry, D.W. Pumplin, S.S. Wasserman, B.D. Tall, K.M. Ketley, and J.B. Kaper. 1991. *Vibrio cholerae* produces a second enterotoxin, which affects tight junctions. *Proc. Natl. Acad. Sci.* **88**: 5242–5246.

Fellay, R., J. Frey, and H. Krisch. 1987. Interposon mutagenesis of soil and water bacteria: A family of DNA fragments designed for in vitro insertional mutagenesis of gram-negative bacteria. *Gene* **52**: 147–154.

Fewson, C.A., R.K. Poole, and C.F. Thurston. 1984. Spectrophotometry in microbiology: Symbols and terminology. Scattered thoughts on opaque problems. *Soc. Gen. Microbiol. Q.* **11**: 87–89.

Field, M. 1980. Intestinal secretion and its stimulation by enterotoxins. In *Cholera and related diarrheas* (ed. O. Ochterlony and J. Holmgren), pp. 46–52. Karger Press, Basel.

Fields, P.I., R.V. Swanson, C.G. Haidaris, and F. Heffron. 1986. Mutants of *Salmonella typhimurium* that cannot survive within macrophages are avirulent. *Proc. Natl. Acad. Sci.* **83**: 5189–5193.

Figurski, D.H. and D.R. Helinski. 1979. Replication of an origin-containing derivative of plasmid RK2 dependent on a plasmid function provided in *trans. Proc. Natl. Acad. Sci.* **76**: 1648–1652.

Finan, T.M., E. Hartwieg, K. LeMieux, K. Bergman, G.C. Walker, and E.R. Signer. 1984. General transduction in *Rhizobium meliloti. J. Bacteriol.* **159**: 120–124.

Fink, G.R. 1988. Notes of a bigamous biologist. *Genetics* **118**: 549–550.

Fink, G.R. and R.G. Martin. 1967. Translation and polarity in the histidine operon. II. Polarity in the histidine operon. *J. Mol. Biol.* **30**: 97–107.

Finlay, B.B. and S. Falkow. 1989. Common themes in microbial pathogenicity. *Microbiol. Rev.* **53**: 210–230.

Finlay, B., S. Ruschkowski, and S. Dedhar. 1991. Cytoskeletal rearrangements accompanying *Salmonella* entry into epithelial cells. *J. Cell Sci.* **99**: 283–296.

Finlay, B., M. Starnbach, C. Francis, B.A.D. Stocker, S. Chatfield, G. Dougan, and S. Falkow. 1988. Identification and characterization of Tn*phoA* mutants of *Salmonella* that are unable to pass through a polarized MDCK epithelial cell monolayer. *Mol. Microbiol.* **2**: 757–766.

Fishov, I. A. Zaritsky, and N.B. Grover. 1995. On microbial states of growth. *Mol. Microbiol.* **15**: 789–794.

Flower, A., R. Osborne, and T. Silhavy. 1995. The allele-specific synthetic lethality of *prlA-prlG* double mutants predicts interactive domains of SecY and SecE. *EMBO J.* **14**: 884–893.

Flynn, J.L. and D.E. Ohman. 1988. Use of a gene replacement cosmid vector for cloning alginate conversion genes from mucoid and nonmucoid *Pseudomonas aeruginosa* strains: *algS* controls expression of *algT. J. Bacteriol.* **170**:

3228–3236.

Foster, P. 1991. *In vivo* mutagenesis. *Methods Enzymol.* **204:** 114–124.

Foster, T.J. 1983. Plasmid-determined resistance to antimicrobial drugs and toxic metal ions in bacteria. *Microbiol. Rev.* **47:** 361–409.

Freese, E., E. Bautz, and E. Freese. 1961. The chemical and mutagenic specificity of hydroxylamine. *Proc. Natl. Acad. Sci.* **47:** 845–855.

Friedberg, E.C., G.C. Walker, and W. Siede. 1995. *DNA repair and mutagenesis.* ASM Press, Washington, D.C.

Friedman, A.M., S. Long, S. Brown, W. Buikema, and F. Ausubel. 1982. Construction of a broad host range cosmid cloning vector and its use in the genetic analysis of *Rhizobium* mutants. *Gene* **18:** 289–296.

Furste, J.P., W. Pansegrau, R. Frank, H. Blocker, P. Scholz, M. Bagdasarian, and E. Lanka. 1987. Molecular cloning of the plasmid RP4 primase region in a multi-host range *tacP* expression vector. *Gene* **48:** 119–131.

Gahring, L., F. Heffron, B. Finlay, and S. Falkow. 1990. Invasion and replication of *Salmonella typhimurium* in animal cells. *Infect. Immun.* **58:** 443–448.

Galan, J.E. and R. Curtiss III. 1989. Cloning and molecular characterization of genes whose products allow *Salmonella typhimurium* to penetrate tissue culture cells. *Proc. Natl. Acad. Sci.* **86:** 6383–6387.

Galan, J.E., V.L. Miller, and D. Portnoy. 1993. Discussion of in vitro and in vivo assays for studying bacterial entry into and survival within eukaryotic cells. *Infect. Agents Dis.* **2:** 288–290.

Gay, P., D. Le Coq, M. Steinmetz, T. Berkelman, and C.I. Kado. 1985. Positive selection procedure for entrapment of insertion sequence elements in gram-negative bacteria. *J. Bacteriol.* **164:** 918–921.

Gelfand, D. 1989. The design and optimization of PCR. In *PCR technology: Principles and applications for DNA amplification* (ed. H. Erlich), pp. 7–16. Stockton Press, New York.

Gelfand, D. and T. White. 1990. Thermostable DNA polymerases. In *PCR protocols: A guide to methods and applications* (ed. M. Innis et al.), pp. 129–141. Academic Press, New York.

Gill, D.M. 1976. The arrangement of subunits in cholera toxin. *Biochemistry* **15:** 1242–1248.

Gill, D.M. and R. Meren. 1978. ADP-ribosylation of membrane proteins catalysed by cholera toxin: Basis of the activation of adenylate cyclase. *Proc. Natl. Acad. Sci.* **75:** 3050–3054.

Ginocchio, C., S. Olmsted, C. Wells, and J. Galan. 1994. Contact with epithelial cells induces formation of surface appendages on *Salmonella typhimurium*. *Cell* **76:** 717–724.

Goldberg, I. and J.J. Mekalanos. 1985. Effect of a *recA* mutation on cholera toxin gene amplification and deletion events. *J. Bacteriol.* **165:** 723–731.

Goldberg, M.B. and P.J. Sansonetti. 1993. Shigella subversion of the cellular cytoskeleton: A strategy for epithelial colonization. *Infect. Immun.* **61:** 4941–4946.

Goldberg, M.B., V.J. DiRita, and S.B. Calderwood. 1990. Identification of an iron-regulated virulence determinant in *Vibrio cholerae*, using Tn*phoA* mutagenesis. *Infect. Immun.* **58:** 55–60.

Goldberg, R.B., R.A. Bender, and S.L. Streicher. 1974. Direct selection for P1-sensitive mutants of enteric bacteria. *J. Bacteriol.* **118:** 810–814.

Goldman, B.S., J.T. Lin, and V. Stewart. 1994. Identification and structure of gene *nasR*, encoding the nitrate- and nitrite-responsive positive regulator of *nas* (nitrate assimilation) operon expression in *Klebsiella pneumoniae* M5al. *J. Bacteriol.* **176:** 5077–5085

Gordon, C.L. and J. King. 1994. Genetic properties of temperature-sensitive folding mutants of the coat protein of phage P22. *Genetics* **136:** 427–438.

Goss, T.J. and R.A. Bender. 1995. The nitrogen assimilation control protein, NAC, is a DNA binding transcription activator in *Klebsiella aerogenes*. *J. Bacteriol.* **177:** 3546–3555.

Graña, D., T. Gardella, and M.M. Susskind. 1988. The effects of mutations in the *ant* promoter of phage P22 depend on context. *Genetics* **120:** 319–327.

Grantham, R. 1974. Amino acid difference formula to help explain protein evolution. *Science* **185:** 862–865.

Griffin, H. and A. Griffin. 1994. *PCR technology. Current innovations.* CRC Press, Boca Raton, Florida.

Groisman, E.A. and M.J. Casadaban. 1986. Mini-Mu bacteriophage with plasmid replicons for in vivo cloning and *lac* gene fusion. *J. Bacteriol.* **168:** 357–364.

———. 1987. Cloning of genes from members of the family *Enterobacteriaceae* with mini-Mu bacteriophage containing plasmid replicons. *J. Bacteriol.* **169:** 687–693.

Groisman, E. and M. Saier. 1990. *Salmonella* virulence: New clues to intramacrophage survival. *Trends Biochem. Sci.* **15:** 30–33.

Groisman, E., P. Fields, and F. Heffron. 1990. Molecular biology of *Salmonella* pathogenesis. In *Molecular biology of bacterial pathogenesis* (ed. B. Iglewski and Clark), pp. 251–272. Academic Press, San Diego.

Guerry, P., J. van Embden, and S. Falkow. 1974. Molecular nature of two non-conjugative plasmids carrying drug resistance genes. *J. Bacteriol.* **117:** 619–630.

Guiney, D.G. and E. Lanka. 1989. Conjugative transfer of IncP plasmids. In *Promiscuous plasmids of gram-negative bacteria* (ed. C.M. Thomas), pp. 27–56. Academic Press, New York.

Guiney, D.G., F.C. Fang, M. Krause, and S. Libby. 1994. Plasmid-mediated virulence genes in non-typhoid *Salmonella* serovars. *FEMS Microbiol. Lett.* **124:** 1–9.

Gulig, P.A. and R. Curtiss. 1987. Plasmid-associated virulence of *Salmonella typhimurium. Infect. Immun.* **55:** 2891–2901.

Guthrie, C., H. Nashimoto, and M. Nomura. 1969. Structure and function of *E. coli* ribosomes. VIII. Cold-sensitive mutants defective in ribosome assembly. *Proc. Natl. Acad. Sci.* **63:** 384–391.

Gutierrez, C. and J.C. Devedjian. 1989. A plasmid facilitating in vitro construction of *phoA* gene fusions in *Escherichia coli. Nucleic Acids Res.* **17:** 3999.

Gutierrez, C., J. Barondess, C. Manoil, and J. Beckwith. 1987. The use of transposon Tn*phoA* to detect genes for cell envelope proteins subject to a common regulatory stimulus. *J. Mol. Biol.* **195:** 289–297.

Guzman, L.-M., D. Belin, M. Carson, and J. Beckwith. 1995. Tight regulation, modulation, and high-level expression by vectors containing the arabinose P_{BAD} promoter. *J. Bacteriol.* **177:** 4121–4130.

Gyllensten, U. 1989. Direct sequencing of in vitro amplified DNA. In *PCR technology: Principles and applications for DNA amplification* (ed. H. Erlich), pp. 45–60. Stockton Press, New York.

Haas, D. and B.W. Halloway. 1976. R-factor variants with enhanced sex factor activity in *Pseudomonas aeruginosa. Mol. Gen. Genet.* **114:** 243–251.

———. 1978. Chromosome mobilization by the R-plasmid R68.45: A tool in *Pseudomonas* genetics. *Mol. Gen. Genet.* **158:** 229–237.

Haas, D. and C. Reimmann. 1989. Use of IncP plasmids in chromosomal genetics of gram-negative bacteria. In *Promiscuous plasmids in gram-negative bacteria* (ed. C.M. Thomas), pp. 185–206. Academic Press, New York.

Haas, D., A. Jann, C. Reimmann, E. Luthi, and T. Leisinger. 1987. Chromosome organization in *Pseudomonas aeruginosa. Antibiot. Chemother.* **39:** 256–263.

Hahn, D.R. and S.R. Maloy. 1986. Regulation of the *put* operon in *Salmonella typhimurium:* Characterization of promoter and operator mutations. *Genetics* **114:** 687–703.

Hamilton, C.M., M. Aldea, B.K. Washburn, P. Babitzke, and S.R. Kushner. 1989. New method for generating deletions and gene replacements in *Escherichia coli. J. Bacteriol.* **171:** 4617–4622.

Hanahan, D. 1983. Studies on transformation of *Escherichia coli* with plasmids. *J. Mol. Biol.* **166:** 557–580.

———. 1987. Mechanisms of DNA transformation. In Escherichia coli *and* Salmonella typhimurium: *Cellular and molecular biology* (ed. F.C. Neidhardt et al.), vol. 2, pp. 1177–1183. American Society for Microbiology, Washington, D.C.

Hanahan, D., J. Jessee, and F. Bloom. 1991. Plasmid transformation of *Escherichia coli* and other bacteria. *Methods Enzymol.* **204:** 63–113.

Hanne, L.F. and R.A. Finkelstein. 1982. Characterization and distribution of the hemagglutinins produced by *Vibrio cholerae. Infect. Immun.* **36:** 209–214.

Hardy, K.G. 1993. *Plasmids: A practical approach.* IRL Press, Geneva.

Harkey, C.W., K.D. Everiss, and K.M. Peterson. 1994. The *Vibrio cholerae* toxin-coregulated-pilus gene *tcpI* encodes a homolog of methyl-accepting chemotaxis proteins. *Infect. Immun.* **62:** 2669–2678.

Harshey, R. 1983. Switch in the transposition products of Mu DNA mediated by proteins: Cointegrates versus simple insertions. *Proc. Natl. Acad. Sci.* **80:** 2012–2016.

Hartman, P.E. and J.R. Roth. 1973. Mechanisms of suppression. *Adv. Genet.* **17:** 1–105.

Hase, C.C., M.E. Bauer, and R.A. Finkelstein. 1994. Genetic characterization of mannose-sensitive (MSHA)-negative mutants of *Vibrio cholerae* derived by Tn5 mutagenesis. *Gene* **150:** 17–25.

Haselkorn, R. 1991. Genetic systems in cyanobacteria. *Methods Enzymol.* **204:** 418–430.

Hatfull, G.F. and G.J. Sarkis. 1993. DNA sequence, structure and gene expression of mycobacteriophage L5: A phage system for mycobacterial genetics. *Mol. Microbiol.* **7:** 395–405.

Heath, J.D., J.D. Perkins, B. Sharma, and G.M. Weinstock. 1992. *Not*I genomic cleavage map of *Escherichia coli* K-12 strain MG1655. *J. Bacteriol.* **174:** 558–567.

Hedges, R.W. and A.E. Jacob. 1974. Transposition of ampicillin resistance from RP4 to other replicons. *Mol. Gen. Genet.* **132:** 31–40.

Hendrix, R.W., J.W. Roberts, F.W. Stahl, and R.A. Weisberg, eds. 1983. *Lambda II*. Cold Spring Harbor Laboratory, Cold Spring Harbor, New York.

Hensel, M., J.E. Shea, C. Gleeson, M.D. Jones, E. Dalton, and D.W. Holden. 1995. Simultaneous identification of bacterial virulence genes by negative selection. *Science* **269:** 400–403.

Herrero, M., V. de Lorenzo, and K.M. Timmis. 1990. Transposon vectors containing non-antibiotic resistance selection markers for cloning and stable chromosomal insertion of foreign genes in gram-negative bacteria. *J. Bacteriol.* **172:** 6557–6567.

Herrington, D.A., R.H. Hall, G.A. Losonsky, J.J. Mekalanos, R.K. Taylor, R.K., and M.M. Levine. 1988. Toxin, toxin-coregulated pili, and the *toxR* regulon are essential for *Vibrio cholerae* pathogenesis in humans. *J. Exp. Med.* **168:** 1487–1492.

Hershfield, V., H. Boyer, C. Yanofsky, M. Lovett, and D. Helinski. 1974. Plasmid colE1 as a molecular vehicle for cloning and amplification of DNA. *Proc. Natl. Acad. Sci.* **71:** 3455–3459.

Higgins, D.E. and V.J. DiRita. 1994. Transcriptional control of *toxT*, a regulatory gene in the toxR regulon of *Vibrio cholerae*. *Mol. Microbiol.* **14:** 17–29.

Higgins, D.E., E. Nazareno, and V.J. DiRita. 1992. The virulence gene activator ToxT from *Vibrio cholerae* is a member of the AraC family of transcriptional activators. *J. Bacteriol.* **174:** 6974–6980.

Hill, C. 1975. Informational suppression of missense mutations. *Cell* **6:** 419–427.

Hirst, T.R., J. Sanchez, J.B. Kaper, S.J.S. Hardy, and J. Holmgren. 1984. Mechansism of toxin secretion by *Vibrio cholerae* investigated in strains harbouring plasmids that encode heat-labile enterotoxins of *Escherichia coli*. *Proc. Natl. Acad. Sci.* **81:** 7752–7756.

Hoffman, C.S. and A. Wright. 1985. Fusions of secreted proteins to alkaline phosphatase: An approach for studying protein secretion. *Proc. Natl. Acad Sci.* **82:** 5107–5111.

Holloway, B.W. 1969. Genetics of *Pseudomonas*. *Bacteriol. Rev.* **33:** 419–443.

Holloway, B. and K.B. Low. 1987. F-prime and R-prime factors. In Escherichia coli *and* Salmonella typhimurium: *Cellular and molecular biology* (ed F.C. Neidhardt et al.), vol. 2, pp. 1145–1153. American Society for Microbiology, Washington, D.C.

Holloway, B.W., U. Römling, and B. Tümmler. 1994. Genomic mapping of *Pseudomonas aeruginosa* PAO. *Microbiology* **140:** 2907–2029.

Hong, J. and B. Ames. 1971. Localized mutagenesis of any specific small region of the bacterial chromosome. *Proc. Natl. Acad. Sci.* **68:** 3158–3162.

Hoppe, I., H.M. Johnston, D. Biek, and J.R. Roth. 1979. A refined map of the *hisG* gene of *Salmonella typhimurium*. *Genetics* **92:** 17–26.

Howe, M.M. 1972. "Genetic studies on bacteriophage Mu." Ph.D. thesis, Massachusetts Institute of Technology, Cambridge.

Hughes, K.T. and J.R. Roth. 1984. Conditionally transposition-defective derivative of Mu d1(Amp Lac). *J. Bacteriol.* **159:** 130–137.

———. 1985. Directed formation of duplications and deletions using Mu d (Ap, lac). *Genetics* **109:** 263–282.

———. 1988. Transitory cis-complementation: A general method for providing transposase to defective transposons. *Genetics* **119:** 9–12.

Hughes, K.T., J.R. Roth, and B.M. Olivera. 1991. A genetic characterization of the *nadC* gene of *Salmonella typhimurium*. *Genetics* **127:** 657–670.

Humphreys, G., G. Willshaw, H. Smith, and E. Anderson. 1976. Mutagenesis of plasmid DNA with hydroxylamine: Isolation of mutants of multi-copy plasmids. *Mol. Gen. Genet.* **145:** 101–108.

Imamoto, F., J. Ito, and C. Yanofsky. 1966. Polarity in the tryptophan operon of *E. coli*. *Cold Spring Harbor Symp. Quant. Biol.* **31:** 235–249.

Innis, M. and D. Gelfand. 1990. Optimization of PCRs. In *PCR protocols: A guide to methods and applications* (ed. M. Innis et al.), pp. 3–12. Academic Press, New York.

Ippen-Ihler, K. and R.A. Skurray. 1993. Genetic organization and transfer-related determinants on the sex factor F and related plasmids. In *Bacterial conjugation* (ed. D.B. Clewell), pp. 23–52. Plenum Press, New York.

Isberg, R.R. 1991. Discrimination between intracellular uptake and surface adhesion of bacterial pathogens. *Science* **252:** 934–938.

Isberg, R.R. and S. Falkow. 1985. A single genetic locus encoded by *Yersinia pseudotuberculosis* permits invasion of cultured animal cells by *Escherichia coli* K-12. *Nature* **317:** 262–264.

Jackson, E.N., F. Laski, and C. Andres. 1982. Bacteriophage P22 mutants that alter the specificity of DNA packaging. *J. Mol. Biol.* **154:** 551–563.

Jarvik, J. and D. Botstein. 1975. Conditional-lethal mutations that suppress genetic defects in morphogenesis by altering structural proteins. *Genetics* **72:** 2738–2742.

Jefferson, R.A. 1989. The *gus* reporter gene system. *Nature* **342:** 837–838.

Jefferson, R.A., S.M. Burgess, and D. Hirsh. 1986. Beta-glucuronidase from *Escherichia coli* as a gene fusion marker. *Proc. Natl. Acad. Sci.* **83:** 8447–8451.

Jeter, R.M., and J.R. Roth. 1987. Cobalamin (vitamin B$_{12}$) biosynthetic genes of *Salmonella typhimurium*. *J. Bacteriol.* **169:** 3189–3198.

Johnston, H.M. and J.R. Roth. 1981. Genetic analysis of the histidine control region of *Salmonella typhimurium*. *J. Mol. Biol.* **145:** 713–734.

Jones, B.D., H.F. Paterson, A. Hall, and S. Falkow. 1993. *Salmonella typhimurium* induces membrane ruffling by a growth factor-receptor-independent mechanism. *Proc. Natl. Acad. Sci.* **90:** 10390–10394.

Jonson, G., J. Holmgren, and A.-M. Svennerholm. 1991. Identification of a mannose-binding pilus on *Vibrio cholerae* El Tor. *Microb. Pathog.* **11:** 433–441.

Joshi, A., V. Baichwal, and G.F.-L. Ames. 1991. Rapid polymerase chain reaction amplification using intact bacterial cells. *BioTechniques* **10:** 42–45.

Kaufman, M.R. and R.K. Taylor. 1994. Identification of bacterial cell-surface virulence determinants with Tn*phoA*. *Methods Enzymol.* **235:** 426–441.

Kaufman, M.R., Shaw, C.E., Jones, I.D., and R.K. Taylor. 1993. Biogenesis and regulation of the *Vibrio cholerae* toxin-coregulated pilus: Analogies to other virulence factor secretory systems. *Gene* **126:** 43–49.

Keen, N.T., D. Kobayashi, and D. Trollinger. 1988. Improved broad-host-range plasmids for DNA cloning in Gram-negative bacteria. *Gene* **70:** 191–197.

Kenyon, C. and G. Walker. 1980. DNA-damaging agents stimulate gene expression at specific loci in *Escherichia coli*. *Proc. Natl. Acad. Sci.* **77:** 2819–2823.

Keusch, G. and D.M. Thea. 1989. The Salmonellae: Typhoid fever and gastroenteritis. In *Mechanisms of microbial disease* (ed. M. Schaechter et al.), pp. 266–275. Williams and Wilkins, Baltimore.

Kleckner, N. 1990. Regulating Tn*10* and IS*10* transposition. *Genetics* **124:** 449–454.

Kleckner, N., J. Bender, and S. Gottesman. 1991. Uses of transposons with emphasis on Tn*10*. *Methods Enzymol.* **204:** 139–180.

Kleckner, N., K. Reichardt, and D. Botstein. 1979a. Inversions and deletions of the *Salmonella typhimurium* chromosome generated by the translocatable tetracycline resistance element Tn*10*. *J. Mol. Biol.* **127:** 89–115.

Kleckner, N., J. Roth, and D. Botstein. 1977. Genetic engineering in vivo using translocatable drug-resistance elements. New methods in bacterial genetics. *J. Mol. Biol.* **116:** 125–159.

Kleckner, N., R.K. Chan, B.-K. Tye, and D. Botstein. 1975. Mutagenesis by insertion of a drug resistance element carrying an inverted repetition. *J. Mol. Biol.* **97:** 561–575.

Kleckner, N., D.A. Steele, K. Reichardt, and D. Botstein. 1979b. Specificity of insertion by the translocatable tetracycline resistance element Tn*10*. *Genetics* **92:** 1023–1040.

Klein, S., K. Lohman, R. Clover, G.C. Walker, and E.R. Signer. 1992. A directional, high-frequency chromosomal mobilization system for genetic mapping of *Rhizobium meliloti*. *J. Bacteriol.* **174:** 324–326.

Klig, L., D. Oxender, and C. Yanofsky. 1988. Second-site revertants of *Escherichia coli trp* repressor mutants. *Genetics* **120:** 651–655.

Kohara, Y., K. Akiyama, and K. Isono. 1987. The physical map of the whole *Escherichia coli* chromosome: Application of a new strategy for rapid analysis and sorting of a large genomic library. *Cell* **50:** 495–508.

Kontomichalou, P., M. Mitani, and R.C. Clowes. 1970. Circular R-factor molecules, controlling penicillinase synthesis, replicating in *Escherichia coli* under either relaxed or stringent control. *J. Bacteriol.* **104:** 33–44.

Krajewska-Grynkiewicz, K. and T. Klopotowski. 1979. Altered linkage values in phage P22-mediated transduction caused by distant deletions or insertions in donor chromosomes. *Mol. Gen. Genet.* **176:** 87–93.

Kukral, A.M., K.L. Strauch, R.A. Maurer, and C.G. Miller. 1987. Genetic analysis in *Salmonella typhimurium* with a small collection of randomly spaced insertions of transposon Tn*10*Δ*16*Δ*17*. *J. Bacteriol.* **169:** 1787–1793.

Kuo, T.-T. and B.A.D. Stocker. 1969. Suppression of proline requirement of *proA* and *proAB* deletion mutants in *Salmonella typhimurium* by mutation to arginine requirement. *J. Bacteriol.* **98:** 593–598.

Lawes, M. and S. Maloy. 1995. Mu*d* SacI, a transposon with strong selectable and counter-selectable markers: Use for rapid mapping of chromosomal mutations in *Salmonella typhimurium. J. Bacteriol.* **177:** 1383–1387.

Lederberg, J. 1948. Detection of fermentative variants with tetrazolium. *J. Bacteriol.* **56:** 695.

———. 1950. The beta-D-galactosidase of *Escherichia coli,* strain K-12. *J. Bacteriol.* **60:** 381–392.

———. 1951. Streptomycin resistance: A genetically recessive mutation. *J. Bacteriol.* **61:** 549–550.

Lederberg, J. and E.M. Lederberg. 1952. Replica plating and indirect selection of bacterial mutants. *J. Bacteriol.* **63:** 399–406.

Lederberg, J. and E.L. Tatum. 1946. Gene recombination in *Escherichia coli. Nature* **158:** 558.

Lederberg, J. and N.D. Zinder. 1948. Concentration of biochemical mutants of bacteria with penicillin. *J. Am. Chem. Soc.* **70:** 4267–4268.

Lee, C.A. and S. Falkow. 1990. The ability of *Salmonella* to enter mammialian cells is affected by bacterial growth state. *Proc. Natl. Acad. Sci.* **87:** 4304–4308.

Lee, C.A., B.D. Jones, and S. Falkow. 1992. Identification of a *Salmonella typhimurium* invasion locus by selection for hyperinvasive mutants. *Proc. Natl. Acad. Sci.* **89:** 1847–1851.

Leemans, J., R. Villarroel, B. Silva, M. Van Montagu, and J. Schell. 1980. Direct repetition of a 1.2Md DNA sequence is involved in site-specific recombination by the P1 plasmid R68. *Gene* **10:** 319–328.

Le Minor, L. and M. Popoff. 1987. Designation of *Salmonella enterica* sp. nov., nom.rev., as the type and only species of the genus *Salmonella. Int. J. Sys. Bacteriol.* **37:** 465–468.

Lennox, E.S. 1955. Transduction of linked genetic characters of the host by bacteriophage P1. *Virology* **1:** 190–206.

Lewin, B. 1994. Preface. In *Genes V.* Oxford University Press, England.

Levine, M. 1957. Mutations in the temperate phage P22 and lysogeny in *Salmonella. Virology* **3:** 22–41.

Levine, M.M., J.B. Kaper, D.A. Herrington, G.A. Losonsky, J.G. Morris, M.L. Clements, R.E. Black, B.D. Tall, and R. Hall. 1988. Volunteer studies of deletion mutants of *Vibrio cholerae* O1 prepared by recombinant techniques. *Infect. Immun.* **56:** 161–167.

Levine, M.M., D. Nalin, J.P. Craig, D. Hoover, E.J. Bergquist, D. Waterman, H.P. Holley, R.B. Hornick, N.F. Pierce, and J.P. Libonati. 1979. Immunity to cholera in man: Relative role of antibacterial versus antitoxic immunity. *Trans. R. Soc. Top. Med. Hyg.* **73:** 3–9.

Lin, J.T., B.S. Goldman, and V. Stewart. 1994. The *nasFEDCBA* operon for nitrate and nitrite assimilation in *Klebsiella pneumoniae* M5al. *J. Bacteriol.* **176:** 2551–2559.

Litwin, C.M. and S.B. Calderwood. 1994. Analysis of the complexity of gene regulation by Fur in *Vibrio cholerae. J. Bacteriol.* **176:** 240–248.

Liu, S.-L. and K.E. Sanderson. 1992. A physical map of the *Salmonella typhimurium* LT2 genome made by using *Xba*I analysis. *J. Bacteriol.* **174:** 1662–1672.

———. 1995. Genomic cleavage map of *Salmonella typhi* Ty2. *J. Bacteriol.* **177:**

5099–5107.

Liu, S.-L., A. Hessel, and K.E. Sanderson. 1993a. The *XbaI-BlnI-CeuI* genomic cleavage map of *Salmonella typhimurium* LT2 determined by double digestion, end labelling, and pulsed-field gel electrophoresis. *J. Bacteriol.* **175:** 4104–4120.

———. 1993b. The *XbaI–BlnI-CeuI* genomic cleavage map of *Salmonella enteritidis* shows an inversion relative to *Salmonella typhimurium* LT2. *Mol. Microbiol.* **10:** 655–664.

Liu, S.-L., A. Hessel, H.-Y.M. Cheng, and K.E. Sanderson. 1994. The *XbaI-BlnI-CeuI* genomic cleavage map of *Salmonella paratyphi* B. *J. Bacteriol.* **176:** 1014–1024.

Lockman, H.A., J.E. Galen, and J.B. Kaper. 1984. *Vibrio cholerae* enterotoxin genes: Nucleotide sequence analysis of DNA encoding ADP-ribosyltransferase. *J. Bacteriol.* **159:** 1086–1089.

Lospalluto, J.J. and R.A. Finkelstein. 1972. Chemical and physical properties of cholera exo-enterotoxin (choleragen) and its spontaneously formed toxoid (choleragenoid). *Biochim. Biophys. Acta.* **257:** 158–166.

Low, K.B. 1972. *E. coli* K12 F ′ factors, old and new. *Bacteriol. Rev.* **36:** 587–607.

———. 1991. Conjugational methods for mapping with Hfr and F-prime strains. *Methods Enzymol.* **204:** 43–62.

Maaløe, O. and N.O. Kjeldgaard. 1966. *Control of macromolecular synthesis.* W.A. Benjamin, New York.

MacConkey, A. 1905. Lactose-fermenting bacteria in feces. *J. Hyg.* **5:** 333–378.

MacPherson, M.J. 1991. *Directed mutagenesis: A practical approach.* IRL Press, Washington, D.C.

Mahan, M.J. and J.R. Roth. 1991. Ability of a bacterial chromosome segment to invert is dictated by included material rather than flanking sequence. *Genetics* **129:** 1021–1032.

Mahan, M.J., J.M. Slauch, and J.J. Mekalanos. 1993. Selection of bacterial virulence genes that are specifically induced in host tissues. *Science* **259:** 686–688.

Maloy, S. 1989. *Experimental techniques in bacterial genetics.* Jones and Bartlett, Boston.

Maloy, S. and W. Nunn. 1981. Selection for loss of tetracycline resistance by *Escherichia col. J. Bacteriol.* **145:** 1110–1112.

Maloy, S.R. and J.R. Roth. 1983. Regulation of proline utilization in *Salmonella typhimurium*: Characterization of *put*::Mu d(Ap, *lac*) operon fusions. *J. Bacteriol.* **154:** 561–568.

Maloy, S. and P. Youderian. 1994. Challenge phage: Dissecting DNA-protein interactions in vivo. *Methods Mol. Genet.* **3:** 205–233.

———. 1995. Genetic approaches for dissecting DNA-protein interaction in vivo. *Am. Biotechnol. Lab.* (in press).

Maloy, S., J. Cronan, Jr., and D. Freifelder. 1994. *Microbial genetics*, 2nd edition, Jones and Bartlett, Boston.

Mandecki, W., K. Krajewska-Grynkiewicz, and T. Klopotowski. 1986. A quantitative model for nonrandom generalized transduction applied to the phage P22-*Salmonella typhimurium* system. *Genetics* **114:** 633–657.

Mandel, G.L., Douglas, R.G., Jr., and J.E. Bennett. 1995. *Principles and practice of infectious diseases.* Churchill Livingstone, New York.

Maniatis, T., E. Fritsch, and J. Sambrook. 1982. *Molecular cloning: A laboratory manual.* Cold Spring Harbor Laboratory, Cold Spring Harbor, New York.

Manoil, C. 1990. Analysis of protein localization by use of gene fusions with complementary properties. *J. Bacteriol.* **172:** 1035–1042.

Manoil, C. and J. Beckwith. 1985. TnphoA: A transposon probe for protein export signals. *Proc. Natl. Acad. Sci.* **82:** 8129–8133.

Manoil, C., J.J. Mekalanos, and J. Beckwith. 1990. Alkaline phosphatase fusions: Sensors of subcellular location. *J. Bacteriol.* **172:** 515–518.

Margolin, W. and M.M. Howe. 1986. Localization and DNA sequence analysis of the *C* gene of bacteriophage Mu, the positive regulator of Mu late transcription. *Nucleic Acids Res.* **14:** 4881–4897.

Martin, M.O. and S.R. Long. 1984. Generalized transduction in *Rhizobium meliloti. J. Bacteriol.* **159:** 125–129.

Martin, R.G., D. Silbert, D. Smith, and H. Whitfield, Jr. 1966. Polarity in the histidine operon. *J. Mol. Biol.* **21:** 357–369.

Martin, S., E. Sodergren, T. Matsuda, and D. Kaiser. 1978. Systematic isolation of transducing phages for *Myxococcus xanthus. Virology* **88:** 44–53.

Masters, M. 1977. The frequency of P1 transduction of the genes of *Escherichia coli* as a function of chromosomal position: Preferential transduction of the origin of replication. *Mol. Gen. Genet.* **155:** 197–202.

Masters, M., B.J. Newman, and C.M. Henry. 1984. Reduction of marker discrimination in transductional recombination. *Mol. Gen. Genet.* **196:** 85–90.

Mathew, M.K., C.L. Smith, and C.R. Cantor. 1988a. High resolution separation and accurate size determination in pulsed field gel electrophoresis. I. DNA size standards and the effect of agarose and temperature. *Biochemistry* **27:** 9204–9210.

———. 1988b. High resolution separation and accurate size determination in pulsed field gel electrophoresis. II. Effect of pulse time and electric field strength, and implications for models of the separation process. *Biochemistry* **27:** 9211–9216.

Mathew, M.K., C.-F. Hui, C.L. Smith, and C.R. Cantor. 1988c. High resolution separation and accurate size determination in pulsed field gel electrophoresis. IV. The influence of DNA topology. *Biochemistry* **27:** 9222–9226.

Mazodier, P., P. Cossart, E. Giraud, and F. Gasser. 1985. Completion of the nucleotide sequence of the central region of Tn5 confirms the presence of three resistance genes. *Nucleic Acids Res.* **13:** 195–205.

McCabe, P. 1990. Production of single-stranded DNA by asymmetric PCR. In *PCR protocols: A guide to methods and applications* (ed. M. Innis et al.), pp. 76–83. Academic Press, New York.

McIver, K.S., E. Kessler, J.C. Olson, and D.E. Ohman. 1995. The elastase propeptide functions as an intramolecular chaperone required for elastase activity and secretion in *Pseudomonas aeruginosa. Mol. Microbiol.* (in press).

McMurry, L., P. Petrucci, and S.B. Levy. 1980. Active efflux of tetracycline encoded by four genetically different tetracycline resistance determinants in

Escherichia coli. Proc. Natl. Acad. Sci. **77:** 3974–3977.

McPherson, M.J. 1991. *Directed mutagenesis.* Oxford University Press, England.

Mecsas, J., P.E. Rouviere, J.W. Erickson, T.J. Donohue, and C.A. Gross. 1993. The activity of σ^E, an *Escherichia coli* heat-inducible σ-factor, is modulated by expression of outer membrane proteins. *Genes Dev.* **7:** 2618–2628.

Mekalanos, J.J. 1983. Duplication and amplification of toxin genes in *Vibrio cholerae. Cell* **35:** 253–263.

————. 1992. Environmental signals controlling the expression of virulence determinants in bacteria. *J. Bacteriol.* **174:** 1–7.

Mekalanos, J.J. and J.C. Sadoff. 1994. Cholera vaccines: Fighting an ancient scourge. *Science* **265:** 1387–1389.

Mekalanos, J.J., R.J. Collier, and W.R. Romig. 1979. Enzymatic activity of cholera toxin. II. Relationships to proteolytic processing, disulfide bond reduction, and subunit composition. *J. Biol. Chem.* **254:** 5855–5861.

Mekalanos, J.J., D.J. Swartz, G.D.N. Pearson, N. Harford, F. Groyne, and M. deWilde. 1983. Cholera toxin genes: Nucleotide sequence, deletion analysis, and vaccine development. *Nature* **306:** 551–557.

Merritt, E.A., S. Sarfaty, F. van den Akker, C. L'Hoir, J.A. Martial, and W.G. Hol. 1994. Crystal structure of cholera toxin B-pentamer bound to receptor GM1 pentasaccharide. *Protein Sci.* **3:**166–175.

Metcalf, W.W. and B.L. Wanner. 1993. Construction of new β-glucuronidase cassettes for making transcriptional fusions and their use with new methods for allele replacement. *Gene* **129:** 17–25.

Michaelis, S., H. Inouye, D. Oliver, and J. Beckwith. 1983. Mutations that alter the signal sequence of alkaline phosphatase in *Escherichia coli. J. Bacteriol.* **154:** 366–374.

Miller, J.F., J.J. Mekalanos, and S. Falkow. 1989. Coordinate regulation and sensory transduction in the control of bacterial virulence. *Science* **243:** 916–922.

Miller, J.H. 1972. *Experiments in molecular genetics.* Cold Spring Harbor Laboratory, Cold Spring Harbor, New York.

————. 1992. *A short course in bacterial genetics.* Cold Spring Harbor Laboratory Press, Cold Spring Harbor, New York.

Miller, S.I., A.M. Kukral, and J.J. Mekalanos. 1989. A two-component regulatory system (*phoP phoQ*) controls *Salmonella typhimurium* virulence. *Proc. Natl. Acad. Sci.* **86:** 5054–5058.

Miller, V.L. and J.J. Mekalanos. 1985. Genetic analysis of the cholera toxin positive regulatory gene *toxR. J. Bacteriol.* **163:** 580–585.

————. 1988. A novel suicide vector and its use in construction of insertion mutations: Osmoregulation of outer membrane proteins and virulence determinants in *Vibrio cholerae* requires *toxR. J. Bacteriol.* **170:** 2575–2583.

Miller, V.L., R.K. Taylor, and J.J. Mekalanos. 1987. Cholera toxin transcriptional activator ToxR is a transmembrane DNA binding protein. *Cell* **48:** 271–279.

Modrich, P. 1989. Methyl-directed DNA mismatch correction. *J. Biol. Chem.* **264:** 6597–6600.

Moir, D., S.E. Stewart, B.C. Osmond, and D. Botstein. 1982. Cold-sensitive cell-division-cycle mutants of yeast: Isolation, properties, and pseudoreversion studies. *Genetics* **100:** 547–563.

Monod, J., A.M. Pappenheimer, Jr., and G. Cohen-Bazire. 1952. La cinétique de la biosynthèse de la β-galactosidase chez *Escherichia coli* considérée comme fonction de la croissance. *Biochim. Biophys. Acta* **9:** 648–660.

Muro-Pastor, A.M. and S. Maloy. 1995. Direct cloning of mutant alleles from the bacterial chromosome into plasmid vectors *in vivo. BioTechniques* **18:** 386–390.

Myers, R. and S. Maloy. 1988. Mutations of *putP* that alter the lithium sensitivity of *Salmonella typhimurium. Mol. Microbiol.* **2:** 749–755.

Nagata, S., C.C. Hyde, and E.W. Miles. 1989. The α subunit of tryptophan synthase. *J. Biol. Chem.* **264:** 6288–6296.

Neal, B.L., P.K. Brown, and P.R. Reeves. 1993. Use of *Salmonella* phage P22 for transduction in *Escherichia coli. J. Bacteriol.* **175:** 7115–7118.

Neidhardt, F.C., P.L. Bloch, and D.F. Smith. 1974. Culture medium for enterobacteria. *J. Bacteriol.* **119:** 736–747.

Neidhardt, F.C., J.L. Ingraham, and M. Schaechter. 1990. *Physiology of the bacterial cell.* Sinauer Associates, Sunderland, Massachusetts.

Newland, J.W., B.A. Green, and R.K. Holmes. 1984. Transposon-mediated mutagenesis and recombination in *Vibrio cholerae. Infect. Immun.* **45:** 428–432.

Newman, B.J. and M. Masters. 1980. The variation with which markers are transduced by phage P1 is primarily a result of discrimination during recombination. *Mol Gen. Genet.* **180:** 585–589.

Nikaido, H. 1994. Prevention of drug access to bacterial targets: Permeability barriers and active efflux. *Science* 382–388.

Nilsson, B. and S. Anderson. 1991. Proper and improper folding of proteins in the cellular environment. *Annu. Rev. Microbiol.* **45:** 607–635.

Nohmi, T., A. Hakura, Y. Nakai, M. Watanabe, S.Y. Murayama, and T. Sofuni. 1991. *Salmonella typhimurium* has two homologous but different *umuDC* operons: Cloning of a new *umuDC*-like operon (*samAB*) present in a 60 megadalton cryptic plasmid of *S. typhimurium. J. Bacteriol.* **173:** 1051–1063.

Norvick, R.P. 1991. Genetic systems in staphylococci. *Methods Enzymol.* **204:** 587–636.

Ogierman, M.A., S. Zabihi, L. Mourtzios, and P.A. Manning. 1993. Genetic organization and sequence of the promoter-distal region of the *tcp* gene cluster of *Vibrio cholerae. Gene* **126:** 51–60.

O'Hoy, K. and V. Krishnapillai. 1987. Recalibration of the *Pseudomonas aeruginosa* strain PAO chromosome map in time units using high-frequency-of-recombination donors. *Genetics* **115:** 611–618.

Oppenheim, D. and C. Yanofsky. 1980. Translational coupling during expression of the tryptophan operon of *Escherichia coli. Genetics* **95:** 785–795.

Orbach, M.J. and E.N. Jackson. 1982. Transfer of chimeric plasmids among *Salmonella typhimurium* strains by P22 transduction. *J. Bacteriol.* **149:** 985–994.

Ornellas, E.P. and B.A.D. Stocker. 1974. Relation of lipopolysaccharide character to P1 sensitivity in *Salmonella typhimurium. Virology* **60:** 491–502.

Ornstein, D. and M. Kashdan. 1985. Sequencing DNA using ^{35}S-labeling: A troubleshooting guide. *BioTechniques* **3:** 476–483.

Osek, J., G. Jonson, A.-M. Svennerholm, and J. Holmgren. 1994. Role of antibodies against biotype-specific *Vibrio cholerae* pili in protection against experimental classical and El Tor cholera. *Infect. Immun.* **62:** 2901–2907.

Ottemann, K.M., V.J. DiRita, and J.J. Mekalanos. 1992. ToxR proteins with substitutions in residues conserved with OmpR fail to activate transcription from the cholera toxin promoter. *J. Bacteriol.* **174:** 6807–6814.

Overbye, L.J., M. Sandkvist, and M. Bagdasarian. 1993. Genes required for extracellular secretion of enterotoxin are clustered in *Vibrio cholerae*. *Gene* **132:** 101–106.

Ozeki, H. 1959. Chromosome fragments participating in transduction in *Salmonella typhimurium*. *Genetics* **44:** 457–470.

Pace, J., M. Hayman, and J. Galan. 1993. Signal transduction and invasion of epithelial cells by *S. typhimurium*. *Cell* **72:** 505–514.

Pardee, A.B., F. Jacob, and J. Monod. 1959. The genetic control and cytoplasmic expression of "inducibility" in the synthesis of β-galactosidase by *E. coli*. *J. Mol. Biol.* **1:** 165–178.

Parkinson, J. and S. Parker. 1979. Interaction of the *cheC* and *cheZ* gene products is required for chemotactic behavior of *Escherichia coli*. *Genetics* **76:** 2890–2394.

Pearson, G.D.N. 1989. "The cholera toxin genetic element: A site-specific transposon." Ph.D thesis, Harvard Medical School, Cambridge, Massachusetts.

Pearson, G.D.N., A. Woods, S.L. Chiang, and J.J. Mekalanos. 1993. CTX genetic element encodes a site-specific recombination system and an intestinal colonization factor. *Proc. Natl. Acad. Sci.* **90:** 3750–3754.

Peek, J.A. and R.K. Taylor. 1992. Characterization of a periplasmic thiol:disulfide interchange protein required for the functional maturation of secreted virulence factors of *Vibrio cholerae*. *Proc. Natl. Acad. Sci.* **89:** 6210–6214.

Pelham, H. 1985. Lithium chloride modification of the rapid alkaline plasmid prep. *Trends Genet.* **1:** 6.

Penfold, R.J. and J.M. Pemberton. 1992. An improved suicide vector for construction of chromosomal insertion mutations in bacteria. *Gene* **118:** 145–146.

Perbal, B. 1988. *A practical guide to molecular cloning*, 2nd ed. John Wiley, New York.

Perkins, J.D., J.D. Heath, B. Sharma, and G.M. Weinstock. 1992. *Sfi*I genomic cleavage map of *Escherichia coli* K-12 strain MG1655. *Nucleic Acids Res.* **20:** 1129–1137.

———. 1993. *Xba*I and *Bln*I genomic cleavage maps of *Escherichia coli* K-12 strain MG1655 and comparative analysis of other strains. *J. Mol. Biol.* **232:** 419–445.

Perry, K.L. and G.C. Walker. 1982. Identification of plasmid (pKM101)-coded proteins involved in mutagenesis and UV resistance. *Nature* **300:** 278–281.

Peterson, K.M. and J. Mekalanos. 1988. Characterization of the *Vibrio cholerae* ToxR regulon: Identification of novel genes involved in intestinal colonization. *Infect. Immun.* **56:** 2822–2829.

Peterson, K., N. Ossanna, A. Thliveris, D. Ennis, and D. Mount. 1988. Derepres sion of specific genes promotes DNA repair and mutagenesis in *Escherichia coli. J. Bacteriol.* **170:** 1–4.

Pfau, J. and R.K. Taylor. 1995. New definition of the ToxR binding site using the challenge phage system to obtain a genetic footprint. *Mol. Microbiol.* (in press).

Pfau, J. and P. Youderian. 1990. Transferring plasmid DNA between different bacterial species with electroporation. *Nucleic Acids Res.* **18:** 6165.

Plum, G. and J.E. Clark-Curtiss. 1994. Induction of *Mycobacterium avium* gene expression foloowing phagocytosis by human macrophages. *Infect. Immun.* **62:** 476–483.

Poteete, A.R. 1988. Bacteriophage P22. In *The bacteriophages*, vol. 2. (ed. R. Calendar), pp. 647–682. Plenum Press, New York.

Poteete, A.R., A.C. Fenton, and K.C. Murphy. 1988. Modulation of *Escherichia coli* RecBCD activity by the bacteriophage λ Gam and P22 Abc functions. *J. Bacteriol.* **170:** 2012–2021.

Priess, H., B. Bräuer, C. Schmidt, and D. Kamp. 1987. Sequence of the left end of Mu. In *Phage Mu* (ed. N. Symonds et al.), pp. 277–296. Cold Spring Harbor Laboratory, Cold Spring Harbor, New York.

Ptashne, M. 1992. *A genetic switch*, 2nd edition. Blackwell Scientific, Cambridge, Massachusetts.

Pugsley, A.P. 1993. The complete general secretory pathway in gram-negative bacteria. *Microbiol. Rev.* **57:** 50–108.

Quandt, J. and M.F. Hynes. 1993. Versatile suicide vectors which allow direct se lection for gene replacement in gram-negative bacteria. *Gene* **127:** 15–21.

Raj, A.S., A.Y. Raj, and H. Schmieger. 1974. Phage genes involved in the forma tion of generalized transducing particles in *Salmonella*-phage P22. *Mol. Gen. Genet.* **135:** 175–184.

Raleigh, E.A., J. Benner, F. Bloom, H.D. Braymer, E. DeCruz, K. Dharmalingam, J. Heitman, M. Noyer Weidner, A. Pickarowicz, P.L. Kretz et al. 1991. Nomenclature relating to restriction of modified DNA in *Escherichia coli. J. Bacteriol.* **173:** 2707–2709.

Ranade, K. and A. Poteete. 1993. Superinfection exclusion (*sieB*) genes of bac teriophages P22 and λ. *J. Bacteriol.* **175:** 4712–4718.

Ratzkin, B. and J. Roth. 1978. Cluster of genes controlling proline degradation in *Salmonella typhimurium. J. Bacteriol.* **133:** 744–754.

Reimmann, C. and D. Haas. 1987. Mode of replicon fusion mediated by the duplicated insertion sequence IS*21* in *Escherichia coli. Genetics* **115:** 619–625.

Reznikoff, W.S. 1993. The Tn*5* transposon. *Annu. Rev. Microbiol.* **47:** 945–963.

Rhine, J.A. and R.K. Taylor. 1994. TcpA pilin sequences and colonization re quirements for O1 and O139 *Vibrio cholerae. Mol. Microbiol.* **13:** 1013– 1020.

Richardson, J.P. 1991. Preventing the synthesis of unused transcripts by Rho fac tor. *Cell* **64:** 1047–1049.

Richardson, K. 1991. Roles of motility and flagellar structure in pathogenicitiy of *Vibrio cholerae*: Analysis of motility mutants in three animal models. *Infect.*

Immun. **59:** 2727–2736.

Ried, J.L. and A. Collmer. 1987. An *nptI-sacB-sacR* cartridge for constructing directed, unmarked mutations in gram-negative bacteria by marker exchange-eviction mutagenesis. *Gene* **57:** 239–246.

Roberts, J.W. 1969. Termination factor for RNA synthesis. *Nature* **224:** 1168–1174.

Roof, D.M. and J.R. Roth. 1988. Ethanolamine utilization in *Salmonella typhimurium.* *J. Bacteriol.* **170:** 3855–3863.

———. 1989. Functions required for vitamin B_{12}-dependent ethanol-amine utilization in *Salmonella typhimurium.* *J. Bacteriol.* **171:** 3316–3323.

Rosenshine, I., S. Ruschkowski, V. Foubister, and B. Finlay. 1994. *Salmonella typhimurium* invasion of epithelial cells: Role of induced host cell tyrosine protein phosphorylation. *Infect. Immun.* **62:** 4969–4974.

Rosset, R. and L. Gorini. 1969. A ribosomal ambiguity mutation. *J. Mol. Biol.* **39:** 95–112.

Roth, J. 1970. Genetic techniques in studies of bacterial metabolism. *Methods Enzymol.* **17:** 1–35.

Roth, J.R. and P.E. Hartman. 1965. Heterogeneity in P22 transducing particles. *Virology* **27:** 297–307.

Rubin, R.A. and S.B. Levy. 1990. Interdomain hybrid Tet proteins confer tetracycline resistance only when they are derived from closely related members of the *tet* gene family. *J. Bacteriol.* **172:** 2303–2312.

Ruvkun, G.B. and F.M. Ausubel. 1981. A general method for site-directed mutagenesis in prokaryotes. *Nature* **289:** 85–88.

Rychlik, W., W. Spencer, and R. Rhodes. 1990. Optimization of the annealing temperature of DNA amplification in vitro. *Nucleic Acids Res.* **18:** 6409–6412.

Salyers, A.A. and D.D. Whitt. 1994. *Bacterial pathogenesis: A molecular approach.* American Society for Microbiology, Washington, D.C.

Sambrook, J., E.F. Fritsch, and T. Maniatis. 1989. *Molecular cloning: A laboratory manual.* Cold Spring Harbor Laboratory Press, Cold Spring Harbor, New York.

Sanderson, K.E. and P.E. Hartman. 1978. Linkage map of *Salmonella typhimurium,* edition V. *Microbiol. Rev.* **42:** 417–519.

Sanderson, K.E. and J.R. Roth. 1988. Linkage map of *Salmonella typhimurium,* edition VII. *Microbiol. Rev.* **52:** 485–532.

Sanderson, K.E. and D.R. Zeigler. 1991. Storing, shipping, and maintaining records on bacterial strains. *Methods Enzymol.* **204:** 248–264.

Sanderson, K.E., A. Hessel, and K. Rudd. 1995. The genetic map of *Salmonella typhimurium,* edition VIII. *Microbiol. Rev.* **59:** 241–303.

Sandkvist, M., V. Morales, and M. Bagdasarian. 1993. A protein required for secretion of cholera toxin through the outer membrane of *Vibrio cholerae.* *Gene* **123:** 81–86.

Sanger, F., S. Nicklen, and A.R. Coulson. 1977. DNA sequencing with chain-terminating inhibitors. *Proc. Natl. Acad. Sci.* **74:** 5463–5467.

Schifferli, D.M. 1995. Use of Tn*phoA* and T7 RNA polymerase as tools to study fimbrial proteins. *Methods Enzymol.* **253:** 242–258.

Schleif, R. 1987. The L-arabinose operon. In Escherichia coli *and* Salmonella typhimurium: *Cellular and molecular biology.* American Society for Microbiology, Washington, D.C.

Schmid, M.B. and J.R. Roth. 1980. Circularization of transduced fragments: A mechanism for adding segments to the bacterial chromosome. *Genetics* **94:** 15–29.

———. 1983. Genetic methods for analysis and manipulation of inversion mutations in bacteria. *Genetics* **105:** 517–537.

Schmieger, H. 1972. Phage P22 mutants with increased or decreased transduction abilities. *Mol. Gen. Genet.* **119:** 75–88.

———. 1982. Packaging signals for phage P22 on the chromosome of *Salmonella typhimurium. Mol. Gen. Genet.* **187:** 516–518.

Schmieger, H. and H. Backhaus. 1976. Altered cotransduction frequencies exhibited by HT-mutants of *Salmonella*-phage P22. *Mol. Gen. Genet.* **143:** 307–309.

Schmieger, H., K.M. Taleghani, A. Meierl, and L. Weiss. 1990. A molecular analysis of terminase cuts in headful packaging of *Salmonella* phage P22. *Mol. Gen. Genet.* **221:** 199–202.

Schwacha, A. and R.A. Bender. 1993. The *nac* (nitrogen assimilation control) gene from *Klebsiella aerogenes. J. Bacteriol.* **175:** 2107–2115.

Selander, R.K., P. Beltran, and N.H. Smith. 1991. Evolutionary genetics of *Salmonella.* In *Evolution at the molecular level* (ed. R.K. Selander et al.), pp. 25–57. Sinauer Associates, Sunderland, Massachusetts.

Shaw, C.E. and R.K. Taylor. 1990. *Vibrio cholerae* O395 *tcpA* pilin gene sequence and comparision of predicted protein structural features to those of type 4 pilins. *Infect. Immun.* **58:** 3040–3049

Shaw, C.E., K.M. Peterson, J.J. Mekalanos, and R.K. Taylor. 1990. Genetic studies of *Vibrio cholerae* TCP pilus biogenesis. In *Adances in research on cholera and related diarrheas* (ed. R.B. Sack and Y. Zinnaka), pp. 51–58. KTK Scientific Publishers, Tokyo, Japan.

Shen, M.M., E.A. Raleigh, and N. Kleckner. 1987. Physical analysis of Tn*10-* and IS*10*-promoted transpositions and rearrangements. *Genetics* **116:** 359–369.

Shigekawa, K. and W. Dower. 1988. Electroporation of eukaryotes and prokaryotes: A general approach to the introduction of macromolecules into cells. *BioTechniques* **6:** 742–751.

Shimkets, L.J., R.E. Gill, and D. Kaiser. 1983. Developmental cell interactions in *Myxococcus xanthus* and the *spoC* locus. *Proc. Natl. Acad. Sci.* **80:** 1406–1410.

Siegel, E.C., S.L. Wain, S.F. Meltzer, M.L. Binion, and J.L. Steinberg. 1982. Mutator mutations in *Escherichia coli* induced by the insertion of phage Mu and the transposable elements Tn*5* and Tn*10. Mutat. Res.* **93:** 25–33.

Silhavy, T.J. and J.R. Beckwith. 1985. Uses of *lac* fusions for the study of biological problems. *Microbiol. Rev.* **49:** 398–418.

Silhavy, T.J., M.L. Berman, and L.W. Enquist. 1984. *Experiments with gene fusions.* Cold Spring Harbor Laboratory, Cold Spring Harbor, New York.

Simon, R. 1984. High frequency mobilization of gram-negative bacterial replicons by the in vitro constructed Tn*5*-Mob transposon. *Mol. Gen.*

Genet. **196:** 413–420.

———. 1989. Transposon mutagenesis in non-enteric gram-negative bacteria. In *Promiscuous plasmids of gram-negative bacteria* (ed. C.M. Thomas), pp. 207–228. Academic Press, Oxford.

Simon, R., U. Priefer, and A. Pühler. 1983. A broad host range mobilization system for *in vivo* genetic engineering: Transposon mutagenesis in gram-negative bacteria. *Bio/Technology* **1:** 784–791.

Simons, R.W., F. Houman, and N. Kleckner. 1987. Improved single and multi-copy *lac*-based cloning vectors for protein and operon fusions. *Gene* **53:** 85–96.

Singer, M., T.A. Baker, G. Schnitzler, S.M. Deischel, M. Goel, W. Dove, K. J. Jaacks, A.D. Grossman, J.W. Erickson, and C.A. Gross. 1989. Collection of strains containing genetically linked alternative antibiotic resistance elements for genetic mapping of *Escherichia coli*. *Microbiol. Rev.* **53:** 1–24.

Sixma, T.K., S.E. Pronk, K.H. Kalk, E.S. Wartna, B.A. van Zanten, B. Witholt, and W.G. Hol. 1991. Crystal structure of a cholera toxin-related heat-labile enterotoxin from *E. coli*. *Nature* **351:** 371–377.

Skorupski, K. and R.K. Taylor. 1995. Positive selection vectors for allelic exchange. *Gene* (in press).

Slaugh, J.M., M.J. Mahan, and J.J. Mekalanos. 1994. *In vivo* expression technology for selection of bacterial genes specifically induced in host tissues. *Methods Enzymol.* **235:** 481–492.

Smith, C.L. and C.R. Cantor. 1987. Purification, specific fragmentation, and separation of large DNA molecules. *Methods Enzymol.* **155:** 449–467.

Smith, C.L. and G. Condemine. 1990. New approaches for physical mapping of small genomes. *J. Bacteriol.* **172:** 1167–1172.

Smith, C., W. Koch, S. Franklin, P. Foster, T. Cebula, and E. Eisenstadt. 1990. Sequence analysis and mapping of the *Salmonella typhimurium* LT2 *umuDC* operon. *J. Bacteriol.* **172:** 4964–4978.

Smith, C.L., S. Klco, T.-Y. Zhang, H. Fang, R. Olivia, D. Wang, M. Bremer, and S. Lawrence. 1993. Analysis of megabase DNA using pulsed-field gel elecrtrophoresis techniques. *Methods Mol. Genet.* **2:** 155–194.

Smith, H.O. and M. Levine. 1967. A phage P22 gene controlling integration of prophage. *Virology* **31:** 207–216.

Spratt, B.G. 1994. Resistance to antibiotics mediated by target alterations. *Science* **264:** 388–393.

Steinmetz, M., D. Le Coq, H.B. Djemia, and P. Gay. 1983. Analyse génétique de *sacB*, gène de structure d'une enzyme sécreteé, la lévane-saccharase de *Bacillus subtilis* Marburg. *Mol. Gen. Genet.* **191:** 138–144.

Steinmetz, M., D. Le Coq, S. Aymerich, G. Gonzy-Tréboul, and P. Gay. 1985. The DNA sequence of the gene for the secreted *Bacillus subtilis* enzyme levansucrase and its genetic control sites. *Mol. Gen. Genet.* **200:** 220–228.

Sternberg, N. and J. Coulby. 1987a. Recognition and cleavage of the bacteriophage P1 packaging site (*pac*). I. Differential processing of the cleaved ends *in vivo*. *J. Mol. Biol.* **194:** 453–468.

———. 1987b. Recognition and cleavage of the bacteriophage P1 packaging site (*pac*). II. Functional limits of *pac* and location of *pac* cleavage termini. *J.*

Mol. Biol. **194:** 469–479.

———. 1990. Cleavage of the bacteriophage P1 packaging site (PAC) is regulated by adenine methylation. *Proc. Natl. Acad. Sci.* **87:** 8070–8074.

Stewart, V. 1982. Requirement of Fnr and NarL functions for nitrate reductase expression in *Escherichia coli* K-12. *J. Bacteriol.* **151:** 1325–1330.

Stewart, V. and J. Parales, Jr. 1988. Identification and expression of genes *narL* and *narX* of the *nar* (nitrate reductase) locus in *Escherichia coli* K-12. *J. Bacteriol.* **170:** 1589–1597.

Stewart, V. and C. Yanofsky. 1986. Role of leader peptide synthesis in tryphophanase operon expression in *Escherichia coli* K-12. *J. Bacteriol.* **167:** 383–386.

Stibitz, S. 1994. Use of conditionally counterselectable suicide vectors for allelic exchange. *Methods Enzymol.* **235:** 458–465.

Stibitz, S. and N.H. Carbonetti. 1994. Hfr mapping of mutations in *Bordetella pertussis* that define a genetic locus involved in virulence gene regulation. *J. Bacteriol.* **176:** 7260–7266.

Stibitz, S., W. Black, and S. Falkow. 1986. The construction of a cloning vector designed for gene replacement in *Bordetella pertussis. Gene* **50:** 133–140.

Stocker, B.A.D. and P. Makela. 1986. Genetic determination of bacterial virulence with special reference to *Salmonella. Current Top. Microbiol. Immunol.* **124:** 153–172.

Straus, N. and R. Zagursky. 1991. In vitro production of large single-stranded templates for DNA sequencing. *Bio/Techniques* **10:** 376–384.

Strom, M.S. and S. Lory. 1993. Structure-function and biogenesis of the type IV pill. *Annu. Rev. Microbiol.* **47:** 567–596.

Summers, D. and H. Withers. 1990. Electrotransfer: Direct transfer of bacterial plasmid DNA by electroporation. *Nucleic Acids Res.* **18:** 2192.

Suskind, S.R. and L.I. Kurek. 1959. On a mechanism of suppressor gene regulation of tryptophan synthesase activity in *Neurospora crassa. Proc. Natl. Acad. Sci.* **45:** 193–196.

Susskind, M.M. and D. Botstein. 1978. Molecular genetics of bacteriophage P22. *Microbiol. Rev.* **42:** 385–413.

Susskind, M. and P. Youderian. 1983. Bacteriophage P22 antirepressor and its control. In *Lambda II* (ed. R.W. Hendrix et al.), pp. 347–363. Cold Spring Harbor Laboratory, Cold Spring Harbor, New York.

Susskind, M., A. Wright, and D. Botstein. 1971. Superinfection exclusion by P22 prophage in lysogens of *Salmonella typhimurium.* II. Genetic evidence for two exclusion systems. *Virology* **45:** 638–652.

Suwanto, A. and S. Kaplan. 1989. Physical and genetic mapping of the *Rhodobacter sphaeroides* 2.4.1 genome: Genome size, fragment identification, and gene localization. *J. Bacteriol.* **171:** 5840–5849.

———. 1992. Chromosome transfer in *Rhodobacter sphaeroides*: Hfr formation and genetic evidence for two unique circular chromosomes. *J. Bacteriol.* **174:** 1135–1145.

Swanson, J. 1982. Colony opacity and protein II compositions of gonococci. *Infect. Immun.* **37:** 359–368.

Swanson, J. and O. Barrera. 1983. Gonococcal pilus subunit size heterogeneity correlates with transitions in colony piliation phenotype, not with changes in colony opacity. *J. Exp. Med.* **158:** 1459–1472.

Symonds, N., A. Toussaint, P. van de Putte, and M.M. Howe, eds. 1987. *Phage Mu.* Cold Spring Harbor Laboratory, Cold Spring Harbor, New York.

Tabor, S. and C. Richardson. 1989a. Effect of manganese ions on the incorporation of dideoxynucleotides by bacteriophage T7 DNA polymerase and *Escherichia coli* DNA polymerase I. *Proc. Natl. Acad. Sci.* **86:** 4076–4080.

———. 1989b. Selective inactivation of the exonuclease activity of bacteriophage T7 DNA polymerase by in vitro mutagenesis. *J. Biol. Chem.* 264: 6447–6458.

Tacket, C.O., G. Losonsky, J.P. Nataro, S.J. Cryz, R. Edelman, A. Fasano, J. Michalski, J.B. Kaper, and M.M. Levine. 1993. Safety and immunogenicity of live oral vaccine candidate CVD110, a delta*ctxA* delta*zot* delta*ace* derivative of El Tor Ogawa *Vibrio cholerae. J. Infect. Dis.* **168:** 1536–1540.

Tai, P.-C., D. Kessler, and J. Ingraham. 1969. Cold-sensitive mutations in *Salmonella typhimurium* which affect ribosome synthesis. *J. Bacteriol.* **97:** 1298–1304.

Taylor, R.K., C. Manoil, and J.J. Mekalanos. 1989. Broad-host-range vectors for delivery of Tn*phoA:* Use in genetic analysis of secreted virulence determinants of *Vibrio cholerae. J. Bacteriol.* **171:** 1870–1878.

Taylor, R.K., V.L. Miller, D.B. Furlong, and J.J. Mekalanos. 1987. Use of *phoA* gene fusions to identify a pilus colonization factor coordinately regulated with cholera toxin. *Proc. Natl. Acad. Sci.* **84:** 2833–2837.

Taylor, R., C. Shaw, K. Peterson, P. Spears, and J. Mekalanos. 1988. Safe, live *Vibrio cholerae* vaccines? *Vaccine* **6:** 151–154.

Tessman, I., S. Liu, and M. Kennedy. 1992. Mechanism of SOS mutagenesis of UV-irradiated DNA: Mostly error-free processing of deaminated cytosine. *Proc. Natl. Acad. Sci.* **89:** 1159–1163.

Thomas, C.M. 1989. *Promiscuous plasmids of gram-negative bacteria.* Academic Press, Oxford.

Trucksis, M., J.E. Galen, J. Michalski, A. Fasano, and J.B. Kaper. 1993. Accessory cholera enterotoxin (Ace), the third member of a *Vibrio cholerae* virulence cassette. *Proc. Natl. Acad. Sci.* **90:** 5267–5271.

Tye, B.-K. 1976. A mutant of phage P22 with randomly permuted DNA. *J. Mol. Biol.* **100:** 421–426.

Tye, B.-K., J.A. Huberman, and D. Botstein. 1974. Non-random circular permutation of phage P22 DNA. *J. Mol. Biol.* **85:** 501–532.

Ullmann, A. 1992. Complementation in β-galactosidase: From protein structure to genetic engineering. BioEssays **14:** 201–205.

Vandeyar, M.A., and S.A. Zahler. 1986. Chromosomal insertions of Tn*917* in *Bacillus subtilis. J. Bacteriol.* **167:** 530–534.

Vincent, A. and S.W. Liebman. 1992. The yeast omnipotent suppressor SUP46 encodes a ribosomal protein which is a functional and structural homolog of the *Escherichia coli* S4 ram protein. *Genetics* **132:** 375–386.

Vogel, H.J. and D.M. Bonner. 1956. Acetyl ornithase of *Escherichia coli:* Partial purification and some properties. *J. Biol. Chem.* **218:** 97–106.

von Heijne, G. 1987. *Sequence analysis in molecular biology. Treasure trove or trivial pursuit.* Academic Press, San Diego.

Waiter, L.A., A.B. Sadosky, and H.A. Shuman. 1994. Mutagenesis of *Legionella pheumophila* using Tn*903* dII/*lacZ*: Identification of a growth-phase-regulated pigmentation gene. *Mol. Microbiol.* **11:** 641–653.

Walker, G.C. 1984. Mutagenesis and inducible responses to deoxyribonucleic acid damage in *Escherichia coli*. *Microbiol. Rev.* **48:** 60–93.

Wall, J.D. and P.D. Harriman. 1974. Phage P1 mutants with altered transducing abilities for *Escherichia coli*. *Virology* **59:** 532–544.

Wang, A. and J.R. Roth. 1988. Activation of silent genes by transposons Tn*5* and Tn*10*. *Genetics* **120:** 875–885.

Wanner, B.L. and R. McSharry. 1982. Phosphate-controlled gene expression in *Eschericia coli* using MudI-directed *lacZ* fusions. *J. Mol. Biol.* **158:** 347–363.

Wanner, B.L., R. Kodaira, and F.C. Neidhardt. 1977. Physiological regulation of a decontrolled *lac* operon. *J. Bacteriol.* **130:** 212–222.

Way, J.C., M.A. Davis, D. Morisato, D.E. Roberts, and N. Kleckner. 1984. New Tn*10* derivatives for transposon mutagenesis and for construction of *lacZ* operon fusions by transposition. *Gene* **32:** 369–379.

Weinstock, G., M. Susskind, and D. Botstein. 1979. Regional specificity of illegitimate recombination by the translocatable ampicillin-resistance element Tn*1* in the genome of phage P22. *Genetics* **92:** 685–710.

Weiss, B.D., M.A. Capage, M. Kessel, and S.A. Benson. 1994. Isolation and characterization of a generalized transducing phage for *Xanthomonas campestris* pv. *campestris*. *J. Bacteriol.* **176:** 3354–3359.

White, P.B. 1926. Further studies of the Salmonella group. *Br. Med. Res. Council Special Rep. Ser.* **103:** 1–160. (As quoted by Selander, R.K., P. Beltran, and W.H. Smith. 1991. Evolutionary genetics of Salmonella. In *Evolution at the molecular level* (ed. R.K. Selander), pp. 25–57. Sinaver Associates, Sunderland, Massachusetts.

Willetts, N. and R. Skurry. 1987. Structure and function of the F factor and mechanism of conjugation. In Escherichia coli *and* Salmonella typhimurium (ed. J.L. Ingraham et al.), pp. 1101–1133. American Society of Microbiology, Washington, D.C.

Wilmes-Riesenberg, M.R. and B.L. Wanner. 1992. Tn*phoA* and Tn*phA'* elements for making and switching fusions for study of transcription, translation, and cell surface localization. *J. Bacteriol.* **174:** 4558–4575.

Woese, C.R. 1994. There must be a prokaryote somewhere: Microbiology's search for itself. *Microbiol. Rev.* **58:** 1–9.

Wolpert, L. and A. Richards. 1988. *A passion for science*, p. 104. Oxford University Press, England.

Wong, K.K. and M. McClelland. 1992a. A *Bln*I restriction map of the *Salmonella typhimurium* LT2 genome. *J. Bacteriol.* **174:** 1656–1661.

———. 1992b. Dissection of the *Salmonella typhimurium* genome by use of a Tn*5* derivative carrying rare restriction sites. *J. Bacteriol.* **174:** 3807–3811.

Wu., T.-H. and M. Marinus. 1994. Dominant negative mutator mutations in the *mutS* gene of *Escherichia coli*. *J. Bacteriol.* **176:** 5393–5400.

Wu, T.T. 1966. A model for three-point analysis of random general transduction. *Genetics* **54:** 405–410.

Yakobson, E. and D.G. Guiney. 1984. Conjugal transfer of bacterial chromosomes mediated by the RK2 plasmid transfer origin cloned into transposon Tn5. *J. Bacteriol.* **160**: 451–453.

Yanisch-Perron, C., J. Vieira, and J. Messing. 1985. Improved M13 phage cloning vectors and host strains: Nucleotide sequences of the M13mp18 and pUC19 vectors. *Gene* **33**: 103–119.

Yanofsky, C. 1971. Tryptophan biosynthesis in *Escherichia coli*. Genetic determination of the proteins involved. *J. Am. Med. Assoc.* **218**: 1026–1035.

———. 1988. Transcription attenuation. *J. Biol. Chem.* **263**: 609–612.

Yanofsky, C. and V. Horn. 1994. Role of regulatory features of the *trp* operon of *Escherichia coli* in mediating a response to a nutritional shift. *J. Bacteriol.* **176**: 6245–6254.

Yanofsky, C. and E.S. Lennox. 1959. Linkage relationship of the genes controlling tryptophan synthesis in *Escherichia coli*. *Virology* **8**: 425–447.

Yanofsky, C., V. Horn, and P. Gollnick. 1991. Physiological studies on tryptophan transport and tryptophanase operon induction in *Escherichia coli*. *J. Bacteriol.* **173**: 6009–6017.

Yanofsky, C., V. Horn, and D. Thorpe. 1964. Protein structure relationships revealed by mutational analysis. *Science* **146**: 1593–1594.

Yanofsky, C., R.L. Kelley, and V. Horn. 1984. Repression is relieved before attenuation in the *trp* operon of *Escherichia coli* as tryptophan starvation becomes increasingly severe. *J. Bacteriol.* **158**: 1018–1024.

Yanofsky, C., V. Horn, M. Bonner, and S. Stasiowski. 1971. Polarity and enzyme functions in mutants of the first three genes of the tryptophan operon of *Escherichia coli*. *Genetics* **69**: 409–433.

Yarmolinsky, M.B. and N. Sternberg. 1988. *Bacteriophage P1*. In *The bacteriophages* (ed. R. Calendar), vol. 1, pp. 291–438. Plenum Press, New York.

Youderian, P., P. Sugiono, K. Brewer, N. Higgins, and T. Elliott. 1988. Packaging specific segments of the *Salmonella* chromosome with locked in Mud-P22 prophages. *Genetics* **118**: 581–592.

Yu, J., H. Webb, and T.R. Hirst. 1992. A homologue of *Escherichia coli* DsbA protein involved in disulfide bond formation is required for enterotoxin biogenesis in *Vibrio cholerae*. *Mol. Microbiol.* **6**: 1949–1958.

Zahrt, T., G. Mora, and S. Maloy. 1994. Inactivation of mismatch repair overcomes the barrier to transduction between *Salmonella typhimurium* and *Salmonella typhi*. *J. Bacteriol.* **176**: 1527–1529.

Zieg, J. and R. Kolter. 1989. The right end of MudI(Ap, *lac*). *Arch. Microbiol.* **153**: 1–6.

Zinder, N.D. 1992. Forty years ago: The discovery of bacterial transduction. *Genetics* **132**: 291–294.

Zinder, N.D. and J. Lederberg. 1952. Genetic exchange in *Salmonella*. *J. Bacteriol.* **64**: 679–699.

Zipser, D., S. Zabell, J. Rothman, T. Grodzicker, H. Wenk, and M. Novitski. 1970. Fine structure of the gradient of polarity in the *Z* gene of the *lac* operon of *Escherichia coli*. *J. Mol. Biol.* **49**: 251–254.

Zurawski, G., D. Elseviers, G.V. Stauffer, and C. Yanofsky. 1978. Translational control of transcription termination at the attenuator of the *Escherichia coli* tryptophan operon. *Proc. Natl. Acad. Sci.* **75**: 5988–5992.

Index